ELECTROCHEMICAL DETECTION TECHNIQUES IN THE APPLIED BIOSCIENCES

Volume 1: Analysis and Clinical Applications

Ellis Horwood books in the
BIOLOGICAL SCIENCES
General Editor: Dr ALAN WISEMAN, University of Surrey, Guildford
Series in
BIOCHEMISTRY AND BIOTECHNOLOGY
Series Editor: Dr ALAN WISEMAN, Senior Lecturer in the Division of
Biochemistry, University of Surrey, Guildford

ELECTROCHEMICAL DETECTION TECHNIQUES IN THE APPLIED BIOSCIENCES

Volume 1:
Analysis and Clinical Applications

Editor:

GUY ALAIN JUNTER
Chargé de Recherche
Laboratoire de Chimie Macromoléculaire
Centre National de la Recherche Scientifique
University of Rouen, France

ELLIS HORWOOD LIMITED
Publishers · Chichester

Halsted Press: a division of
JOHN WILEY & SONS
New York · Chichester · Brisbane · Toronto

CHEMISTRY

。3180335

~~ENGINEERING~~

First published in 1988 by
ELLIS HORWOOD LIMITED
Market Cross House, Cooper Street,
Chichester, West Sussex, PO19 1EB, England
The publisher's colophon is reproduced from James Gillison's drawing of the ancient Market Cross, Chichester.

Distributors:

Australia and New Zealand:
JACARANDA WILEY LIMITED
GPO Box 859, Brisbane, Queensland 4001, Australia

Canada:
JOHN WILEY & SONS CANADA LIMITED
22 Worcester Road, Rexdale, Ontario, Canada

Europe and Africa:
JOHN WILEY & SONS LIMITED
Baffins Lane, Chichester, West Sussex, England

North and South America and the rest of the world:
Halsted Press: a division of
JOHN WILEY & SONS
605 Third Avenue, New York, NY 10158, USA

South-East Asia
JOHN WILEY & SONS (SEA) PTE LIMITED
37 Jalan Pemimpin # 05–04
Block B, Union Industrial Building, Singapore 2057

Indian Subcontinent
WILEY EASTERN LIMITED
4835/24 Ansari Road
Daryaganj, New Delhi 110002, India

© **1988 G. A. Junter/Ellis Horwood Limited**

British Library Cataloguing in Publication Data
Electrochemical detection techniques in the applied biosciences.
Vol. 1: Analysis and clinical applications
1. Biochemistry. Chemical analysis.
Electrochemical techniques. Applications.
I. Junter, Guy Alain, *1950–*
574.19′285

Library of Congress CIP available

ISBN 0–7458–0494–2 (Ellis Horwood Limited)
ISBN 0–21179–2 (Halsted Press)

Typeset in Times by Ellis Horwood Limited
Printed in Great Britain by Hartnolls, Bodmin

Table of contents

Part 1 Clinical Applications

Contributing authors (to Volumes 1 and 2)

P. Blond, Institut de Chimie et Physique Industrielles de Lyon (ICPI), Service de Chimie Organique et Cinétique, Lyon, France.

S. L Brooks, PHLS Centre for Applied Microbiology and Research, Microbial Technology Laboratory, Porton Down, Salisbury, England.

G. Charrière, Laboratoire Municipal d'Hygiène, Le Havre, France.

D. J. Clarke, PHLS Centre for Applied Microbiology and Research, Microbial Technology Laboratory, Porton Down, Salisbury, England.

H. El Yamani, Laboratoire de Biotechnologie-Enzymes, Ecole Nationale Supérieure des Mines de Saint-Etienne, Saint-Etienne, France.

G.-A. Junter, Laboratoire de Chimie Macromoléculaire, Unité Associée au C.N.R.S. n° 500, Université de Rouen, Mont-Saint-Aignan, France.

J.-M. Kauffmann, Institut de Pharmacie, Université Libre de Bruxelles (ULB), Bruxelles, Belgium.

D. Michel, Institut de Chimie et Physique Industrielles de Lyon (ICPI), Service de Chimie Organique et Cinétique, Lyon, France.

P. Nabet, Laboratoire Central de Chimie, Centre Hospitalier Régional de Nancy, Nancy, France.

G. J. Patriarche, Institut de Pharmacie, Université Libre de Bruxelles (ULB), Bruxelles, Belgium.

M. Porthault, Laboratoire des Sciences Analytiques, Université Claude Bernard (Lyon I), Villeurbanne, France.

J.-L. Romette, Laboratoire de Technologie Enzymatique, Université de Technologie de Compiègne (UTC), Compiègne, France.

J.-P. Schwing, Laboratoire de Chimie Physique et Electrochimie, Ecole Européenne des Hautes Etudes des Industries Chimiques de Strasbourg, Strasbourg, France.

J. L. Taboureau, Institut de Chimie et Physique Industrielles de Lyon, Service de Chimie Organique et Cinétique, Lyon, France.

C. Tran-Minh, Laboratoire de Biotechnologie-Enzymes, Ecole Nationale Supérieure des Mines de Saint-Etienne, France.

J. C. Vire, Institut de Pharmacie, Université Libre de Bruxelles, Belgique.

W. Weppner, Max-Planck-Institut für Festkörperforschung, Stuttgart, West Germany.

Acknowledgement

The Editor is greatly indebted to Mrs D. Moscato for valuable revision of the English language; he also wants to thank Mrs S. Burton for secretarial assistance and his co-workers of the Microbial Technology Group for their comprehensive support.

Guy-Alain Junter

Part 1

Analysis and clinical applications

Part I

Introduction to Part 1

J. P. Schwing

Electrochemically-based biodeterminations *in vivo* or *in vitro* are performed with different kinds of electrodes, each electrode being characterized by an electrochemical reaction involving at least an electron or an ion exchange.

The electrode response is usually measured as a potential V (potentiometry) or as a current density I plotted vs the potential of the so-called 'working electrode'. The potential of the working electrode is usually given with respect to the constant (reference) potential of the Saturated Calomel Electrode (SCE: $+ 0.2415$ volt vs NHE, i.e. the Normal Hydrogen Electrode). The shape and the interpretation of the curves $I = f(V)$ or $V = \phi(I)$ depend on the variation of the controlled parameter, V or I, with time. A very great number of techniques can thus be devised, such as potentiometry at zero current, voltammetry with linear potential sweep, cyclic voltammetry, chronopotentiometry, etc. For a mercury drop electrode or for a rotating platinum electrode, different methods are also defined by considering the time dependence of the voltage or current applied to these electrodes. For a mercury drop electrode some of the most popular methods are direct current (d.c.) and alternating current (a.c.) polarography, pulse polarography, and tast polarography. Theoretical and practical details of these methods can be found in specialized books devoted to electrochemistry, e.g. those given in references [1–5]. For the sake of simplicity, only two important techniques, which are widely applied in clinical analysis, will be briefly considered in this introduction: zero-current potentiometry and voltammetry.

1 ZERO-CURRENT POTENTIOMETRY

1.1 Inert platinum or gold electrodes

For an inert platinum or gold electrode in equilibrium with an electrochemical redox system of the type:

$$\alpha \text{Ox} + m\text{H}^+ + ne^- \rightleftharpoons \beta \text{ Red} + q \text{ H}_2\text{O} \ ,$$

the electrode potential with respect to the NHE is:

$$E_h = E_h^\circ + (RT/nF) \ln[(a_{ox}^\alpha a_{H^+}^m)/(a_{red}^\beta a_{H_2O}^g)]$$

where E_h° is the standard redox potential of the electrode or electrochemical system, a_{ox} and a_{red} are the activities of the oxidized and of the reduced form of the redox couple, respectively, and a_{H^+} and a_{H_2O} are the activities of hydrogen ions and of water, respectively. For dilute solutions the activity of water is taken as unity.

Single redox systems are characterized by their standard reduction potential at 25°C with respect to the NHE. If the solution contains several redox systems the resulting redox potential is called a 'mixed potential' and lies between the potentials of the separated systems. This is very often the case for complex mixtures such as those found in biological systems.

Examples of redox potential measurements using noble metal electrodes in complex biological media (i.e. microbial culture broths) are presented by Junter in Chapter 1 (§1.5).

1.2   Metal–metal oxide electrodes

The Sb/Sb_2O_3 electrode can be prepared as a small needle electrode [1], always covered, through spontaneous air oxidation, with a thin Sb_2O_3 film. According to the electrochemical equilibrium:

$$Sb_2O_3 + 6H^+ + 6e^- \rightleftharpoons 2Sb + 3H_2O \ ,$$

the reduction potential of this electrode is:

$$E_h = E_h^\circ + (RT/F) \ln(a_{H^+})$$

$$Sb_2O_3/Sb$$

This electrode has been used as a pH indicator electrode in biological media [1], especially for measurements on small samples (1 drop of solution). Examples of application are reported by Romette in Chapter 1 (§1.4) (pH-based enzyme electrodes for *in vitro* determinations) and by Michel & Blond in Chapter 2 (§2.2) (pH microelectrodes for *in vivo* measurements).

1.3   Ion-selective electrodes

Ion-selective electrodes are mainly based on membrane potentials (or Donnan potentials) [4,6] using semiconductors, ion-exchange membranes, or liquid membranes.

The most frequently used membrane electrode is the glass electrode [1] whose sensitive surface of Si-OH groups can exchange protons with the test solution. Very small glass electrodes have been devised and can be used for pH measurements in drops or other small samples, as used in microbiology.

Applications of pH and other ion-selective electrodes in clinical chemistry are discussed by Nabet in Chapter 1 (§1.3), whereas Michel & Blond in Chapter 2 (§2.2) deal with ion-selective microelectrodes, their characteristics and their use for *in vivo* determination of biologically important cations.

1.4 Enzyme electrodes

An enzyme electrode is usually an electrode comprising a thin inert film whose active surface is covered with chemically bound enzymes. Behind this film an electrode, sensitive to a reactant or a product of the enzymatic reaction, allows the diffusion of the reactants or the products through the film to be measured. As an example, urea can be catalyzed by urease to produce CO_2 and NH_3 according to:

$$CO(NH_2)_2 + H_2O \xrightarrow{\text{urease}} CO_2 + 2NH_3 \ .$$

In this case the sensing electrode may be a pH-indicating glass electrode. Many examples of potentiometric biosensors associating different types of transducing electrodes with various biocatalysts are reported by Romette in Chapter 1 (§1.4).

2 ELECTROCHEMICAL TECHNIQUES WITH CURRENT FLOW

Stationary (or rotating) platinum electrodes can be used to measure the current flow due to the electrochemical reduction or oxidation of a biochemical substrate.

Glucose in blood [6], for example, can be determined by using the following enzymatic reaction:

$$\text{Glucose} + O_2 \xrightarrow[\text{oxidase}]{\text{glucose}} H_2O_2 + \text{gluconic acid}$$

The product H_2O_2 is determined by amperometry at a platinum electrode.

Many other examples relying on the same principles have led to satisfactory working electrodes for a great number of substrates whose reactions are catalyzed by one or more enzymes fixed on a membrane, covering a suitable detection electrode for one of the reactants or products of the enzymatic reaction (see Chapter 1 (§1.4)).

The dropping mercury electrode is used in polarography mainly for the reduction, at a convenient potential, of many substrates important in biochemistry. Nicotinamide derivatives, purine and pyrimidine derivatives, flavin compounds, quinone and catecholamine systems have been studied and characterized by d.c., a.c. and pulse polarography, single sweep and cyclic voltammetry (Ref. [3], vol. 6). The discussion of the nature of the electrochemical reactions corresponding to polarographic waves is not always straightforward and complementary experiments using constant potential electrolysis (to determine the number of electrons exchanged) or voltammetry (single sweep or cyclic) are often performed to arrive at a better understanding of the direct polarographic curves. A good knowledge of these electrochemical methods is necessary for the interpretation of the data obtained with biochemical compounds.

In the first section of this book (Chapter 1 (§1.1)), Viré *et al.* give an extensive review of clinical analyses *in vitro* by voltammetric and polarographic techniques, whereas Taboureau & Blond in Chapter 2 (§2.1) deal with *in vivo* voltammetric determinations of neurotransmitters using carbon fibre electrodes. In Chapter 1 (§1.2), Porthault focuses on amperometric, polarographic and coulometric detection of drugs and biological compounds after separation with high-performance liquid chromatography.

REFERENCES

[1] Ives, D. J. G. & Janz, G.J. (1961) *Reference electrodes*. New York, Academic Press.
[2] Koryta, J. (1982) *Ions, electrodes and membranes*. New York, John Wiley.
[3] Bard, A. J. (ed.) *Electroanalytical Chemistry*. New York, Marcel Dekker (14 volumes published since 1966).
[4] Moody, G. J. & Thomas, J.D.R. (1971) *Selective ion-sensitive electrodes*. Watford, Merrow Publishing Company.
[5] Headridge, J. B. (1969) *Electrochemical techniques for inorganic chemists*. New York, Academic Press.
[6] Koryta, J. (1975) *Ion-selective electrodes*. New York, Cambridge University Press.

1

In vitro determinations

1.1 CONVENTIONAL AND NOVEL ELECTROCHEMICAL DETECTION TECHNIQUES IN CLINICAL ANALYSIS

J. C. Viré, J. M. Kauffmann, and G. J. Patriarche

If electroanalytical techniques and particularly polarography were applied relatively early to problems of pharmaceutical analysis, the first examples being described in the 1930s and 1940s, it is due to the fact that the samples were well defined, containing only one active compound in relatively large amounts. Furthermore, numerous physiologically active compounds were shown to be electroactive.

The situation is now quite different, the requirements of pharmaceutical analysis becoming more demanding, not only for the different controls performed from the synthesis to the marketing of a given compound, but also for information concerning the absorption, distribution, biotransformation, and elimination of that compound. Such information may only be obtained by the analysis of complex media such as biological fluids or tissues after administration of the drug. On the other hand, very potent compounds are now being developed, resulting in a decrease, to the milligram or submilligram level, of the quantities required to maintain clinical effectiveness. Similarly, the identification of pharmacologically important compounds present in body fluids at very low levels, the knowledge of their mode of action, and their evaluation in biological media have become of great interest in detecting human diseases. These requirements imply that highly selective analytical methods will be used, reaching routinely the nanogram level of sensitivity. The appropriate technique, frequently spectrophotometric, chromatographic, radiometric or electrochemical methods, will be selected according to the sample matrix or the nature of the compound to be analysed, and also the sensitivity of the technique with regard to the investigated molecule. While the two former methods are often preferred by clinical chemists owing to the ease of their use, the electrochemical approach will be a choice method for solving certain analytically oriented clinical problems.

1 Treatment of the sample

As illustrated in this chapter, the most commonly analysed samples are such biological fluids as blood and urine and different tissues in the field of particular diseases or toxicology.

Although electroanalytical techniques offer, at least in some cases, the possibility of a direct investigation of the drug in biological samples, it is usually necessary to separate the compound to be analysed first.

The first step in the treatment of blood is to discard the red cells from the serum to prevent the presence of haemoglobin or its degradation products which may interfere in subsequent analysis.

The separation of the compound from its biological matrix may be performed by using several procedures. Ultracentrifugation or filtration methods may be employed to remove the interference of molecules of high molecular mass, such as proteins, which may inhibit the electrode response. A precipitation of the proteins using an adequate precipitating agent may also be performed. In such cases, care must be taken to avoid a loss of the drugs, some of them being strongly bound to the proteins.

The most commonly used procedure involves solvent extraction of blood or urine. The choice of the solvent is critical in order to eliminate interferences of naturally occurring electroactive compounds of the biological matrix. Another important factor in such a procedure is the pH at which the extraction is carried out. A judicious choice of this parameter may contribute to the elimination of some interfering molecules but also may cause separation of metabolites from the parent drug when using multiple extractions, back-extractions, or 'clean-up' procedures. This subject has been discussed in detail by de Silva [1].

If these techniques are unable to resolve one compound from another, the concentrated solvent extract may be spread on to a thin layer chromatography plate, and the separation will thus be achieved. The scraped area is then eluted with an appropriate solvent and diluted with the supporting electrolyte.

Some difficult cases may also be solved by using HPLC. This technique, associated with ultraviolet, fluorescence, or amperometric detection, has been shown to be useful and powerful in measuring low levels of drugs in blood or urine (see Chapter 1(§1.2) of this book).

Trace metal analysis remains of particular interest. Some biological materials can concentrate certain elements, and the precise knowledge of their concentration may be helpful in medical diagnosis. The biological matrix may complex some metals in varying strengths, causing a change in the free metal ions concentration or a shift of the reduction potential. This justifies a pretreatment of the sample. In numerous cases, the mineralization of blood consists of digestion in a strongly acid medium at a high temperature. Several mixtures have been proposed, the most common being the nitric–perchloric–sulfuric acid mixture which is left in contact with the sample for two hours between 200°C and 300°C [2,3]. Generally, these methods result in a complete destruction of the sample and ensure excellent reproducibility. Although this procedure presents a supplementary contamination risk, the residual current obtained is improved. However, the residual current is not as well defined as when using low temperature ashing of the sample in an oxygen plasma, but this technique generally requires a longer operation time.

The direct and rapid determination of numerous metals by anodic stripping techniques is possible by adding the sample to a reagent, which results in haemolysis of the erythrocytes and the setting free of the metal from its intracellular binding sites. This reagent, Metexchange ($CrCl_3$, 6 H_2O: 10.7 g/kg; $(CH_3COO)_2Ca$, H_2O: 14.3 g/kg; Hg^{2+}: 28 mg/kg), serves as the supporting electrolyte. The mineralization procedure, however, gives rise to a lower residual current.

2 Electrochemical techniques

Although several electroanalytical methods, namely potentiometric and ampero-metric titrations, constant current and potential coulometry, ion-selective, enzyme, bacterial, and tissular electrodes, and numerous voltammetric techniques have been used in pharmaceutical analysis, some of them may be immediately rejected for quantitative analysis of biological material owing to their lack of sensitivity and/or selectivity in such complex media.

The facility of inexpensive titration techniques has been restricted by their low sensitivity to the measurement of drugs in dosage forms and bulk materials.

More interesting are the coulometric techniques, mainly the constant current methods, which have also been applied to pharmaceutical analysis with the advantage over volumetry that they eliminate problems of reagent stability and standardization. The sensitivity is closely related to the nature and the rate of the chemical reaction between the sample and the electrogenerated reagent, but, in favourable cases and using microcells, microgram levels of organic or several tens of nanograms of inorganic compounds may be determined. However, very little attention has been paid to these techniques in clinical analysis owing to their low selectivity.

Constant potential methods are more generally associated with fundamental electroanalysis in order to determine the number of electrons transferred in an electrode process. The interested reader is advised to consult specific literature [4–12].

Potentiometric methods at present play a significant role in biological and clinical analysis with the introduction of new ion-selective indicator electrodes (ISEs). The ability to make direct measurements in complex samples and the fact that they only require relatively inexpensive equipment make ISE-based techniques attractive in many disciplines, particularly in clinical chemistry where a selective, accurate, and easy-to-use analytical system is always of great interest. The growing success of ISEs in the biomedical field is reflected by the increasing number of papers which appear in the literature and are frequently reviewed [13–15]. Most of these detectors are still under investigation in order to improve their analytical performances and to be adapted to routine analysis, but some of them, like Cl^-, Na^+, K^+, Li^+, Ca^{2+}, CO_2, or pH detectors, are now used in clinical laboratories, in automated flow analysis systems, or for single measurements. A detailed overview of the subject will be developed in Chapter 1 (§1.3).

A further step in the search for highly selective sensors has been achieved with the development of biocatalytic membrane electrodes [16–21]. The selectivity of enzymatic reactions linked to the sensitivity of an electrochemical sensor, mainly gas-sensing electrodes which are themselves highly selective, may produce a powerful tool for the investigation of biologically active compounds and will be of particular interest for biochemists. Just as for ISE detectors, considerable research activity in this area of electrochemical biosensors is observed from the abundant literature, and this field will be covered in Chapter 1 (§1.4).

Although some of these methods exhibit a great selectivity, their sensitivity allows their use in biological fluids only for high drug levels, following high-dose administration. The determination of drug concentrations resulting from therapeutic doses may be performed by using modern voltammetric techniques which demonstrate a large linear dynamic range and reach a very low detection limit. Several requirements are to be considered, however. The drug to be analysed must be either directly reducible at the mercury electrode or oxidizable at a solid electrode. Many organic compounds behave in this way, but if this is not the case, a derivatization procedure may be applied to give rise to an electroactive product. This procedure includes nitration, nitrosation, addition, substitution, hydrolysis, oxidation, or complex formation.

Before starting trace analysis in complex media, the voltammetric response of the investigated drug must be well defined. Thus, preliminary studies of the compound are often required, using the conventional direct current (d.c.) technique which is the most suitable for identifying the nature of the recorded wave and studying the influence of several parameters such as pH, nature and concentration of the supporting electrolyte, or of an organic solvent added to dissolve the drug. The knowledge of the oxidation or reduction mechanism may also be helpful in defining the optimal operating parameters.

For most reduction processes, the working electrode is a hanging mercury drop electrode (HMDE) or, in the particular case of polarography, the dropping mercury electrode (DME). In addition to the high hydrogen overvoltage of this metal, the mercury electrode exhibits a low charging current and a good reproducibility, even in the dropping mode where each new clean surface is practically unaffected by previous electrolysis. In order to decrease the capacitive current due to the growing of the drop, the total current may be sampled at the end of the drop life. New polarographic analysers are now equipped with a static mercury drop electrode (SMDE) [22–24]. After the required volume of mercury has been delivered, the surface of the drop remains constant during the sampling of the current. These stands are fully automated and are controlled by the instrument. The main limitation of the mercury electrode is that this metal dissolves anodically in many solutions in the vicinity of +0.1 V to +0.4 V, particularly in the presence of complexing or precipitating agents of mercury (I) or (II). This means that oxidation processes occurring at more positive potentials must be analysed with solid electrodes.

If the accessible potential range is limited on the negative side by the reduction of the supporting electrolyte or of the solvent, usually hydrogen evolution in protic solvents, on the positive side the potential range will be limited by oxidation of the solvent or by oxidation of the electrode material itself to form soluble metal ions or metal oxides.

Platinum and gold are the most commonly used solid metallic electrodes, but they have small overvoltages for hydrogen evolution; this is especially true for platinum which also readily adsorbs hydrogen, giving rise to adsorption–desorption peaks. This feature is observed to a lesser extent on gold. Both metals also exhibit oxidation peaks at sufficiently positive potentials, corresponding to the formation of oxide films which modify the electrode surface during the scan. Reproducible 'active' surfaces may be prepared by using a chemical or an electrochemical pretreatment. In order to obtain better inertness within larger available potential ranges, several different forms of carbon have been used to make satisfactory electrodes. One of the most popular working electrodes is the vitreous or glassy carbon electrode, made of very pure material, highly resistant to chemical attack, and gas impermeable [25,26]. It offers a larger hydrogen overvoltage than platinum and shows good reversibility of a redox couple such as hexacyanoferrate (III/II). This electrode, however, suffers from a high residual current. Its surface is cleaned by polishing with metallographic paper or alumina powder.

Also well known is the carbon paste electrode which is a mixture of spectrographic-grade graphite with Nujol [27]. Its residual current is very low in the positive potential range, but higher in the negative range, owing to the reduction of the oxygen entrapped in the paste. This electrode shows a fair reproducibility and a lower reversibility than the glassy carbon electrode but it is easier to use, the surface being renewed by extruding and cutting out a thin slice of paste. This sensor cannot be used in non-aqueous solvents which disintegrate the paste. Several other dispersing materials have been proposed such as polymers [28,29] or silicone rubber [30].

A porous graphite rod has also been used after its impregnation by molten wax or paraffin in order to avoid a penetration in the electrode by the solution or by oxygen. This treatment improves reproducibility. Several other carbon-based electrodes have also been proposed [30–34].

Solid electrodes may also be used as rotating disk electrodes which improve sensitivity by a well-controlled increase of the mass transport.

Conventional d.c. polarography, with an inherent detection limit of about 10^{-5} M for which the capacitive current becomes comparable with the Faradic current, is not sensitive enough to determine drugs in body fluids after therapeutic doses. On the other hand, if exceptionally high levels are to be analysed, the resolution of the method is relatively poor owing to the wave form of the d.c. polarographic response.

Fast scan or linear sweep polarography [35], sometimes called 'cathode ray' polarography, corresponds to the application of a rapid voltage sweep either to an HMDE or during the last third of the life of a drop in the DME mode. This results in an increase in the recorded current. The technique improves sensitivity up to 10^{-6} or 10^{-7} M, depending on the electrode process, and has found an application in the determination of several compounds. In addition, the asymmetrical peak wave form is easier to measure than the d.c. wave. Nevertheless, measurements of peaks at potentials 0.1–0.3 V more negative than the preceding reduction peak present difficulties because they must be measured by using the descending current of the preceding peak as baseline.

The same technique where the voltage sweep direction is reversed at the end of the scan is called 'cyclic voltammetry' [35]. This is an excellent tool for the investigation of mechanisms of chemical reactions following an electron uptake, but it has been of little analytical use.

Alternating current (a.c.) polarography [36], where a small alternating voltage (2–10 mV) is superimposed on a linearly increasing d.c. potential ramp, improves selectivity because of the peak-shaped response, but the technique has been found to be more sensitive for reversible (10^{-6} M) than for irreversible processes (10^{-4}–10^{-5} M). As this latter case is more frequently encountered in organic polarography and as a.c. techniques are very sensitive to adsorption phenomena, they are no longer applied to biological analysis.

The most popular electroanalytical method used to solve biopharmaceutical oriented problems remains differential pulse (d.p.) polarography [37–41] which uses voltage pulses of constant amplitude superimposed on a gradually increasing ramp of d.c. potential. The current is measured shortly before the application of the pulse and during the last quarter of the pulse life. The difference between these two currents is then plotted as a function of the applied potential. The current–voltage curves are peak-shaped and often symmetrical, and the peak height is a linear function of concentration. This technique offers the best resolution between adjacent peaks, and the detection limit is improved (10^{-7}–10^{-8} M) owing to the lower background current. This increases the signal-to-noise ratio, and the resulting current may be considerably amplified. However, the technique suffers from being time-consuming since slow scan rates must be employed (1–5 mV s^{-1}).

The use of reliable modern electronics for potential control, current measurement, and accurate timing in the millisecond range, and the development of microprocessor-controlled instruments, have given rise to the development of a new powerful pulse technique called 'square-wave voltammetry' (SWV). The waveform consists of a symmetrical square wave superimposed on a staircase, one full period of the square wave occurring for each period of the staircase. The recorded current is the difference between the currents sampled during the last portion of the forward and reverse half-cycles of the square wave.

This technique offers several advantages. Sensitivity is at least as good as, and usually better than, that of d.p. voltammetry. Using comparable parameters, the ratio of the peak current of the square-wave voltammogram to that of the d.p. voltammogram would result in a 30% increase in sensitivity. However, applying the usual values of the operating parameters of the square-wave mode, a ten-fold increase of sensitivity may be reached.

Scan rates are also increased a hundred-fold compared to those of the d.p. mode. Such high scan rates diminish the quantity of compound electrolysed and, owing to the capacity of the computer, it is possible to perform several scans in a short time under the same experimental conditions and to average the results in order to increase the signal-to-noise ratio and so improve the detection limit. An additional advantage of the fast scan capability of square-wave voltammetry is the ability to record repetitively the total voltammogram as a function of time. This allows the study of the kinetics of chemical reactions or the detection of any change in the voltammogram of a compound eluted from the chromatographic column in HPLC or in flow injection analysis. If used as a detection system in these two methods, it should be noted that the square-wave current is insensitive to currents arising from convective mass transport and is thus insensitive to fluctuations in flow rate.

If the pulse techniques with sampling of the current at the end of the pulse life are able to eliminate the capacitive current resulting from the application of the pulse, any currents that are generally independent of potential are rejected by using the square-wave mode. This feature gives rise to a zero current in the limiting current region, providing a very low background current.

Square-wave voltammetry, based on the work of Barker in the 1950s [42], but commercialized only at the beginning of the 1980s because of its dependence on microcomputers, appears to be a powerful analytical tool. While its principles [43–45] and a large part of its theoretical treatment have been described [43,46,47], few applications have been published [48–51], particularly in the biomedical field, but it is beyond doubt that this technique will soon supplant d.p. voltammetry in the trace analysis area.

3 Voltammetric techniques with preconcentration step

Owing to the very low levels to be determined in biological analysis, it is useful to apply, whenever possible, a preconcentration step which will increase the amount of electroactive compound at the electrode. The method was first developed when using an HMDE for amalgamable metals. Preconcentration is performed at a potential selected in the diffusion current region for a length of time depending on the solution concentration and under controlled stirring. After a rest period, the potential is scanned towards less negative values in order to release the amalgamated metal. The technique, called 'anodic stripping voltammetry' (ASV), is able to detect levels as low as 10^{-10} mole per liter [52,53]. Several waveforms such as alternating voltage, differential pulse, or square-wave can be applied during the stripping step to improve sensitivity.

With regard to direct voltammetry, stripping techniques exhibit additional sources of selectivity because the deposition potential may be so selected that certain interferences occurring at more negative potentials are avoided, but also because the preconcentration step may be performed in the medium to be analysed, whereas the stripping step can be realized in a more appropriate supporting electrolyte. However, the HMDE is not mechanically stable enough to allow such medium exchanges. This problem can be solved by use of the mercury film electrode where a thin film of mercury is electrodeposited on to the surface of a glassy carbon electrode. This thin film is mechanically resistant and exhibits higher sensitivity which is due to the increased surface-to-volume ratio of the mercury layer.

Several anions, but also organic compounds, can be determined by using cathodic stripping voltammetry. The deposition step is performed at anodic potentials, and the analyte forms a complex with the electrode material or with a component added to the electrolyte. The deposit must be only slightly soluble and must adhere strongly to the electrode. After the rest period, the preconcentrated material, which must be reducible, is redissolved electrochemically from the electrode by a negative-directed potential scan.

Another method allowing preconcentration of organic compounds has been developed during the last three or four years [54–56]. Numerous organic derivatives exhibit adsorption phenomena which are generally considered to have adverse effects on voltammetric studies. In adsorptive stripping analysis, however, this adsorption process is purposely used as a preconcentration step. As a result, many important substances possessing surface-active properties are easily measurable at the subnanomolar concentration level.

Because adsorption is highly dependent on several parameters such as the nature of the solvent, the electrode material, pH, ionic strength, and temperature, these parameters will also influence the amount of analyte accumulated on the electrode surface. The voltammetric response will also depend on potential and time accumulation.

Mercury or solid electrodes can be used, the former being preferred owing to its lower background current and reproducible surface area.

Adsorptive stripping voltammetry (ASV) needs the same requirements but offers the same advantages as the stripping modes with electrolytic preconcentration. However, selectivity can be improved by a proper choice of the accumulation potential. Here, metals which are not amalgamable may be accumulated at the electrode as organometallic complexes, and the latter could be reduced or oxidized. As the investigated concentrations are low, aqueous solutions may be used, increasing the adsorption capability of most organic compounds by using pulse modes during the stripping step, highly dilute electrolytes can be used, minimizing some interferences and increasing the stripping current for certain analytes.

These advantages, associated with those of conventional voltammetric techniques, make adsorptive stripping analysis a powerful analytical tool which will find, in the near future, numerous applications in various fields of trace analysis. However, the presence of other surface-active species in the sample solution may dramatically interfere in adsorptive stripping measurements. These species can affect the accumulation via a competitive coverage of the electrode surface, resulting in a depletion of the stripping peak. The presence of halide ions, which exhibit specific adsorption, can result in similar effects. These problems, which mainly concern biological analysis, may be minimized by various approaches. Modest peak depressions should be corrected by using the standard addition method, shorter accumulation times, or with a proper choice of the accumulation potential. More severe peak depressions require preliminary chromatographic separation of the interfering surfactants.

Anodic, cathodic, and adsorptive stripping voltammetric techniques are considered to be the most powerful electroanalytical methods for investigating heavy metals in complex media, but increasing attention is being paid to potentiometric stripping analysis (PSA) [57–60]. Similarly to ASV, the high sensitivity of PSA is obtained by an electrolytic preconcentration step. If the working electrode, the rotating glassy carbon electrode being the most popular, is held potentiometrically at a sufficiently reductive potential in a deaerated sample containing mercury (II) ions and mercury soluble metal ions, a mercury film containing the reduced metals will be deposited on to the electrode. However, if ASV applies a positive-directed potential scan to reoxidize the amalgamed metals, these are chemically reoxidized by the excess mercury (II) ions of the solution in the PSA technique, and the time needed for oxidation is monitored instead of the oxidation current. The experimental recorded curve, potential vs time, consists of a normal redox titration curve.

In addition to high sensitivity, PSA offers several other advantages over voltammetric techniques:

- Time can be measured with greater accuracy than current.

- The lengths of the stripping plateaux are independent of the electrode surface area.

- PSA is less sensitive to the presence of low concentrations of surface-active agents, which are frequently encountered in biological samples. As no current passes through the working electrode during stripping, the technique is insensitive to interferences from electroactive substances which are in most cases a limiting factor in stripping voltammetry, because of the high background current.

- To obtain the same resolution as PSA, the d.p. mode must be applied to ASV, but in this case the total stripping time will be of the order of several minutes.
- The method is well suited to analysis without loss of resolution of solutions in which the concentrations of the different metals vary considerably, e.g. by several orders of magnitude. This is because it is the rate of oxidation which is constant rather than the rate of potential increase, as in ASV.
- The technique can be used in samples containing only small concentrations of supporting electrolyte.
- Finally, the required instrumentation is very simple and inexpensive.

An instrument modification which consists of signal enhancement by employing a scheme involving multiple stripping and re-reduction of the preconcentrated analytes has been introduced [61]. This increases the sensitivity of the technique, called 'differential potentiometric stripping analysis' (DPSA). The feasibility of determining trace elements by anodic preconcentration and subsequent reductive PSA has also been demonstrated [62]. As the analyte concentration is sufficiently high (above approximately 25 μg/l for the common heavy metals), the chemical reoxidation step can be more rapidly performed by the dissolved oxygen contained in the non-deaerated sample [63].

4   Applications

As has been pointed out, voltammetric techniques are well suited to the determination of organic or inorganic compounds in biological media owing to their selectivity, accuracy, and precision. However, direct measurement of drugs in biological fluids without any pretreatment of the sample are restricted to high concentration levels, and most analyses require separation procedures. This explains the increasing popularity of chromatographic techniques, mainly HPLC, associated with amperometric detection (see Chapter1 (§1.2)). This technique allows a rapid and effective separation of the compound to be determined from the biological components, but also the differentiation of metabolites from the parent drug. In the organic analytical field the sensitivity of voltammetric techniques is often limited to the 10^{-6}–10^{-7} M concentration range. This parameter has recently been improved by the use of an adsorptive preconcentration step which increases the sensitivity by two orders of magnitude [64].

4.1   *Psychotropic compounds*

As electroanalytical techniques are selected according to the electrochemical properties of the drug to be analysed, the drug level in the body fluid, and the type of analytical problems involved [65,66], and as the azomethine bond of the benzodiazepine derivatives is easily reduced, these compounds have been extensively investigated, mainly by d.p. polarography. Although some of their metabolites exhibit particular electrochemical behaviour, most of them must be separated by extraction procedures. Such problems have often been reviewed [67–73].

Diazepam has been determined in urine [74]. An extraction procedure allows the study of its metabolization by rapid scan a.c. polarography [75]. The stability of nordiazepam has been tested in blood and urine during storage by d.p. polarography after its separation on Amberlite XAD-2 [76].

The polarographic assay of bromazepam extracted from blood with ether in a pH 7 phosphate buffer containing 5% of dimethylformamide allows levels as low as 0.05 μg/ml to be determined [77]. With a 0.5 ml capacity microcell, a detection limit of 10–20 ng/ml can be reached [78]. Extraction of bromazepan can also be performed using a benzene–methylene chloride (9:1)† mixture [79].

The stability of dipotassium clorazepate has been evaluated at 37°C in whole blood [80]. Flurazepam and its metabolites N-1-hydroxyethyl- and N-1-desalkyl-derivatives exhibit differences in their acid–base characteristics. They can be extracted selectively by a judicious choice of solvents and pH. This procedure allows 70 ng of each compound per ml of plasma to be determined. The sensitivity is improved down to 30 ng/ml plasma for the total benzodiazepine content [81].

The new class of imidazo-1,4-benzodiazepines has also been investigated [82]. Nimetazepam has been recovered from whole blood and serum, either directly or after extraction with a mixture of ethyl acetate–toluene (1:1). The linearity of the polarographic response is observed between 30 and 150 μg/ml in a pH 3.8 Britton Robinson buffer [83]. Flunitrazepam [84], clonazepam [84,85], and nitrazepam [84], which also possess a nitro group, have been analysed.

A number of benzodiazepines cannot be differenciated as the reduction of the azomethine bond always occurs in the same potential range, so even if substituents fixed on to the central ring are quite different, oxidation processes concerning these substituents may often give rise to a more convenient separation of these compounds. The electro-oxidation of sixteen benzodiazepines and some of their metabolites at a glassy carbon electrode has been reviewed by Smyth & Ivaska [86].

Other psychotropic drugs have also been investigated.

Ciclazindol has been recovered from plasma and urine after extraction with toluene at pH 12.3. Using a 10% methanolic 0.01 M tetraethylammonium chloride supporting electrolyte, an 80% mean recovery is observed for plasma containing 0.5 to 5.0 μg/ml with a variation coefficient of 5.5 % [87].

Considerable attention has been paid to chlorpromazine which has been directly monitored in urine by d.p. voltammetry using a 23 μl capacity thin-layer cell equipped with a wax-impregnated graphite electrode. A linear calibration curve is obtained in the $4.8 \times 10^{-8} - 2.4 \times 10^{-4}$ M concentration range with 97% recovery. Direct measurement in plasma is only possible in a $2.4 \times 10^{-5} - 4.8 \times 10^{-4}$ M concentration range with 89% recovery [88]. Since chlorpromazine is adsorbed at a carbon paste electrode, sensitivity has been enhanced with regard to the conventional voltammetric measurement by imposing a preconcentration step before scanning the potential. When this procedure is applied in a flow injection system, a higher selectivity is demonstrated by the determination of chlorpromazine in the presence of a hundred-fold excess of non-adsorbable solution species with similar redox potentials [89,90]. This compound has been shown to be accumulated at the carbon paste electrode by adsorption at the surface of the electrode, but also by a dissolution process into the pasting liquid [91,92].

---

† Here, and in the following, dilutions are indicated in vol./vol.

The problems of co-adsorption of electroinactive surfactants or adsorbable electroactive molecules have been minimized by coating a glassy carbon electrode with a cellulose acetate membrane [93]. Similarly, anionic and neutral interferences have been diminished by using a glassy carbon electrode coated with Nafion film for the determination of cationic neurotransmitters in urine [94]. Adsorptive stripping voltammetry has also been applied to the determination of promethazine, dietha-zine, fluphenazine, trifluoperazine, and clozapine. It has been demonstrated that the compounds are both adsorbed at the surface and extracted into the wax-impregnated graphite working electrode. A 5×10^{-8} M detection limit is observed in urine. A membrane-covered electrode is used for plasma measurements [95].

The possible determination of the tricyclic antidepressants imipramine, trimipra-mine, and desipramine in blood and urine at a glassy carbon electrode has been demonstrated. A medium exchange procedure is required to eliminate electroactive interferences at the oxidation potentials of the compounds [96].

4.2 Antiseptics, antimicrobial agents, antibiotics

A d.p. polarographic procedure has been developed for the determination of nitroimidazole compounds after their extraction from protein-free plasma or urine with ethyl acetate. The overall recovery for plasma is about $55 \pm 3.0\%$ with a detection limit of 0.1 μg/ml [97].

A direct evaluation of urinary nitrofurantoin can be performed after a ten-fold dilution of the sample with 0.5 M borate buffer (pH 12.1). The height of the reduction d.c. wave obtained at a rotating platinum electrode increases linearly with concentrations between 0.5 and 15 μg/ml [98]. The same compound can be analysed by using a mixture of pyridine–formic acid–0.1 M tetramethylammonium chloride (2:1:12) (pH 4.3) as supporting electrolyte. A 3 to 60 μg/ml concentration range can be exploited [99].

The d.p. polarographic assay of nalidixic acid in urine requires a previous extraction with chloroform. A linear relationship is observed between 4×10^{-7} and 1×10^{-4} M, and 99% recovery with a relative standard deviation of 11 % is reported [100].

A d.c. polarographic method has been proposed to determine ampicillin in serum and urine after acid hydrolysis in 1 M hydrochloric acid which also serves as supporting electrolyte. Two waves are exhibited, the first being linear with concen-tration down to 2.8×10^{-6} M [101].

Piperacillin is directly analysed in urine and plasma by adding a 0.3 M hydro-chloric acid supporting electrolyte with detection limits of 9×10^{-6} M and 3.7×10^{-5} M, respectively. Sensitivity is improved to 6.3×10^{-7} M and 1.9×10^{-6} M after acid hydrolysis and extraction with ethyl acetate [102].

Streptomycin and related antibiotics are also directly determined in urine diluted 1:20 with 0.01 M sodium hydroxyde down to 1×10^{-6} M. Lower concentrations, however, require an extraction procedure [103].

4.3 Analgesics, anti-inflammatory drugs

The d.p. voltammetric assay of acetaminophen based on the oxidation of the phenolic moiety at a carbon paste electrode has been proposed. Dilution of the plasma with an equal volume of 0.5 M phosphate buffer at pH 8.0 allows determinations between 20 and 400 $\mu g/ml$ [104]. The drug may also be extracted with ethyl acetate and injected in a flow system passing through a thin-layer cell equipped with a silicone–graphite–grease electrode. Linearity of the amperometric response is observed between 10 and 300 $\mu g/ml$ [105]. The drug has also been derivatized in a protein-free plasma with sodium nitrite. The resulting *o*-nitroderivative gives rise to a 6-electron a.c. polarographic peak allowing the determination of 0.6 $\mu g/ml$ plasma [106]. Alternating current polarography has also been applied to the assay of indomethacin extracted from serum. The residue is hydrolyzed with sodium hydroxyde and the indole derivative formed is nitrosated to an indole–nitrosamine electroactive compound. Levels as low as 0.3 $\mu g/ml$ serum can be analysed [107].

Piroxicam has been determined in urine after daily oral absorption of 20 mg of the drug. The sample is diluted ten-fold and buffered at pH 4.0. A detection limit of 1.5 $\mu g/ml$ urine is reported [108].

4.4   Antitumor drugs

Differential pulse polarography allows the determination of plasmatic carmustine and other *N*-alkyl-*N*-nitrosourea at levels below 1 $\mu g/ml$ in a 0.2 M citric acid medium after extraction with *n*-pentane. Calibration curves of carmustine are linear in the $5 \times 10^{-8} - 5 \times 10^{-7}$ M concentration range [109].

5-Fluorouracil, which can form mercury salts, is analysed by d.p. cathodic stripping voltammetry in a pH 7.8 borax–perchloric acid buffer at the hanging mercury drop electrode. Extraction is carried out with diethyl ether and propanol (4:1) in order to discard uric acid which seriously interferes with detection. The detection limit was found to be 5×10^{-6} M with a mean recovery of 43% [110]. 5-Fluorouracil, as well as chlorambucil, can also be determined by an adsorptive stripping voltammetric technique [111].

1,2,3,4-Tetrahydrocarbazole, used as a model compound for indole alkaloids, has been recovered from spiked urines after adsorptive preconcentration on to a carbon paste electrode. The stripping step is performed in a pH 9.4 ammonia buffer. After preconcentration for 10 min, a detection limit near 1.4×10^{-8} M is obtained. Urinary vinblastine, for example, has been determined down to 1.5×10^{-8} M in a pH 4.5 acetate buffer [112]. Vinblastine and vincristine also exhibit a catalytic wave in a pH 9.3 ammonia buffer, allowing their determination between 5 ng and 5 $\mu g/ml$ with a relative standard deviation of 7.3% for urine and 8.6% for plasma and recoveries ranging from 90 to 96%. The technique is claimed to be a thousand-fold more sensitive than the anodic oxidation [113].

Differential pulse polarography has been applied to the determination of methotrexate in spiked plasma and urine after dilution of the sample in 0.1 M sodium hydroxyde or 0.2 M hydrochloric acid. The same sensitivity and accuracy are observed for both supporting electrolytes in a 4 to 40 ppm concentration range [114]. An adsorptive preconcentration of methotrexate at a hanging mercury drop electrode improves sensitivity. Dilution of urine samples (1:4) with a pH 2.5 phosphate buffer allows determinations at the 5×10^{-7} M concentration level with a 10-second deposition time [115].

Adsorptive preconcentration of adriamycin in diluted urine at a carbon paste electrode followed by a d.p. stripping step after medium exchange allows this anthracycline compound to be determined in a 1×10^{-7} to 1×10^{-5} M concentration range with a 1×10^{-8} M detection limit [116]. Doxorubicine in urine has been demonstrated to be easily resolved from the uric acid response by using a flow injection procedure with a carbon paste electrode. Linearity of the response is established between 1×10^{-9} and 1×10^{-6} M in a 0.2 M acetate buffer [117]. A modified commercial instrument has been used to perform d.p. polarographic measurements of mitomycin C in blood and urine. As the compound is previously separated on Amberlite XAD-2, a detection limit of 25 ng/ml can be reached. Without any pretreatment, a 200 ng/ml detection limit is observed. The calibration curves are linear up to 10 μg/ml [118].

The N-oxide group of indicine-N-oxide allows the d.p. polarographic analysis of the compound after extraction of the lyophilized residue in a pH 4.63 buffer. Recovery of the drug is $88 \pm 7\%$ in the 1-20 μg/ml range [119].

Platinum derivatives are often used as cancerostatic drugs. Differential pulse polarography is well suited to the determination of total platinum in urine, plasma, or tissues after administration of *cis*-dichlorodiamminoplatinum. The sample is calcined at 700°C and the residue is dissolved in a 0.1 M KCL–0.1 M ethylene diamine supporting electrolyte. The platinum–ethylene diamine complex is reduced at -1.65 V vs SCE. Linearity is observed between 50 and 1600 ng/ml for blood and urine and between 100 and 3200 ng/ml for tissues [120]. The metal can also be extracted as a diethyldithiocarbamate complex with chloroform, the residue being digested with aqua regia. The resulting hexachloroplatinic acid is determined in a 0.05 M sulfuric acid solution after deposition of the metal on to a rotating glassy carbon electrode at -0.2 V vs SCE. The recorded current corresponds to the oxidation of the deposited platinum without stripping. 0.1 to 100 ppm of platinum can be determined [121].

4.5   *Blood pressure regulators, cardiotonics*

Therapeutic levels of sodium nitoprusside from 1×10^{-5} to 8×10^{-5} M are determined in human serum by using the a.c. polarographic response of the reducible nitroso group. Serum is deproteinated with perchloric acid and diluted ten-fold with a pH 9.5 borate buffer. The detection limit is fixed at 1.5×10^{-6} M in serum [122]. Using a 3-ml sample treated with perchloric acid and ferrocyanide, d.p. polarography allows determinations of this molecule in plasma, serum or blood between 30 and 1000 ng/ml with a detection limit of 15 ng/ml [123].

Norpace (diisopyramide phosphate) can be oxidized at a glassy carbon electrode in a pH 8.0 Britton Robinson buffer. The peak increases linearly with concentrations between 1 and 5 μg/ml. The molecule is extracted from alkalinized plasma with diethyl ether. Recovery is $81 \pm 3.0\%$ [124].

Cardiac glycosides such as digoxin, digitoxin, and digitoxigenin adsorb strongly on to the hanging mercury drop electrode. They can be determined in urine after a twenty-fold dilution with 0.05 M sodium hydroxyde at the 10^{-7} M concentration level. Differential pulse voltammetry is performed after a 1 minute preconcentration step at -0.9 V vs Ag/AgCl,KCLs [125].

Nicardipine may undergo either reduction of its nitro group or oxidation of its dihydropyridine moiety. Using d.p. voltammetry at a hanging mercury drop electrode, the drug is determined down to 2×10^{-8} M in human urine diluted 1:4 with a pH 7.4 phosphate buffer. However, as flow measurements are more suited to routine analysis, the oxidative detection of the molecule may be realized on carbon electrodes with a detection limit of 1.3×10^{-7} M, corresponding to 1.4 ng in the 20 μl injected volume [126].

4.6 Other compounds
Therapeutic levels of plasmatic theophyllin (10–20 μg/ml) are monitored at a carbon paste electrode by d.p. voltammetry. Plasma buffered at pH10 is extracted with a chloroform/2-propanol mixture. The organic solvent is back-extracted with 0.1 M sodium hydroxyde. The aqueous phase is buffered at pH 7.4 and analysed [127].

Disodium cromoglycate is extracted from urine into chloroform with tri-*n*-butylbenzylphosphonium chloride, forming an ion-pair. It can also be extracted as free cromoclycic acid into ethyl acetate from 2 M hydrochloric acid solutions. Differential pulse polarography exhibits a detection limit of 0.5 μg/ml [128].

The first d.c. wave developed by the diuretic triamterene (2,4,7-triamino-6-phenyl pteridine) in 10% acetic acid has been used to determine its urinary excretion. Linear calibration curves have been established between 10 and 100 μg/ml with a 1 μg/ml detection limit. Freeze-dried samples are extracted with a pyridine-water (9:1) mixture [129]. Alternating current polarography has been applied to the direct determination of another diuretic drug, tienilic acid [(thienoyl-2)-4-dichloro-2,3-phenoxyacetic acid] in urine. A pH 8.0 phosphate buffer provides a linear response between 10 and 500 μg/ml [130].

Differential pulse polarography is quite useful for determining K vitamins owing to their quinonic structure. Vitamin K_1 has been monitored after intravenous administration of 20 mg. Plasma is extracted with chloroform in the presence of zeolite, and the residue is dissolved in an ethanol–0.5 M acetate buffer (9:1) mixture. The peak observed at -0.58 V vs SCE increases linearly in the 0.1–1.2 μg/ml concentration range. Recovery is 72.2% [131]. Using the same procedure, the simultaneous determination of ubiquinone–50 has been reported [132]. A methanol–0.2 M borate buffer (9.1) supporting electrolyte allows the assay of plasmatic vitamin K_3 in a 0.6–10 μM concentration range with a 0.3 μM detection limit [133].

Cimetidine has been determined in urine samples mixed 1:1 with 0.1 M hydrochloric acid, using square-wave adsorptive stripping voltammetry. Linear calibration plots were obtained up to 2 μg/ml with a detection limit of 1 ng/ml. Serum can also be analysed but requires a pretreatment to remove surfactants which strongly decrease the peak [134].

A linear scan adsorptive stripping procedure has been developed for the direct measurement of thiourea in urine diluted with an equal volume of 0.5 M sodium perchlorate solution at a hanging mercury drop electrode. A 60-second accumulation time allows levels as low as 1 ng/ml of the compound to be detected [135]. Thioamide drugs have been recovered from plasma and urine down to 2×10^{-8} M using cathodic stripping voltammetry at the hanging mercury drop electrode. The sample is diluted 1:1 with a pH 4.78 Britton Robinson buffer. Naturally-occurring sulphur compounds do not interfere [136,137].

Several heterocyclic *N*-oxide and azo-compounds have been determined in urine by normal and d.p. polarography after an extraction procedure. Detection limits of about 5×10^{-5} M and 5×10^{-6} M, respectively, have been reported. Nitro-derivatives can be analysed without previous separation. The sample is diluted in a pH 12 Britton Robinson buffer [138].

4.7   Toxic compounds

The thiourea structure is often included in pesticides. Cathodic stripping voltammetry is able to detect these compounds down to 1 ng/ml in urine samples diluted 1:1 with 1 M sodium hydroxyde [139]. The d.c. polarographic determination of nitrophenol metabolites of organophosphorus pesticides in the urine [140] and blood [141] of poisoned animals has also been described. No previous extraction procedure is required. Paraquat, which is frequently used as a herbicide, can be monitored in serum and urine after dilution of the samples with a pH 7 buffer and adding gelatin and *n*-butanol as antifoam. Detection limits were found to be 0.03 μg/ml serum and 0.05 μg/ml urine by using d.p. polarography [142].

N-nitroso-N-methylaniline, which is a carcinogenic compound, can be detected in whole blood and serum by a d.p. polarographic technique. The sample mixed with a phosphate buffer is passed through a Sep Pak C 18 cartridge and eluted with methanol. An aliquot is diluted 1:9 with a 0.1 M perchloric acid–sodium perchlorate supporting electrolyte. A detection limit of 0.01 ppm is reported with recoveries of 82–84% for whole blood and 88–90% for serum [143].

4.8   Naturally occurring compounds

Uric acid is known to adsorb rapidly on a carbon paste electrode. A preconcentration step in the sample followed by a medium exchange procedure improves the d.p. voltammetric response with respect to sensitivity and selectivity, owing to the interference of ascorbic acid in direct voltammetric measurements. The technique has been applied to serum and urine samples, but a flow injection system is preferred to the medium exchange, avoiding a loss of weakly adsorbed species during the transfer of the electrode [144]. A simple electrochemical method requiring no reagent or sample pretreatment has been proposed to analyse uric acid in serum. The molecule undergoes oxidation at a carbon paste electrode, a constant potential of +0.64 V vs Ag/AgCl,KCls being applied by a digital potentiostat. The latter is previously calibrated with standard solutions in a pH 7 Mac Ilvaine buffer. Measurements are directly performed in 0.5 ml untreated serum samples. The results are in good agreement with those obtained by the phosphotungstate or the uricase methods [145].

The total cholic acid content, free and conjugated, of human bile is determined by d.p. polarography after dilution of the sample with water (1:24) and adding aliquots to a pH 5.25 acetate buffer. Linearity is observed between 2×10^{-6} and 1×10^{-4} M with a relative standard deviation of 2.4 to 8.9% [146].

Free bilirubin exhibits an irreversible reduction peak at the hanging mercury drop electrode. Serum is mixed with methanol (1:5), and 50 μl of the supernatant is diluted with 10 ml of 0.1 M tris(hydroxymethyl)methylamine (pH 8.5). The d.p. voltammetric response increases linearly with concentration between 0 and 1 μg/ml [147].

Vanillylmandelic acid (VMA) is a major metabolite of epinephrine. It is able to form a complex with molybden (VI) which exhibits, in the presence of bromate ions, a catalytic maximum wave in a 0.1 M formic acid medium. Differential pulse polarography has been applied to the determination of VMA in urine after extraction with ethyl acetate. Calibration plots are linear in the $2\times10^{-7}-1\times10^{-6}$ M concentration range with a detection limit of 7×10^{-9} M [148].

A d.p. polarographic method based on the interaction of serum albumin with potassium titanium (IV) oxalate allows determinations of this protein in a continuous flow system at a fixed potential of -0.68 V vs. SCE. Titanium concentration (6×10^{-2} M) and pH (phosphate-citrate buffer, pH 4.89) must be strictly controlled. Linearity of the polarographic response is obtained between 50 and 500 μg/ml [149].

Inorganic blood and urine components are also of particular interest.

Bromide ion levels have been evaluated in blood and urine after oxidation to bromate with hypochlorite. An amplification process is applied: bromate is reduced to bromine with an excess of bromide; the bromine formed is extracted with chloroform and then reduced to bromide. These steps can be repeated, allowing determinations at the nanomolar level in 15% perchloric acid by a.c. polarography. Recoveries between 98.7% and 99.8% are reported [150].

The sulfide released in 0.2 M sodium hydroxyde from the disulfide bonds of proteins can be measured in 0.2 M sodium hydroxyde on a mercury pool using cathodic stripping voltammetry. A $2\times10^{-7}-8\times10^{-7}$ M concentration range can be analysed with a 3-minute deposition time. No interferences from cysteine, cystine, sulfite, or thiosulfate are reported but copper (II) and lead ions seriously interfere [151,152].

A cathodic stripping procedure has also been developed to determine thiocyanate ions down to 2×10^{-8} M and copper (II) ions down to 1×10^{-8} M in pure solutions. The method is based on the electrolytic accumulation of copper (I) thiocyanate at the surface of a hanging mercury drop electrode followed by its cathodic stripping. One per cent of urine sample is added to a 0.1 M perchloric acid supporting electrolyte. About 1×10^{-5} M thiocyanate are found in urine. Concentrations of about 2×10^{-3} M are generally detected in saliva which must be diluted before analysis [153].

It can be concluded, from recent literature, that electrochemical methods remain of great interest for solving clinical analysis problems. However, to avoid time-consuming separation procedures, it is preferable to apply direct voltammetric techniques to measurements of concentrations above 1 μg/ml. Such techniques are able to detect lower levels, but a separation or purification procedure must be previously carried out to increase the signal-to-noise ratio. Therefore, an increasing number of clinical problems are now solved by the powerful HPLC separation technique, although an electrochemical detection step is still used [154–157] (see also Chapter 1(§1.2) of this book). Simultaneously, flow injection analysis with electrochemical detection remains of considerable interest [157–161]. However, increasing attention is being paid to adsorptive stripping voltammetry of organic compounds, owing to its sensitivity and selectivity [54,55,64].

5   Metal analysis

A large part of analytical literature is always devoted to trace metal analysis, and stripping voltammetry remains a choice method for the determination of several representative cations [54,162–167], as will be illustrated below. However, several interferences may affect the accuracy of the measurements, and here potentiometric stripping analysis is a more accurate, sensitive, versatile, and inexpensive method for metal ions analysis [57–63,164].

5.1 Direct voltammetric analysis

Very few examples are devoted to voltammetric assays of metals without a precon-centration step.

The analysis of arsenic in blood and urine includes a mineralization of the sample with a nitric–sulfuric–perchloric acid mixture (5:3:2), the reduction of As(V) to As(III) with hydrazine sulfate, and the extraction of As(III) with diethylammonium diethyldithiocarbamate in chloroform. The residue is dissolved in a 6 M sulfuric acid supporting electrolyte and analysed using d.p. polarography. A detection limit of 5×10^{-8} M is reported [168].

Molybdenum contained in plasma and urine has been determined in a pH 2.2 potassium nitrate–sulfuric acid medium by cathode ray polarography. After dry ashing of the sample, extraction of the acidified residue is carried out with pentyl acetate. Molybdenum is back-extracted with the supporting electrolyte [169]. This metal can also be extracted from bovine liver after wet digestion as its Mo-8-hydroxyquinolinate complex with dichloromethane containing 0.1 M tetrabutyl-ammonium perchlorate. The organic layer is submitted to d.c. and d.p. polarography [170].

5.2 Voltammetric stripping analysis

Numerous papers deal with the analysis of lead and cadmium. This is largely because of their toxicity [171], but they are also chosen as representative elements for the development of a new procedure or to compare different mineralization methods, depending on the nature of the sample. Sulfuric acid as the ashing acid at 500°C was found to be the most suitable for tissues [172], while a mixture of nitric, perchloric, and sulfuric acids in a pressurized device is recommended for blood [173]. It has been demonstrated that irradation techniques give comparable results with conventional acid digestion. Required irradiation doses were 4–6 h for UV irradiation and 20–30 Mrad for γ-irradiation [174]. The influence of twelve mineralization procedures has been reviewed. The risks of ignoring the sources of contribution to the blank values have been pointed out [175]. Several assays have been performed by using Metex-change as the decomplexing agent and supporting electrolyte [176–178]. Lumaton, which contains a quaternary ammonium hydroxyde in isopropanol, has also been proposed as the digesting agent [179,180]. To improve the baseline of the minera-lized sample, some metals have been extracted after complexation in an organic solvent [181,182]. The same procedure has been used to discard interfering metals [183]. Separation by ion-exchange chromatography has also been proposed [184, 185]. It is not always necessary to extract interfering elements from the solution to be analysed. In some cases, they can be complexed in order to shift their reduction potential. This procedure is generally used to mask the response of lead and cadmium with EDTA, allowing the determination of thallium which is not com-plexed [186,187]. An attempt has been made to determine lead and cadmium in urine without any pretreatment. Despite a higher background current, the peak heights have been enhanced by working at temperatures between 40°C and 60°C [188].

Some metals which cannot be amalgamated are complexed into an organometallic compound which is adsorbed at the surface of the hanging mercury drop electrode. Cobalt and nickel are usually accumulated as their dimethylglyoximate complex [189–191].

To avoid loss of metal, a digestion procedure including the immobilization of mercury (II) ions with thioacetamide in sulfide form has been proposed. The sulfide is released in a further step with nitric acid and hydrogen peroxide [192].

Several elements have to be chemically treated after digestion of the sample to acquire electroactivity. Antimony [193] and arsenic [194] have been reduced with hydrazine sulfate and copper (I) respectively to Sb(III) and As(III) before anodic stripping measurement.

Selenium has been extensively studied [195,196]. After mineralization, hydrochloric acid is generally used to convert all the selenium to the Se(IV) form which is electroactive [197–199]. In a multi-element analysis including the determination of selenium, the electroinactivity of Se(VI) can be turned to account to avoid its interference by oxidizing this element with potassium permanganate, the excess of the latter being reduced with hydrazine sulfate [200].

A multi-element analysis of a single biological sample allows the determination of Se, Pb, Cd, Cu, Zn, Ni, Co. After adding hydrochloric acid to the digestion residue to convert selenium to Se(IV), this element is determined by d.p. cathodic stripping voltammetry. Copper, lead, and cadmium are then analysed by d.p. anodic stripping voltammetry or d.p. polarography, depending on their concentration. Ammonia buffer is then added for the assay of zinc. This buffer is suitable for the determination of nickel and cobalt as their dimethylglyoximate complexes by d.p. adsorptive stripping voltammetry. Arsenic can also be analysed if selenium, which interferes, is absent [201].

Several examples of trace metal analysis by voltammetric stripping methods are summarized in Table 2 (Table 1 indicates the symbols used in Tables 2 and 3).

5.3 *Potentiometric stripping analysis*

As well as voltammetric stripping techniques, potentiometric stripping analysis has been applied to the simultaneous determination of zinc, lead, cadmium and copper in biological samples [202]. If a simple potentiostat is used, solutions must be deoxygenated to avoid reoxidation by dissolved oxygen which increases the reoxidation rate [203]. However, computerized devices, able to measure times in the millisecond range, allow the deoxygenation step to be left out and offer a shorter analysis time [204,205].

Potentiometric stripping detection has also been applied to flow injection systems [206–208].

Table 3 summarizes the data of several assays using potentiometric stripping analysis.

Acknowledgements

Thanks are expressed to the 'Fonds National de la Recherche Scientifique' (FNRS Belgium) for help to one of us (G. J. Patriarche) and to the SPPC (Belgium Politic Research, ARC), contract No. 86/91-89.

Table 1 — Symbols used in Tables 2 and 3

HMDE	Hanging Mercury Drop Electrode
MFE	Mercury Film Electrode (mainly on glassy carbon electrode)
GCE	Glassy Carbon Electrode
AuE	Gold Electrode
PtE	Platinum Electrode
CMGrE	Composite Mercury Graphite Electrode
WIGrE	Wax Impregnated Graphite Electrode
GrSMFE	Graphite Spray Mercury Film Electrode
GCAuFE	Glassy Carbon Gold Film Electrode
PGrTAuFE	Pyrolytic Graphite Tube Gold Film Electrode
RGCE	Rotating Glassy Carbon Electrode
RAuDE	Rotating Gold Disk Electrode
RGCMFE	Rotating Glassy Carbon Mercury Film Electrode
RGCAuFE	Rotating Glassy Carbon Gold Film Electrode
LSASV	Linear Scan Anodic Stripping Voltammetry
DPASV	Differential Pulse Anodic Stripping Voltammetry
DPCSV	Differential Pulse Cathodic Stripping Voltammetry
ACASV	Alternating Current Anodic Stripping Voltammetry
DPAdsSV	Differential Pulse Adsorptive Stripping Voltammetry
PSA	Potentiometric Stripping Analysis
FIPSA	Flow Injection Potentiometric Stripping Analysis
ComPSA	Computerized Potentiometric Stripping Analysis
AAS	Atomic Absorption Spectrometry
DMG	Dimethylglyoxime
Lin.	Linearity range
d.l.	Detection limit
r.s.d.	Relative standard deviation

Table 2 — Applications of voltammetric stripping analysis

Sample	Sample pretreatment	Working electrode	Determined elements	Supporting electrolyte	Methods and comments	Refs
Blood	Acid digestion H_2SO_4-$HClO_4$ (1:5) 200°C	HMDE	Pb	Digested sample diluted with water	DPASV 0–4 μM in blood	[209]
Blood	Acid digestion HNO_3-$HClO_4$-H_2SO_4 (24:24:1)	WIGrE	Pb	Acetate buffer pH 4.6 + Hg^{2+} 20 μg/ml	DPASV Thin-layer cell (60 μl) ng/ml level	[210]
Blood	No pretreatment	CMGrE	Pb	Blood+Metexchange	LSASV Use of 100 μl of blood Correlation with AAS	[176]
Blood	No pretreatment	MFE	Cd	Blood+Metexchange	LSASV d.l.:4 nM EDTA interference eliminated by adding $NiCl_3$ Use of 500 μl of blood	[177]
Blood Serum	Acid digestion HNO_3-$HClO_4$ (5:2)	HMDE	Se	+HCL→Se(IV) Diluted with water	DPCSV Lin.: 0.1–12 ppb d.l.: 0.1 ppb†	[199]
Urine	Acid digestion HNO_3 5h 120°C	HMDE	Pb, Cd	Acetate buffer pH 4.9	DPASV ppb level	[211]
Urine	Acid digestion HNO_3-$HClO_4$ (1:2) 210°C	MFE	Pb, Cd	Acetate buffer pH 4.5	DPASV d.l.: Pb: ≤ 2 μg/l Cd: ≤ 0.3 μg/l Use of 1 ml of sample	[212]
Urine	Acid digestion with thioacetamide (see text)	RAuDE	Hg	KCl 1 M	ACASV Recovery ≥ 97% d.l.: ppb level	[192]
Urine	No pretreatment	HMDE	Pb, Cd		DPASV Working temperature: 40–60°C Mineralization if copper to be determined.	[188]
Urine	0.01 M HCl Chromatogr. on Sep Pak C18 Filtration on 0.45 μm filter	RGCMFE	Pb	Directly transferred in the cell	DPASV Lin.: 10–400 ng/ml	[185]

Sample	Pretreatment	Element	Electrolyte	Details	Ref.
Blood Urine	Acid digestion HNO_3-$HClO_4$,H_2SO_4 (23:23:1)	As	+Cu (I) in HCl 105°C → As (III); HCl 7 M	LSASV Lin.: 5–500 ng	[194, 213]
Blood Urine	Acid digestion HNO_3-$HClO_4$-H_2SO_4 (24:24:1)	Sb	+hydrazine→Sb (III) 1 M hydroxyl-ammonium chloride — 6 M HCl 1:1	LSASV Lin.: 10–600 ng Interference from Cu, Zn, Fe.	[193]
Blood Urine	Chromatography on Amberlite IRA-400	Bi	H_2SO_4-KCl	DPASV D.l.: 5 ng/ml	[184]
Blood Urine	No pretreatment	Bi	Blood+metexchange	DPASV d.l.: 5 ng/ml 50-fold excess of Cu tolerated	[178]
Blood Urine	Acid digestion HNO_3-H_2SO_4 (5:1)	Co	Ammonia buffer pH 7.5+DMG	DPAdsSV d.l.: blood: 0.8 ng/ml urine: 0.2 ng/ml	[190]
Urine Saliva	No pretreatment	Tl	Acetate buffer pH 4.7 +2×10^{-3} M EDTA	DPASV d.l.: urine: 0.02 ppb saliva: 0.2 ppb	[187]
Human teeth	HNO_3 in Teflon bomb 2h 200°C	Cu, Pb, Cd, Zn.	HNO_3+water+acetate pH 0.8–1.0	DPASV Sensitivity: 3–7 $\mu A/\mu M$ Electrolysis time: 30 s–4 min	[214]
Human hair	HNO_3 and fused alkali metal nitrates	Cu, Pb, Cd, Zn.	Acetate buffer pH 4.7	DPASV Recoveries: 91–101% r.s.d.: 2.6–13%	[215]
Human bones	HNO_3-$HClO_4$ (1:1)	Pd, Cd.	Digested sample diluted with water	DPASV Same results as AAS	[216]
Blood Teeth	HNO_3-$HClO_4$-H_2SO_4 (2:2:1) HNO_3-$HClO_4$ (5:2)	Cu, Pb, Cd, Zn.	Digested sample + 0.25% HNO_3 (Cu, Pb, Cd) +0.3 M CH_3COONH_4 (Zn)	DPASV d.l.: 1 ppb	[217]
Blood Urine Bones	Acid digestion $HClO_4$,H_2SO_4	Bi	0.2–0.5 M HCl +5×10^{-5} M $AuCl_3$	DPASV Cu and Hg extracted as EDTA complexes with chloroform Lin.: 10^{-6}–10^{-9} M d.l.: urine: 0.1 ng/ml blood: 0.6 ng/ml bones: 30 ng/g	[183]
Blood Urine	Acid digestion $HClO_4$·H_2SO_4 (2:1)	Bi	0.2–0.5 M HCl +5×10^{-5} M $AuCl_3$	DPASV Bi extracted as diethyldithio-	[181]

Table 2 — Applications of voltammetric stripping analysis — continued

Sample	Sample pretreatment	Working electrode	Determined elements	Supporting electrolyte	Methods and comments	Refs
Bones					carbamate complex with CHCl$_3$. Lin., d.l.: same as ref [182]	
Blood Urine Hair	Acid digestion HNO$_3$-H$_2$SO$_4$-H$_2$O$_2$ 150°–200°C	HMDE	Ni	Ammonia buffer pH 9.2+DMG.	DPAdsSV No d.l. in biological media d.l. in water: 1 ng/ml	[189]
Bovine liver	Digestion with Lumaton	HMDE	Se		DPCSV	[180]
Bovine liver	Acid digestion HNO$_3$-HClO$_4$ (1:1)	RGCAuFe	Se	Digested sample+water +0.1 M HCl+1×10^{-6} M Au(III)	DPCSV Lin.: up to 1×10^{-6} M d.l.:~2×10^{-9} M	[197]
Bovine liver	Acid digestion HNO$_3$+Mg(NO$_3$)$_2$ and dry ashing 500°C 30 min.	HMDE	Se, Cu, Pb, Cd.	Ashed sample + water + HCl up to pH 1	Se: DPCSV others: DPASV After Se det.: +KMnO$_4$ (Se(IV)→Se(VI))+ hydrazine d.l.: Se: 0.1 µg/l; Cu: 0.5 µg/l Pb, Cd: 0.01 µg/l	[200]
Bovine liver	Dry ashing 450°C 8h	HMDE	Co, Ni.	Ashed sample dissolved with HCl, +ammonia buffer + DMG	DPAdsSV d.l.: Co: 0.01 µg/g Ni: 0.02 µg/g Influence of DMG concentration	[191]
Bovine liver	Acid digestion HNO$_3$-HClO$_4$-H$_2$SO$_4$ (23:23:1)	PGrTAuFE	As	+Cu(I) in HCl, 105°C→As(III); 7 M HCl	LSASV Lin.: 0.01–1.0 µg	[219]
Urine Bovine liver	Acid digestion HNO$_3$-H$_2$SO$_4$ (5:1)	HMDE	Se	Digested sample + water + HCl	DPCSV d.l.: ng/ml level Comparative study of wet digestion methods	[196]
Blood, urine, feces, teeth, bones, hair.	Acid digestion in Parr bomb, 2–7 h (urine: no pretreatment)	HMDE	Bi, Cu, Pb, Cd, Zn. Tl	Acetate pH 6 Acetate pH 6+EDTA	DPASV Use of 10 µl liquid sample and 10–1000 mg tissues	[186]

Sample	Preparation	Electrode	Elements	Medium	Technique / results	Ref.
Calf liver, mussels, leaves	Mineralized sample extracted with salicylaldoxime in 1,2-dichloroethane.	HMDE	Cu	DMF+5% $HClO_4$	LSASV Lin.: 0.1–4.5 µg/ml d.l.: 15 ng/ml	[182]
Animal muscles Bovine liver Blood	Acid digestion HNO_3-H_2SO_4 (5:2) If Se>0.1 µg/l: anion exchange separation	(1) HMDE (2) RAuDE	Se	Digested sample+HCl	(1) DPCSV peak at −0.5 V (2) DPASV peak at +0.45 V d.l.:<1 ng/ml for 1) and (2) better reproducibility with (1)	[198]
Animal tissues	Digestion with Lumaton	HMDE	Sn, Pb, Cu	1 M HCl in methanol	ACASV d.l.:1 ppb	[179]
Meat Organs from cows calves, bulls	Acid digestion 200°C HNO_3-$HClO_4$ (2:1)	HMDE	Cu, Pb, Cd, Zn. Co, Ni.	pH 2	DPASV 0.5 g samples DPAdsSV	[218]
Animal muscles Bovine liver Urine	Acid digestion HNO_3-H_2SO_4	HMDE	Se, Cd, Pb, Cu, Zn, Ni, Co.	pH 9 + DMG (1) +HCL (2) idem (3) + ammonia buffer (4) + DMG	(1) DPCSV (Se) (2) DPASV (Cd, Pb, Cu) (3) DPASV (Zn) (4) DPAdsSV (Ni, Co) d.l.: ng/ml level As analysed if Se absent	[201]

†Parts per billion (i.e. USA billion, 10^9).

Table 3 — Applications of potentiometric stripping analysis

Sample	Sample pretreatment	Working electrode	Determined elements	Supporting electrolyte	Methods and comments	Refs
Whole blood serum	No pretreatment	RGCE	Pb, Cd.	Sample -0.5 M HCl (9:1)$+Hg^{2+}$ 75 mg/l	ComPSA d.l.: 25 nM (r.s.d. ≤4%) No deaeration	[205]
Whole blood	No pretreatment	Carbon fibre electrode	Pb, Cd		ComFIPSA d.l.: 25 nM; 12 samples/h	[207]
Urine	No pretreatment	MFE	Pb, Cd	Sample+HCl up to 0.5 M	ComPSA d.l.: 1 nM; no deaeration	[204]
Urine	No pretreatment	MFE	Pb	Acidified sample+ Triton X-100$+Hg^{2+}$	PSA d.l.: 1 ng/ml	[203]
Urine	Acid digestion (HNO_3)	AuE	Hg	1 M NaBr+0.1 M HCl +Cr (VI)	ComFIPSA d.l.: 5 nM (15 min depos, time)	[220]
Bovine liver	(1) NHO_3 (2) HNO_3–$HClO_4$ (comparison)	MFE	Cu, Pb, Cd, Zn.	0.06 M HCl$+Hg^{2+}$ (Cu, Pb, Cd) 0.06 M HCl+0.18 M acetic acid+Ga(III) $+Hg^{2+}$ (Zn)	ComPSA Comparison with AAS Theoretical values found	[202]
Bovine liver	HNO_3 in pressurized container 110°C overnight	PtE GCE	Mn	Digested sample+ace-tate buffer pH 4.5	Reductive PSA with hydroquinone. Theoretical values found (10 μg/g)	[62]
Horse kidney	Acid digestion	GCE	Pb, Cd.	Digested sample directly injected	FIPSA d.l.: 50 μg/l 100 μl samples Use of a wall-jet cell	[208]

References

[1] de Silva, J. A. F. (1970) In: Swarbrick, J. (ed.) *Current concepts in the pharmaceutical sciences: biopharmaceutics*. Philadelphia, Lea and Febiger p. 203.

[2] Peter, F. & Reynolds, R. G. (1976) *Anal. Chem.* **48**, 2041–2042.

[3] Deangelis, T. P., Bond, R. E., Brooks, E. E. & Heineman, W. R. (1977) *Anal. Chem.* **49**, 1792–1797.

[4] Lingane, J. J. (1970) *Electroanalytical chemistry*. 2nd ed. New York, Interscience.

[5] Wilson, C. L. & Wilson, D. W. (eds) (1975) *Comprehensive Analytical Chemistry: Coulometric Analysis*. Vol. II D, Amsterdam, Elsevier.

[6] Patriarche, G. J. (1964) *Contribution à l'analyse coulométrique*. Paris, Arscia-Maloine.

[7] Patriarche, G. J., Chateau-Gosselin, M., Vandenbalck, J. L. & Zuman, P. (1979) In: Bard, A. J. (ed.) *Electroanalytical chemistry*. Vol. 11. New York, M. Dekker p. 141–289.

[8] Nürnberg, H. W. (1974) *Electroanalytical chemistry*. New York, J. Wiley.

[9] Curran, D. J. (1984) In: Kissinger, P. T. & Heineman, W. R. (eds) *Laboratory techniques in electroanalytical chemistry*. New York, M. Dekker p. 539–568.

[10] Janata, J. & Mark, H. B., Jr. (1969) In: Bard, A. J. (ed.) *Electroanalytical chemistry*. Vol. 3. New York, M. Dekker.

[11] Bard, A. J. & Santhanam, K. S. V. (1970) In: Bard, A. J. (ed.) *Electroanalytical chemistry*. Vol. 4. New York, M. Dekker.

[12] Harrar, J. E. (1975) In: Bard, A. J. (ed.) *Electroanalytical chemistry*. Vol. 8. New York, M. Dekker.

[13] Meyerhoff, M. E. & Fraticelli, Y. M. (1982) *Anal. Chem.* **54**, 27R–44R.

[14] Davis, J. E., Solsky, R. L., Giering, L. & Malhotra, S. (1983) *Anal. Chem.* **55**, 202R–214R.

[15] Arnold, M. A. & Meyerhoff, M. E. (1984) *Anal. Chem.* **56**, 20R–48R.

[16] Guilbault, G. G. (1984) *Analytical uses of immobilized enzymes*. New York, M. Dekker.

[17] Savory, J., Bertholf, R. L., Boyd, J. C., Bruns, D. E., Felder, R. A., Lovell, M., Shipe, J. R., Willis, M. R., Czaban, J. D., Coffey, K. F. & O'Connell, K. M. (1986) *Anal. Chim. Acta* **180**, 99–136.

[18] Czaban, J. D. (1985) *Anal. Chem.* **57**, 345A–356A.

[19] Carr, P. W. & Bowers, L. D. (1980) *Immobilized enzymes in analytical and clinical chemistry*. New York, Wiley–Interscience.

[20] Bowers, L. D. (1986) *Anal. Chem.* **58**, 513A–530A.

[21] Guilbault, G. G. & Kauffmann, J. M. (1987) *Biotechnol. Appl. Biochem.*, **9**, 95–113.

[22] Peterson, W. M. (1979) *Am. Lab. (Fairfield, Conn.)* **11**, 69–78.

[23] Bond, A. M. & Jones, R. D. (1980) *Anal. Chim. Acta* **121**, 1–11.

[24] Anderson, J. E., Bond, A. M. & Jones, R. D. (1981) *Anal. Chem.* **53**, 1016–1020.

[25] Plock, C. E. (1968) *J. Electroanal. Chem.* **18**, 289–293.

[26] Zittel, H. E. & Miller, F. J. (1965) *Anal. Chem.* **37**, 200–203.

[27] Adams, R. N. (1958) *Anal. Chem.* **30**, 1576.

[28] Swofford, H. S., Jr. & Carman III, R. L. (1966) *Anal. Chem.* **38**, 966–969.

[29] Prete, M. P., Kauffmann, J. M., Viré, J. C., Patriarche, G. J., Debye, B. & Geuskens, G. (1984) *Anal. Lett.* **17**, 1391–1401.

[30] Pungor, E., Szepesuary, E. & Havas, J. (1968) *Anal. Lett.* **1**, 213–221.

[31] Heineman, W. R. & Kissinger, P. T. (1980) *Anal. Chem.* **52**, 138R–151R.

[32] Sawyer, D. T. & Roberts, J. L., Jr. (1974) *Experimental electrochemistry for chemists*. New York, Wiley–Interscience.

[33] Kauffmann, J. M., Laudet, A. & Patriarche, G. J. (1982) *Talanta* **29**, 1077–1082.

[34] Patriarche, G. J., Kauffmann, J. M. & Viré, J. C. (1983) *J. Pharm. Biomed. Anal.* **1**, 469–474.

[35] Parker, V. D. (1986) In: Bard, A. J. (ed.) *Electroanalytical chemistry*. Vol. 14. New York, M. Dekker.

[36] Smith, D. E. (1966) In: Bard, A. J. (ed.) *Electroanalytical chemistry*. Vol. 1. New York, M. Dekker.

[37] Christie, J. H. & Osteryoung, R. A. (1974) *J. Electroanal. Chem.* **49**, 301–311.

[38] Flato, J. B. (1972) *Anal. Chem.* **44**, 75A–87A.

[39] Osteryoung, J. G. & Hasebe, K. (1976) *Rev. Polarogr. Jap.* **22**, 1–25.

[40] Heijne, G. J. M. & Vanderlinden, W. E. (1976) *Anal. Chim. Acta* **82**, 231–243.

[41] Bircke, R. L. (1978) *Anal. Chem.* **50**, 1489–1496.

[42] Barker, G. C. (1958) *Anal. Chim. Acta* **18**, 118–128.

[43] Osteryoung, J. G. & O'Dea, J. J. (1986) In: Bard, A. J. (ed.) *Electroanalytical chemistry*. Vol. 14. New York, M. Dekker.

[44] Yarnitzky, C., Osteryoung, R. A. & Osteryoung, J. G. (1980) *Anal. Chem.* **52**, 1174–1178.

[45] Osteryoung, J. G. & Osteryoung, R. A. (1985) *Anal. Chem.* **57**, 101A–110A.

[46] Christie, J. H., Turner, J. A. & Osteryoung, R. A. (1977) *Anal. Chem.* **49**, 1899–1903.

[47] Turner, J. A., Christie, J. H., Vucovic, M. & Osteryoung, R. A. (1977) *Anal. Chem.* **49**, 1904–1908.

[48] Stojek, Z. & Osteryoung, J. G. (1981) *Anal. Chem.* **53**, 847–851.

[49] Webber, A., Shah, M. & Osteryoung, J. G. (1983) *Anal. Chim. Acta* **154**, 105–109.

[50] Webber, A., Shah, M. & Osteryoung, J. G. (1984) *Anal. Chim. Acta* **157**, 1–16.

[51] Webber, A. & Osteryoung, J. G. (1984) *Anal. Chim. Acta* **157**, 17–29.

[52] Barendrecht, E. (1967) In: Bard, A. J. (ed.) *Electroanalytical chemistry*. Vol. 2. New York, M. Dekker.

[53] Vydra, F. Stulik, K. & Julakova, E. (1976) *Electrochemical stripping analysis*. Chichester, Ellis Horwood Ltd.

[54] Wang, J. (1985) *Stripping analysis: principles, instrumentation and applications*. Deerfield Beach, Verlag Chemie.

[55] Wang, J. (1985) *Intern. Lab.* **17**, 41–76.

[56] Wang, J. Luo, D. B. Farias, P. A. M. & Mahmoud, J. S. (1985) *Anal. Chem.* **57**, 158–162.
[57] Jagner, D. & Graneli, A. (1976) *Anal. Chim. Acta* **83**, 19–26.
[58] Jagner, D. & Aren, K. (1978) *Anal. Chim. Acta* **100**, 375–388.
[59] Jagner, D. (1978) *Anal. Chem.* **50**, 1924–1929.
[60] Jagner, D. (1982) *Analyst* **107**, 593–599.
[61] Kryger, L. (1980) *Anal. Chim. Acta* **120**, 19–30.
[62] Christensen, J. K. & Kryger, L. (1980) *Anal. Chim. Acta* **118**, 53–64.
[63] Jagner, D. (1979) *Anal. Chem.* **51**, 342–345.
[64] Smyth, W. F. (1986) In: Smyth, M.R. & Vos, J.G. (eds) *Electrochemistry, sensors and analysis. Analytical chemistry symposia series.* Vol. 25. Amsterdam, Elsevier p.29–36.
[65] Pungor, E., Feher, Z., Nagy, G., Lindner, E. & Toth, K. (1982) *Anal. Proc. (London)* **19**, 79–82.
[66] Smyth, M. R. & Smyth, W. F. (1976) *Meth. Dev. Biochem.* **5**, 15–24.
[67] Brooks, M. A. (1980) In: Smyth, W. F. (ed.) *Electroanalysis in hygiene, environmental, clinical and pharmaceutical chemistry. Analytical chemistry symposia series.* Vol. 2. Amsterdam, Elsevier p.287–298.
[68] Brooks, M. A. (1983) *Bioelectrochem. Bioenerg.* **10**, 37–55.
[69] Brooks, M. A. (1984) In: Kissinger, P. T. & Heineman, W.R. (eds) *Laboratory techniques in electroanalytical chemistry.* New York, M. Dekker p.569–609.
[70] Smyth, W. F. (1982) *Anal. Proc. (London)* **19**, 82–86.
[71] Brooks, M. A., de Silva, J. A. F. & Hackman, M. R. (1973) *Am. Lab.* **5**, 23–33.
[72] Brooks, M. A. & de Silva, J. A. F. (1975) *Talanta* **22**, 849–860.
[73] Clifford, J. M. & Smyth, W. F. (1977) *Proc. Anal. Div. Chem. Soc.* **14**, 325–329.
[74] Ellaithy, M. M., Volke, J. & Manousek, O. (1977) *Talanta* **24**, 137–140.
[75] Dugal, R., Caillé, G. & Cooper, S. F. (1973) *Union Méd. Canada* **102**, 2491–2497.
[76] Meeles, M. T. H. A., Dreyer–Van der Glas, S. M. & De Jong, H. J. (1977) *Pharm. Weekbl.* **112**, 501–508.
[77] Sengun, F. I. & Oelschläger, H. (1975) *Arch. Pharm. (Weinheim)* **308**, 720–723.
[78] Brooks, M. A. & Hackman, M. R. (1975) *Anal. Chem.* **47**, 2059–2062.
[79] de Silva, J. A. F., Berkersky, I., Brooks, M. A., Weinfeld, R. E., Glover, W. & Puglisi, C.V. (1974) *J. Pharm. Sci.* **63**, 1440–1445.
[80] Abruzzo, C. W., Brooks, M. A., Cotler, S. & Kaplan, S. A. (1976) *J. Pharmacol. Biopharm.* **4**, 29–41.
[81] Clifford, J. M., Smyth, M. R. & Smyth, W. F. (1974) *Fres. Z. Anal. Chem.* **272**, 198–201.
[82] Puglisi, C. V., Meyer, J. C., D'Arconte, L., Brooks, M. A. & de Silva, J. A. F. (1978) *J. Chromatogr.* **145**, 81–96.
[83] Kobiela, A. (1977) *Pharmazie* **32**, 693–694.
[84] de Silva, J. A. F., Puglisi, C. V. & Munno, N. (1974) *J. Pharm. Sci.* **63**, 520–527.

[85] Kobelia–Krzyzanowska, A. (1976) *Pharmazie* **31**, 649–650.

[86] Smyth, W. F. & Ivaska, A. (1985) *Analyst* **110**, 1377–1379.

[87] Chan, H. K. & Fogg, A. G. (1978) *Anal. Chim. Acta* **98**, 101–109.

[88] Jarbawi, T. B., Heineman, W. R. & Patriarche, G. J. (1981) *Anal. Chim. Acta* **126**, 57–64.

[89] Wang, J. & Freiha, B. A. (1983) *Anal. Chem.* **55**, 1285–1288.

[90] Wang, J. & Freiha, B. A. (1983) *Anal. Chim. Acta* **148**, 79–85.

[91] Wang, J., Deshmukh, B. K. & Bonakdar, M. (1985) *J. Electroanal. Chem.* **194**, 339–353.

[92] Wang, J. & Freiha, B. A. (1984) *Anal. Chem.* **56**, 849–852.

[93] Wang, J., Bonakdar, M. & Pack, M. M. (1987) *Anal. Chim. Acta* **192**, 215–223.

[94] Wang, J., Tuzhi, P. & Golden, T. (1987) *Anal. Chim. Acta* **194**, 129–138.

[95] Jarbawi, T. B. & Heineman, W. R. (1986) *Anal. Chim. Acta* **186**, 11–19.

[96] Wang, J., Bonakdar, M. & Morgan, C. (1986) *Anal. Chem.* **58**, 1024–1028.

[97] Brooks, M. A., D'Arconte, L. & de Silva, J. A. F. (1976) *J. Pharm. Sci.* **65**, 112–114.

[98] Mason, W. D. & Sandmann, B. (1976) *J. Pharm. Sci.* **65**, 599–601.

[99] Morales, A., Toral, M. I. & Richter, P. (1984) *Analyst* **109**, 633–636.

[100] Van Oort, W. J., Sorel, R. H. A., Brussee, D., Schulman, S.G., Zuman, P. & Den Hartigh, J. (1983) *Anal. Chim. Acta* **149**, 175–191.

[101] Schroder, S., Bonow, A., Nschel, H. & Horn, G. (1978) *Pharmazie* **33**, 432–434.

[102] Schroder, S., Voigt, R., Horn, G., Sestakova, I. & Skarka, P. (1985) *Pharmazie* **40**, 333–336.

[103] Wang, J. & Mahmoud, J. S. (1986) *Anal. Chim. Acta* **186**, 31–38.

[104] Munson, J. W. & Abdine, H. (1978) *J. Pharm. Sci.* **67**, 1775–1776.

[105] Falkowski, A. & Wei, R. (1981) *Anal. Lett.* **14**, 1003–1012.

[106] Alkayer, M., Vallon, J. J., Pégon, Y. & Bichon, C. (1981) *Anal. Chim. Acta* **124**, 113–119.

[107] Alkayer, M., Vallon, J. J., Pégon, Y. & Bichon, C. (1981) *Anal. Lett.* **14**, 1047–1070.

[108] Viré, J. C., Kauffmann, J. M., Braun, J. & Patriarche, G. J. (1985) *J. Pharm. Belg.* **40**, 133–138.

[109] Bartosek, I., Daniel, S. & Sykora, S. (1978) *J. Pharm. Sci.* **67**, 1160–1163.

[110] Miranda Ordieres, A. J., Garcia Gutierrez, M. J., Costa Garcia, A. & Tunon Blanco, P. (1987) *Analyst* **112**, 243–246.

[111] Wang, J., Lin, M. S. & Villa, V. (1987) *Analyst* **112**, 247–251.

[112] Wang, J. & Bonakdar, M. (1985) *Anal. Lett.* **18**, 2569–2579.

[113] Temizer, A. (1986) *Talanta* **33**, 791–794.

[114] Ellaithy, M. M., El-Tarras, M. F., Tadros, N.B. & Amer, M.M. (1982) *Anal. Lett.* **15**, 981–988.

[115] Wang, J., Tuzhi, P., Lin, M. S. & Tapia, T. (1986) *Talanta* **33**, 707–712.

[116] Chaney, E. N., Jr. & Baldwin, R. P. (1982) *Anal. Chem.* **54**, 2556–2560.

[117] Chaney, E. N., Jr. & Baldwin, R. P. (1985) *Anal. Chim. Acta* **176**, 105–112.

[118] Van Bennekom, W. P., Tjaden, U. R., De Bruijn, E. A. & Van Oosterom, A.T. (1984) *Anal. Chim. Acta* **156**, 289–294.

[119] McComish, M., Bodek, I. & Branfman, A. R. (1980) *J. Pharm. Sci.* **69**, 727–729.
[120] Brabec, V., Vrana, O. & Kleinwächter, V. (1983) *Coll. Czech. Chem. Commun.* **48**, 2903–2908.
[121] Schmid, G. M. & Atherton, D. R. (1986) *Anal. Chem.* **58**, 1956–1959.
[122] Alkayer, M., Vallon, J. J. & Pégon, Y. (1981) *Anal. Lett.* **14**, 399–414.
[123] Leeuwenkamp, O. R., Van Der Mark, E. J., Jousma, H., Van Bennekom, W. P. & Bult, A. (1984) *Anal. Chim. Acta* **166**, 129–140.
[124] Burmicz, J. S., Smyth, W. F., Smyth, M. R. & Palmer, R. F. (1981) *Analyst* **106**, 802–805.
[125] Wang, J., Mahmoud, J. S. & Farias, P. A. M. (1985) *Analyst* **110**, 855–859.
[126] Wang, J., Deshmukh, B. K. & Bonakdar, M. (1985) *Anal. Lett.* **18**, 1087–1102.
[127] Munson, J. W. & Abdine, H. (1978) *Talanta* **25**, 221–222.
[128] Fogg, A. G. & Fayad, N. (1978) *Anal. Chim. Acta* **102**, 205–210.
[129] Pellerin, F., Letavernier, J. F. & Chanon, N. (1977) *Analysis* **5**, 19–22.
[130] Pellerin, F., Letavernier, J. F., Chanon, N. & Lezian, T. (1977) *Eur. J. Med. Chem.* **12**, 477–482.
[131] Hart, J. P., Nahir, A. M., Chayen, J. & Catterall, A. (1982) *Anal. Chim. Acta* **144**, 267–271.
[132] Hart, J. P. & Catterall, A. (1980) In: Smyth, W.F. (ed.) *Electroanalysis in hygiene, environmental, clinical and pharmaceutical chemistry. Analytical chemistry symposia series.* Vol. 2. Amsterdam, Elsevier p.145–153.
[133] Akman, S. A., Kusu, F., Takamura, K., Chlebowski, R. & Block, J. (1984) *Anal. Biochem.* **141**, 488–493.
[134] Webber, A., Shah, M. & Osteryoung, J. G. (1983) *Anal. Chim. Acta* **154**, 105–119.
[135] Stara, V. & Kopanika, M. (1984) *Anal. Chim. Acta* **159**, 105–110.
[136] Davidson, I. E. & Smyth, W. F. (1977) *Anal. Chem.* **49**, 1195–1198.
[137] Smyth, W. F. (1980) In: Smyth, W. F. (ed.) *Electroanalysis in hygiene, environmental, clinical and pharmaceutical chemistry. Analytical chemistry symposia series.* Vol. 2. Amsterdam, Elsevier p.271–286.
[138] Vaneesorn, Y. & Smyth, W. F. (1980) In: Smyth, W. F. (ed.) *Electroanalysis in hygiene, environmental, clinical and pharmaceutical chemistry. Analytical chemistry symposia series.* Vol. 2. Amsterdam, Elsevier p.299–308.
[139] Smyth, M. R. & Osteryoung, J. G. (1977) *Anal. Chem.* **49**, 2310–2314.
[140] Zietek, M. (1979) *Mikrochim. Acta (Wien)* **II**, 75–83.
[141] Zietek, M. (1976) *Mikrochim. Acta (Wien)* **II**, 549–557.
[142] Franke, G., Pietrulla, W. & Preussner, K. (1979) *Fres. Z. Anal. Chem.* **298**, 38–42.
[143] Pylypiw, H. M., Jr. & Harrington, G. W. (1981) *Anal. Chem.* **53**, 2365–2367.
[144] Wang, J. & Freiha, B. A. (1984) *Bioelectrochem. Bioenerg.* **12**, 225–234.
[145] Park, G., Adams, R. N. & White, W. R. (1972) *Anal. Lett.* **5**, 887–896.
[146] Ferri, T., Campanella, L. & Deangelis, G. (1984) *Analyst* **109**, 923–925.
[147] Koch, T. R. & Akingbe, O. O. (1981) *Clin. Chem.* **27**, 1295–1299.
[148] Hasebe, K., Kakizaki, T. & Yoshida, H. (1987) *Anal. Chem.* **59**, 373–376.
[149] Alexander, P. W. & Shah, M. H. (1980) *Anal. Chem.* **52**, 1896–1900.

[150] Vallon, J. J., Pégon, Y. & Accominotti, M. (1980) *Anal. Chim. Acta* **120**, 65–74.

[151] Florence, T. M. (1979) *J. Electroanal. Chem.* **97**, 237–255.

[152] Florence, T. M. (1978) *Anal. Lett.* **11**, 913–924.

[153] Bilewicz, R. & Kublik, Z. (1981) *Anal. Chim. Acta* **123**, 201–212.

[154] Lunte, C. E., Kissinger, P. T. & Shoup, R. E. (1985) *Anal. Chem.* **57**, 1541–1546.

[155] Musch, G., De Smet, M. & Massart, D. L. (1985) *J. Chromatogr.* **348**, 97–110.

[156] Hart, J. P. (1986) In: Smyth, M. R. & Vos, J. G. (eds) *Electrochemistry, sensors and analysis. Analytical chemistry symposia series.* Vol. 25. Amsterdam, Elsevier p.355–366.

[157] Kissinger, P. T. (1984) In: Kissinger, P. T. & Heineman, W. R. (eds) *Laboratory techniques in electroanalytical chemistry.* New York, M. Dekker p.611–636.

[158] Frenzel, W., Welter, G. & Brätter, P. (1986) In: Smyth, M. R. & Vos, J. G. (eds) *Electrochemistry, sensors and analysis. Analytical chemistry symposia series.* Vol. 25. Amsterdam, Elsevier p.77–82.

[159] Tougas, T. P., Jannetti, J. M. & Collier, W. G. (1985) *Anal. Chem.* **57**, 1377–1381.

[160] Toth, K., Nagy, G., Feher, Z., Horvai, G. & Pungor, E. (1980) *Anal. Chim. Acta* **114**, 45–58.

[161] Pungor, E., Feher, Z., Nagy, G., Toth, K., Horvai, G. & Gratzl, M. (1979) *Anal. Chim. Acta* **109**, 1–24.

[162] Heineman, W. R., Mark, H. B., Jr., Wise, J. A. & Roston, D. A. (1984) In: Kissinger, P. T. & Heineman, W. R. (eds.) *Laboratory techniques in electroanalytical chemistry.* New York, M. Dekker p.499–538.

[163] Wang, J. (1982) *J. Electroanal. Chem.* **139**, 225–232.

[164] Florence, T. M. (1984) *J. Electroanal. Chem.* **168**, 207–218.

[165] Viré, J. C., Kauffmann, J. M. & Patriarche, G. J. (1982) *J. Pharm. Belg.* **37**, 442–462.

[166] Nürnberg, H. W. (1985) *Fres. Z. Anal. Chem.* **320**, 741–743.

[167] Veillon, C., Lewis, S. A., Patterson, K. Y., Wolf, W. R., Harnly, J. M., Versieck, J., Van Ballenberghe, L., Cornelis, R. & O'Haver, T. C. (1985) *Anal. Chem.* **57**, 2106–2109.

[168] Rat, J. C., Burnel, D., Hutin, M. F. & Kolla, F. (1982) *Ann. Fals. Exp. Chim.* **75**, 265–272.

[169] Christian, G. D. & Patriarche, G. J. (1979) *Analyst* **104**, 680–683.

[170] Nagaosa, Y. & Kobayashi, K. (1984) *Talanta* **31**, 593–596.

[171] Boeckx, R. L. (1986) *Anal. Chem.* **58**, 274A–288A.

[172] Adeloju, S. B., Bond, A. M. & Noble, M. L. (1984) *Anal. Chim. Acta* **161**, 303–314.

[173] Oehme, M. & Lund, W. (1979) *Fres. Z. Anal. Chem.* **298**, 260–268.

[174] Batley, G. E. & Farrar, Y. J. (1978) *Anal. Chim. Acta* **99**, 283–292.

[175] De Ruig, W. G. (1981) *Mikrochim. Acta (Wien)* **II**, 199–206.

[176] Morrell, G. & Giridhar, G. (1976) *Clin. Chem.* **22**, 221–224.

[177] Christensen, J. M. & Angelo, H. (1978) *Scand. J. Clin. Lab. Invest.* **38**, 655–662.

[178] Bantjes, N. V. & Hamilton, D. W. (1985) *S–Afr. Tydskr. Chem.* **38**, 21–24.
[179] Dogan, S., Nembrini, G. & Haerdi, W. (1981) *Anal. Chim. Acta* **130**, 385–390.
[180] Ahmad, R. B., Hill, J. O. & Magee, R. J. (1983) *Analyst* **108**, 835–839.
[181] Hutin, M. F., Burnel, D. & Deparis, D. (1982) *Ann. Fals. Exp. Chim.* **75**, 319–325.
[182] Aznarez, J., Vidal, J. C. & Rabadan, J. M. (1986) *Analyst* **111**, 619–624.
[183] Hutin, M. F., Burnel, D., Netter, P., Faure, G. & Gaucher, A. (1981) *Ann. Med. Nancy Est* **20**, 495–498.
[184] Kauffmann, J. M., Patriarche, G. J. & Christian, G. D. (1981) *Anal. Lett.* **14**, 1209–1220.
[185] Bond, A. M. & Reust, J. B. (1984) *Anal. Chim. Acta* **162**, 389–392.
[186] Kinard, J. T. (1977) *Anal. Lett.* **10**, 1147–1161.
[187] Kauffmann, J.M., Montenez, T., Vandenbalck, J. L. & Patriarche, G. J.(1984) *Mikrochim. Acta (Wien)* **I**, 95–105.
[188] Lund, W. & Eriksen, R. (1979) *Anal. Chim. Acta* **107**, 37–46.
[189] Pihlar, B., Valenta, P. & Nürnberg, H. W. (1981) *Fres. Z. Anal. Chem.* **307**, 337–346.
[190] Heinrich, R. & Angerer, J. (1984) *Int. J. Environ. Anal. Chem.* **16**, 305–314.
[191] Adeloju, S. B., Bond, A. M. & Briggs, M. H. (1984) *Anal. Chim. Acta* **164**, 181–194.
[192] Leu, M. & Seiler, H. (1985) *Fres. Z. Anal. Chem.* **321**, 479–482.
[193] Costantini, S., Giordano, R., Rizzica, M. & Benedetti, F. (1985) *Analyst* **110**, 1355–1359.
[194] Davis, P. H., Berlandi, F. J., Dulude, G. R., Griffin, R. M., Matson, W. R. & Zink, E. W. (1978) *J. Amer. Ind. Hyg. Assoc.* **39**, 480–490.
[195] Raptis, S. E., Kaiser, G. & Tölg, G. (1983) *Fres. Z. Anal. Chem.* **316**, 105–123.
[196] Adeloju, S. B., Bond, A. M. & Briggs, M. H. (1984) *Anal. Chem.* **56**, 2397–2401.
[197] Posey, R. S. & Andrews, R. W. (1981) *Anal. Chim. Acta* **124**, 107–112.
[198] Adeloju, S. B., Bond, A. M., Briggs, M. H. & Hughes, H. C. (1983) *Anal.Chem.* **55**, 2076–2082.
[199] Huang, B. X., Zhang, H. C., Pu, G. G., Yin, F., Zheng, S. C. & Yang, H. (1985) *Anal. Lett.* **18**, 279–285.
[200] Adeloju, S. B., Bond, A. M. & Hughes, H. C. (1983) *Anal. Chim. Acta* **148**, 59–69.
[201] Adeloju, S. B., Bond, A. M. & Briggs, M. H. (1985) *Anal. Chem.* **57**, 1386–1390.
[202] Danielsson, L., Jagner, D., Josefson, M. & Westerlund, S. (1981) *Anal. Chim. Acta* **127**, 147–156.
[203] Jagner, D., Danielsson, L. & Aren, K. (1979) *Anal. Chim. Acta* **106**, 15–21.
[204] Jagner, D., Josefson, M. & Westerlund, S. (1981) *Anal. Chim. Acta* **128**, 155–161.
[205] Jagner, D., Josefson, M., Westerlund, S. & Aren, K. (1981) *Anal. Chem.* **53**, 1406–1410.
[206] Jagner, D. & Aren, K. (1982) *Anal. Chim. Acta* **141**, 157–162.

[207] Almestrand, L., Jagner, D. & Renman, L. (1987) *Anal. Chim. Acta* **193**, 71–79.
[208] Frenzel, W. & Brätter, P. (1986) *Anal. Chim. Acta* **179**, 389–398.
[209] Martin, M., Pelletier, M. & Haerdi, W. (1980) *Mitt. Geg. Lebensmittel Unters. Hyg.* **71**, 260–262.
[210] Deangelis, T. P., Bond, R. E., Brooks, E. E. & Heineman, W. R. (1977) *Anal. Chem.* **49**, 1792–1797.
[211] Onar, A. N. & Temizer, A. (1987) *Analyst* **112**, 227–229.
[212] Golimowski, J., Valenta, P., Stoeppler, M. & Nürnberg, H. W. (1979) *Talanta* **26**, 649–656.
[213] Davis, P. H., Dulude, G. R., Griffin, R. M., Matson, W. R. & Zink, E. W. (1978) *Anal. Chem.* **50**, 137–143.
[214] Oehme, M., Lund, W. & Jonsen, J. (1978) *Anal. Chim. Acta* **100**, 389–398.
[215] Chittleborough, G. & Steel, B. J. (1980) *Anal. Chim. Acta* **119**, 235–241.
[216] Simon, J. & Liese, T. (1981) *Fres. Z. Anal. Chem.* **309**, 383–385.
[217] Khandekar, R. N. & Mishra, U. C. (1984) *Fres. Z. Anal. Chem.* **319**, 577–580.
[218] Narres, H. D., Valenta, P. & Nürberg, H. W. (1984) *Fres. Z. Anal. Chem.* **317**, 484–485.
[219] Win Lee, S. & Meranger, J. C. (1981) *Anal. Chem.* **53**, 130–131.
[220] Jagner, D., Josefson, M. & Aren, K. (1982) *Anal. Chim. Acta* **141**, 147–156.

1.2 HIGH-PERFORMANCE LIQUID CHROMATOGRAPHY WITH ELECTROCHEMICAL DETECTION

M. Porthault

Modern or high-performance liquid chromatography (HPLC) has become a very successful analytical tool in recent years. The increasing popularity of this technique is due to its high analytical efficiency, and it is now approaching the level of sophistication achieved in gas chromatography, for which it can be regarded as a true alternative in situations where low vapour pressure and/or thermal instability prohibit the use of gas chromatography. It is not surprising, therefore, that HPLC has found a particularly enthusiastic acceptance in pharmaceutical and clinical analysis. Instrumental developments, have followed, and there are a large number of reliable chromatographic systems on the market. Nevertheless, sample preparation in the analysis of complex samples must be improved, and perfectly stable stationary phases and very sensitive and selective detection must be obtained. Consequently, many research groups are focusing their efforts on these areas.

1 General aspects of HPLC

1.1 Main parameters in HPLC†

In a chromatographic analysis, each separated component 'i' is characterized by its residence time on the column, which corresponds to the retention time t_{R_i}. On a chromatogram giving the elution profile in terms of time, the position of the summit of peak 'i' coincides with t_{R_i}

The characterization of the component retention is also expressed by the capacity factor k_i' according to:

$$k_i' = (t_{R_i} - t_o) \tag{1}$$

where t_0 is the retention time of the unretained species. The k' value range is usually 1 to 10.

The column efficiency of a length L is measured by the number of theoretical plates N. The higher N, the smaller the height equivalent to a theoretical plate H (HETP), since:

$$N = L/H . \tag{2}$$

Using the chromatogram, it is possible to calculate N from the following equation:

$$N = 16 \, (t_R/W)^2 \tag{3}$$

where W is the peak base width. For the component 'i':

† The chromatographic symbols used are summarized in Table 1.

Table 1 — List of chromatographic symbols

α	selectivity factor
H	height equivalent to a theoretical plate (HETP)
k'	capacity factor of solute
k'_i	capacity factor of solute i
L	column length
N	number of theoretical plates of the column
R_s	resolution of two peaks
σ	standard deviation of the Gaussian distribution represented by the chromatographic peak
σ_c	column standard deviation
σ_c^2	column variance
σ_{ec}^2	extra-column variance
σ_{tubing}^2	connection variance
$\sigma_{V_{cell}}^2$	cell volume variance
σ_τ^2	response time variance
t_o	residence time of an unretrained species
t_R	retention time of solute
τ	response time of the detector
u	eluent velocity
V_{cell}	detection cell volume
V_R	retention volume of the solute
W	peak base width
W_i	base width of peak i

$$N_i = 16 \, (t_R/W_i)^2 \qquad (4)$$

where W_i is the base width of peak 'i'.

In a first approximation, a chromatographic peak can be considered as a Gaussian peak; consequently, its width W represents four standard deviations (4σ):

$$W = 4\sigma \qquad (5)$$

The smaller W, the higher N, that is, efficiency.

The separation of two peaks 1 and 2 is mainly characterized by the selectivity α which is deduced from the ratio k'_2/k'_1 ($k'_2 > k'_1$); the first condition for the separation of two solutes is $\alpha > 1$.

This separation is also illustrated by the resolution R_s. In practice, R_s can be calculated from the following equation:

$$R_s = 2(t_{R_2} - t_{R_1})/(W_1 + W_2) \qquad (6)$$

where t_{R_1} and t_{R_2} are th. retention times and W_1 and W_2 the peak base widths of peaks

1 and 2, respectively. R_s presents the advantage of being related to three fundamental parameters: selectivity α, retention, expressed by the capacity factor k', and efficiency, expressed as N. Equation (6) can be written theoretically according to:

$$R_s = \frac{1}{4}\frac{\alpha - 1}{\alpha}\frac{k_2'}{1 + k_2'}N^{1/2} \;.$$

(7)

These three factors, α, k', N, contributing to R_s can be optimized independently. The fundamental contribution, selectivity, can be optimized by changing the nature of the stationary and mobile phases, and the capacity factor k' can be optimized by varying the mobile phase composition. Efficiency can be improved by increasing the column length (see equation (2)), decreasing the particle size of packing material (explained in another section), and by optimizing the flow rate u of the mobile phase:

$$u = l/t_o$$

(8)

The first equation relating plate height to eluent velocity was proposed by Van Deemter *et al.* [1]. Since then, other improved versions have been formulated by several authors [2,3]. However, for purposes of simplicity it is preferable to use the Van Deemter equation as follows:

$$H = A + B/u + Cu$$

(9)

This equation shows the main dynamic contributions to the spreading of the peak through the column during the solute transfer:

- parameter A is an eddy diffusion term independent of flow rate,
- parameter B is a longitudinal diffusion term,
- parameter C is a mass transfer term between mobile and stationary phases and within the phases.

By using packings with particles of small regular size, the A and C terms are diminished; consequently good efficiency can be obtained if the velocity is correctly chosen. In practice, it is important to note that if the diameter of the particles is decreased, the column resistance to eluent flow is increased. A compromise may be necessary. In modern HPLC the mobile phase is delivered by convenient pumping systems under a relatively high pressure (usually between 50 and 300 atm) in order to overcome the column resistance to the eluent. Under these conditions, analyses with short retention times are obtained.

The columns are packed with small particles, usually 5 or 10 μm in diameter, essentially in order to make sufficiently quick transfers between the mobile and stationary phases and consequently to obtain an efficient column at commonly used velocities of mobile phase (near 0.1 cm s^{-1}) and chromatographic peaks with low base width.

Then for the detection problem, note that each peak has a low volume, and if the solute band flows through the detector cell in a short time, some limitations for the detectors must also be respected, as is shown in the following discussion.

1.2 Extra-column peak broadening (or band broadening)

It has been seen that the plate number N (or the height equivalent to a theoretical plate $H = L/N$) is a measurement of the amount of broadening of a solute band as it moves down the column.

Under these conditions, at the bottom of the column, band width W corresponds to four column standard deviations ($W = 4\sigma_c$). In fact, for the recorded peak width, the total contribution to band broadening must be considered, i.e. chromatographic column and extra-column contributions. Because of the additivity of variances (σ^2), these effects are usually discussed in terms of the variances concerning several contributions. Therefore, the total variance characterizing the chromatographic peak σ_{total}^2 is the sum of several contributions according to:

$$\sigma_{total}^2 = \sigma_c^2 + \Sigma\sigma_{ec}^2 \tag{10}$$

where σ_c^2 and $\Sigma\sigma_{ec}^2$ are random variances due to dispersion through the column and out of the column, respectively.

The main extra-column contributions are due to the injection, the connective tubing, and the detector. Each extra-column band broadening is often presented as a fraction of the column band broadening:

$$\sigma_{ec}^2 = \theta^2\sigma_c^2 \ . \tag{11}$$

From equations (3) and (5), we obtain

$$\sigma^2 = t_R^2/N \tag{12}$$

or

$$\sigma^2 = V_R^2/N \tag{13}$$

in time or volume units, respectively. In (13), V_R is the retention volume of the solute, obtained by multiplying t_R by the volumic flow rate of the mobile phase.

By combining the relations (10), (11) and (12), we have

$$N_{obs} = N_c(1 + \Sigma\theta^2) \tag{14}$$

where N_c and N_{obs} are the column plate numbers and the plate numbers measured from the chromatogram, respectively. Consequently, with a phase system and a given column, because of the extra-column effects, a loss of the resolution R_s is observed: the observed resolution $R_{s_{obs}}$ is inferior to the column resolution R_{s_c}.

From equations (7) and (14), we have

$$R_{s_{obs}}/R_{s_c} = (\sigma_c^2/\sigma_{total}^2)^{1/2} \tag{15}$$

$$= (1 + \Sigma\theta^2)^{-1/2} \tag{16}$$

A loss of 5% is acceptable; this implies that $\Sigma\theta^2$ is equal to 0.1.

This conclusion means that the environment of the column, e.g. injector, connective tubing, detector, must be adapted geometrically and electronically (if necessary), if the column performances are not to be destroyed.

The detector contributions are, mainly, connection tubing, detection cell volume V_{cell}, and response time τ: by considering the variances we can write

$$\sigma_{detection}^2 = \sigma_{tubing}^2 + \sigma_{V_{cell}}^2 + \sigma_{\tau}^2 \tag{17}$$

Equations are available to determine these different variances in relation to main significant parameters: tubing length and diameter, V_{cell}, τ . . .

Therefore, by imposing a maximum tolerated resolution loss (i.e. a value for $\Sigma\theta^2$) the suitable sizes of connective tubing, V_{cell}, τ can be evaluated. Some practical appropriate values which comply with these limitations are given in Table 1, with $\theta^2 = 0.1$ and N corresponding to $k' = 1$.

These characteristics of the instrumentation are not always encountered on the market. For strict applications poor results are often obtained with a very good column if insufficient attention is paid to extra-column effects.

1.3 Main modes of separations

The chromatographic behaviour of various solutes depends on theoretical physico-chemical properties and the chosen stationary and mobile phases (see selectivity α and capacity factor k'). Various separation modes result from the variety of stationary phases used in liquid chromatography. Each mode has a predominant mechanism of solute distribution between the two phases used.

In HPLC, four main modes are often distinguished [4]: adsorption, partition, ion-exchange, and size-exclusion chromatography.

Adsorption chromatography can be used for separating non-polar or fairly polar organic molecules. Usually silica is the stationary phase. It is used with an organic mobile phase (e.g. 2,2,4 trimethylpentane-dichloromethane) whose polarity is adjusted. The eluotropic series of various solvents established by Snyder [5] is very useful for adjusting the compostion of the mobile phase. Nowadays, this mode is often substituted by reversed-phase chromatography.

In fact, 'partition chromatography', which initially corresponded to liquid–liquid partition chromatography, where the separation was governed by the difference of solute solubility between the stationary and mobile phases, is now always applied with bonded phases. In this study, applications with apolar bonded phases and polar mobile phases are most often encountered. This mode is usually known as 'reversed-phase chromatography'. The most common method uses alkyl-bonded silica in conjunction with a hydroorganic eluent such as methanol or acetonitrile/water mixtures. The polar molecules are eluted more quickly than weak polar molecules because of the apolar nature of the packing surface.

Ion-exchange chromatography is well known. The stationary phase, called 'ion exchanger', contains ions capable of being exchanged with ionic solutes, and the mobile phase is aqueous. Usually, the separation is controlled by adjusting the two main parameters, i.e. the ionic strength of the eluent and the pH of the buffer solution. In HPLC, ion exchangers are chemically-bonded silicas or HPLC polymeric ion-exchange resins. This mode can only be used for ionizable solutes. It is important to note that it has now become possible to separate solutes by using other modes. Two common possibilities are encountered, both based on the suppression of the ionised state.

Firstly, a simple ionisation retrogradation can be used. In this case, it is sufficient to adjust pH in order to obtain the molecular form of the ionizable solute. The chromatographic problem becomes the same as the separation of polar molecules, which is usually achieved by reversed-phase chromatography. Secondly, the ionized state can be suppressed by adding a convenient counter-ion to the mobile phase. The counter-ion possesses an electric charge opposite to the solute and permits the formation of ion pairs. This second method is usually operated as reversed-phase ion-pair chromatography. The stationary phase is an apolar chemically-bonded silica, and the eluent an aqueous or possibly hydroorganic phase containing the counter-ion.

The first general approach is to consider that the ion pairs are partitioned into the stationary phase and the free ions eluted by the mobile phase. In practice, the main parameters for adjusting the separation are the pH of the mobile phase, the ratio of the organic component to the aqueous component, and the concentration and the nature of the counter-ion.

The fourth mode is size-exclusion chromatography, where the separation is governed by the size of the molecules. The small molecules enter all the porous particles of the stationary phase and are the last species to be eluted, but the other large molecules are excluded from the particles of stationary phase and thus they are eluted before the small molecules. This chromatography mode is interesting for organic polymers and biopolymers.

The main subject of this chapter, HPLC with electrochemical detection, involves ion-exchange chromatography, essentially reversed-phase chromatography with classical molecular solutes or equivalent species after suppression of the ionized state as in ion-pairing chromatography. When using an electrochemical detection mode, aqueous or hydroorganic mobile phases are very convenient.

1.4 Main detection techniques

The four best-known detectors in HPLC are UV–visible absorbance, fluorescence, electrochemical and refractometric detectors.

The UV–visible absorbance detector is still by far the most widely used in liquid chromatography. In this case, when monochromatic UV or visible radiation is used, the absorbance of light due to the presence of solute in the mobile phase obeys Beer's law. At present, there is a wide range of reliable photometric detectors, from single wavelength detectors to UV–visible photodiode array detectors. A good detection limit and a reasonable dynamic range can often be obtained, and, if the mobile phase is conveniently chosen, the response is unaffected by the eluent composition. Nevertheless, not all molecules possess a sufficiently strong UV chromophore for satisfactory detection, and it is necessary to take the incompatibility with certain eluents into account.

The use of fluorescence detectors has become increasingly widespread in HPLC, particularly in biochemical and clinical applications. This mode detects the specific fluorescence energy emitted by the analysed solute after excitation by UV radiation of a well-chosen wavelength. In the adsorptiometry detector range, relatively reliable fluorimetric detectors are available and the fluorimetric detection systems often give useful selectivity in trace analysis work. However, not all molecules are detectable with this detection mode.

The refractive index detector is mainly used in a differential mode. The magnitude of the response is related to the difference in refractive index between the pure solute and the solvent. This detector mode offers the advantage of allowing determination of all molecular types, and it can be considered as a universal detector. However, it is relatively limited in sensitivity, because the refractive index is subject to changes in temperature, pressure, and eluent composition.

Although recent interesting improvements have been realized in the performance of these various detection modes, with a tendency to apply HPLC to increasingly more complex mixtures, particularly in the analysis of biological samples, other improvements and other available detection techniques are still expected. From this viewpoint, the developments of electrochemical detectors are very interesting. At present, the electrochemical detection technique is the fourth well-known mode for monitoring various solutes after HPLC separations.

1.5 Micro-HPLC

At present, 'micro-high-performance liquid chromatography' (micro-HPLC) is HPLC with column diameters of less than 2 mm. This relatively recent development involves several different techniques according to the column characteristics. The main types are micro-HPLC with packed microbore columns and micro-HPLC with open tubular capillary. The latter technique is due mainly to the work of Novotny & Ishii [6] and is still in the research stage. But the first two methods have been dealt with in numerous works, and commercial instrumentation is already available.

The technique with microbore columns where the stainless steel columns generally used have an inner diameter of 1 mm is a direct extension of the conventional technique [7]. The requirements are: pumps with a low flow rate (10 to 100 μl/min is sufficient), detectors with a small volume (less than 1 μl) and a very low value of the term σ_{ec}^2 taking into account the low value of σ_c^2 (Table 2). The main advantages are a low consumption of eluent and a higher sensitivity because of a lower chromatographic dilution. Also good results can be obtained with a lower amount of sample injected. This aspect can be interesting in bioanalysis.

The technique with packed microcolumns devised by Ishii *et al.* [8,9] employs even smaller columns than those mentioned above. Classically, in this technique polytetrafluoroethylene columns have an inner diameter of 0.5 mm and are 10–30 cm long; so they need miniaturized instrumentation for supplying the eluent and for sample injection. Furthermore, the cell volume of the detector must be less than 0.2–0.3 μl.

Table 2 — Appropriate cell volume and response time of various columns

Column	Classical	Microbore	Short
Length (cm)	25	25	3.2
Inner diameter (mm)	4.6	1	4.6
Dead volume (ml)	3.3	0.15	0.4
Linear velocity of mobile phase (cm s^{-1})	0.1	0.1	0.32
Plate number N	15 000	15 000	5000
Column variance σ_c^2 (μl^2)	3000	6	≈ 130
Maximal cell volume V_{cell} (μl)	17	0.8	3.6
Maximum response time τ (s)	1.3	1.3	≈ 0.1

Over the past seven years micro-HPLC has developed significantly, and this is necessarily of interest to analysts and bioanalysts. Nevertheless, the main applications of HPLC are currently carried out with conventional HPLC.

2   General aspects of electrochemical detection techniques in HPLC

In the field of HPLC, electrochemical detection techniques (EDTs) are playing an increasingly larger role. In the last decade, liquid chromatography with EDT has developed from a specialized technique for catecholamines. Many recently-published papers deal with various substances which are electroactive and the choice of commercially available electrochemical detectors for HPLC instrumentation now offers apparatus in this rapidly developing field. The main merits attributed to this technique are its high sensitivity and selectivity; these detectors are also relatively inexpensive.

For the purposes of this technique, the solutes must be electroactive within the working parameters of the detector, and the mobile phase must be adequately conductive. This last condition is easily achieved by using aqueous or hydroorganic mobile phases as in reversed-phase or ion-exchange and ion-pairing chromatography. Fortunately, this favourable situation is often encountered in the field of fundamental and applied biosciences.

In this presentation, electrochemical detection in liquid chromatography will be limited to its most common meaning, that is, to all techniques in which the detection principle is the transfer of electric charges between the monitored substance in the column effluent and an electric conductor referred to as a 'working electrode'.

This transfer of electric charges is due to the oxidation or reduction of the substance to be detected while keeping the working electrode at a constant and well-chosen potential with respect to a reference electrode. The amount of transferred charge corresponds to the current which is measured. The electrochemical reaction involving the equal and opposite transfer of charge at the counter or auxiliary electrode is not taken into consideration in studying the behaviour of the detection cell. The only components flowing through the cell which are detected are those undergoing an electrochemical reaction at the chosen potential for the working electrode.

Usually, the electrochemical detector can be operated in one of two modes, amperometric or coulometric. For the amperometric mode, only a fraction of the analyte is converted electrochemically, and, for the coulometric mode, the analyte is fully converted while flowing through the detector cell. The first technique is by far the most widely used. Other modes are occasionally mentioned, for example potentiometry and conductimetric detection, in which the charge transfer process at the electrodes is secondary. However, these modes are often not mentioned in the HPLC–EDTs association.

Thus, our discussion will deal only with the first two modes mentioned above.

2.1 *Amperometric detectors*

Amperometric measure, at a fixed voltage of the working electrode, the current produced by the redox reaction of the electroactive solute as it passes in front of the electrode. There is a concentration gradient in the immediate region surrounding the electrode surface. The rate of diffusion of solute to the electrode surface through the mobile phase limits the current produced.

The geometrical configuration of the detector cell, the electrodes chosen, and the flow rate of mobile phase all affect the electrochemical detector response. Numerous papers have discussed these problems [10–13], and several theoretical equations for describing the current response have been proposed. Given the different possible assumptions on the flow regime in the cell and different approximate diffusion models, this study has been complicated by the inappropriate use of theory for steady-state amperometric detectors to describe response under liquid chromatography conditions. The recent work by Weber [14], Elbicki *et al.* [15], and Moldoveanu & Anderson [16] is particularly helpful in developing an efficient approach.

Furthermore, as mentioned previously, the combination of the HPLC column and an electrochemical detector makes it necessary to minimize the extra-column peak broadening due to the detector. This constraint means that the detector must satisfy a number of requirements: a low-volume electrochemical cell, connective tubing with low dead volume, and a low response time (see Table 2). To satisfy these requirements, various detector cell types have been proposed [17–19]. For amperometric detection, three cell models with solid working electrodes are encountered: 'thin-layer', 'wall-jet' and 'tubular' cells, and a fourth type with dropping or hanging mercury electrodes.

Among these four types, it was found [15,20] that a wall-jet cell behaves almost like a classical thin-layer cell in numerous applications. The tubular electrode cell has not yet been used in commercially available apparatus. Mercury electrode cells are less frequently employed than the first two types. They can, however, be of interest for reductive reaction.

2.2 *Cells with solid working electrodes*

Solid working electrodes are mainly used in anodic oxidation. Their great advantage is the freedom of choice in the geometry of the cell and electrode. Fig. 1 indicates the

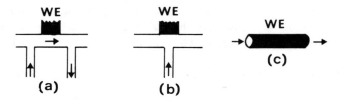

Fig. 1 — Cells with solid working electrodes. WE = working electrode; (a) = thin-layer cell; (b) wall-jet cell; (c) = tubular cell; → = mobile-phase flowing.

schematic configurations of three solid working electrodes (WE) in thin-layer (a), wall-jet (b), and tubular (c) cells, respectively.

In model (a), thin-layer cell, the mobile phase flows parallel to the solid electrode surface. Generally, a rectangular channel is formed.

In model (b), the mobile phase emerges from the inlet as a jet of fluid which impinges perpendicularly onto the electrode surface ('the wall'), then flows radially over it. Since it has been established [15,20] that the mass transfer behaviour can be (approximately) considered as in the normal thin-layer cell, given the appropriate parameters the name 'centrally-injected thin-layer cell' has been suggested for model (b).

In model (c), the eluent flows parallel to the electrode surface.

2.3 Working electrode materials
Carbon paste was made popular for some time by Kissinger *et al.* [17,21]. This material is, however, relatively difficult to handle [22] and is unsuitable for use with non-aqueous solvents and high flow rates. Noble metal electrodes, such as those made of platinum or gold, have also been employed [23]; nevertheless, their use is limited.

Glassy carbon is the most commonly employed material. Variability in performance can be observed because of carbon–oxygen functionalities on the surface and variations in the surface area [24]. Some carbons are more compact at the surface and hence offer a smaller surface area. This compactness depends to a large extent on the manufacturing process. On the other hand, the history of the electrode [25] can influence its behaviour. It is better to apply periodically an electrode-cleaning procedure by mechanical, chemical or electrochemical means, or with a combination of these methods [25]. Despite these problems, glassy carbon appears to be the best material available for working electrodes.

2.4 Noise
The sensitivity of electrochemical detectors, one of their principal adavantages, is a function of several variables. In the case of amperometric detectors, the ratio of produced current to concentration present in the cell depends on electrode area, eluent velocity, and cell dimensions. But the signal-to-noise ratio is the quantity which is of most interest. Noise has recently been studied by Van Rooijen & Poppe [26] and Morgan & Weber [13]. They take into account the various noise sources, the

stability of the reference electrode, impedance changes of the working electrode (due to adsorption–desorption processes and formation of small gas bubbles on the surface, flow pulsations, . . .), and electronic equipment.

Finally, Morgan & Weber [13] offer, for low-frequency noise with glassy carbon electrodes at moderate oxidation potentials, a linear relationship between the noise and the surface area of the A electrode. With A of the order of 10^{-3} cm^2, if i_{total} is the total current through the cell, the noise, characterized by $\sigma^2_{i_{\text{total}}}$, is given by the equation

$$\sigma^2_{i_{\text{total}}} \approx 10^{-20} A^2 + 10^{-26} \tag{18}$$

where i_{total} and A are expressed in amperes and cm^2, respectively. Under these conditions, a peak-to-peak noise of 0.1 nA cm^{-2} should be obtained; experimentally, interesting values in the 0.1–1 nA cm^{-2} range can be found.

Smaller electrode areas give lower noise levels; nevertheless, with microelectrodes ($A < 10^{-3}$ cm^2) the impedance noise contribution increases and with larger electrodes the electronic noise contribution can become dominant.

2.5 Signal
Several equations have been established to describe the electrical current corresponding to the electrochemical reaction at a solid working electrode in amperometric detectors for HPLC [10,11,15,27]. Different results have arisen, essentially from different assumptions on the flow regime in the detector cell and different approximate diffusion models.

Prabhu & Anderson [11] appear to take liquid chromatographic conditions into account in greater detail. However, the discrepancy between theory and experimental results is generally sufficiently low to show up the most suitable equation.

Table 3 shows the main equations describing the current responses for different types of cell geometry, according to the paper from Elbicki et al. [15].

These equations have been respectively proposed for:

- tubular electrode cells with fully developed laminar flow (I),
- thin-layer cells with laminar flow parallel to the electrode surface and rectangular channel (II,III) depending on the state of development of the flow (for a typical channel thin-layer cell, generally equation III applies).
- wall-jet electrode cells (IV,V) (an initial criterion is the ratio of electrode diameter to stream diameter).

All these equations show the influence of geometric parameters and of the flow rate of mobile phase.

An interesting paper [33] was published by Hanekamp & Van Nieuwkerk, attempting to generalize these equations in a form applicable to all types of detector cell. To describe the diffusion phenomena in the detector cell, the modified Reynolds number Re_x ($Re_x = Ul\nu^{-1}$, where l characterizes the length of the electrode) and the Schmidt number Sc ($Sc = \nu D^{-1}$) were introduced. Finally, the limiting current for solid-state electrode detectors was obtained with the following equation:

Table 3 — Current equations in amperometric detectors

Equations		References
(I)	$i = 1.61nFC(DA/r)^{2/3}U^{1/3}$	[28,29]
(II)	$i = 0.68nFCD^{2/3}v^{-1/6}(A/b)^{1/3}U^{1/2}$	[30]
(III)	$i = 1.47nFC(DA/b)^{2/3}U^{1/3}$	[29]
(IV)	$i = 0.903nFCD^{2/3}v^{1/6}A^{3/4}u^{1/2}$	[31]
(V)	$i = 0.898nFCD^{2/3}v^{-5/12}a^{-1/2}A^{3/8}U^{3/4}$	[32]

Symbols:

a	diameter of input conduit (cm)
A	electrode area (cm^2)
b	channel height or thickness (cm)
C	concentration (mM)
D	diffusion coefficient (cm^{2}s^{-1})
F	Faraday
v	kinetic viscosity (cm^{2}s^{-1})
n	electron number
r	radius of tubular electrode
u	velocity (cm s^{-1})
U	volume flow rate (cm^{3}s^{-1})

$$i = knFCD(Sc)^{1/3}Y \, (Re_x)^{\beta} \tag{19}$$

where k is a dimensionless parameter depending on the cell geometry, Y is characteristic of the electrode width, and the exponent β of Re_x depends on the cell geometry as indicated in Table 4.

Table 4 — Parameters k, Y, β of generalized equation for detector

Type of cell	k	Y	β
Tubular	8.0	$l^{1/3}r^{1/3}$	1/3
Thin-layer	0.8	b	1/2
Wall-jet	1.2	a	3/4

l = length of tubular electrode
a, b, r: same meaning as in Table 2

If the sensitivity of the response is considered in terms of electrode surface A, two influences must be examined: the increasing of noise with A and the increasing of the reaction current with A exponent, ranging between 1/3 and 3/4. Also, the highest

sensitivity would be obtained with small electrodes. In conventional conditions, the
optimal electrode area would be $0.2 \, \text{mm}^2$, and a detection limit of $c. \, 5 \times 10^{-11} \, \text{M}$
should appear. On the other hand, it is difficult to claim that the wall-jet design is
more sensitive than thin-layer design or vice versa. The influence of the flow rate of
mobile phase, under certain hydrodynamic conditions, is represented in each
equation in Table 3 by the term U with an exponent β (values of β: see Table 4).
However, if the contribution of the chromatographic peak broadening is also taken
into account in terms of U as was done by Prabhu & Anderson [11], we have a better
approach of the influence of the mobile phase on the current value. The current is
modified by a term which varies with $1/[(1 + k')H^{1/2}]$; H is the HETP which depends
on the flow rate as in equation (9).

Another aspect which is partly interpreted from the previous current equations is
the influence of the mobile phase composition. If the eluent composition varies, the
dielectric properties, the electric conductivity, the viscosity (v) of eluent charge, the
diffusion coefficient (D), and the electrochemical reaction characteristics of the
solute are modified. Also there is a significant problem in using amperometric
detection in gradient elution HPLC. Gunasingham et al. [27] mention a possibility
for minimizing the fluctuation current during a gradient run by using a large volume
wall-jet detector without significantly affecting the extra-column band broadening.

2.6 Cell with dropping mercury electrode

As mentioned previously, most practical amperometric detectors operate with solid
electrodes and respond to anodically oxidizable compounds. However, many other
substances can be reduced electrochemically and thus can be monitored after the
chromatographic column by a polarographic detector where the mercury electrode
permits easy access to a wide cathodic range with a periodically renewed surface.

In spite of its efficiency, this technique has a major disadvantage: it requires the
use of mercury. Nevertheless, this type of cell is encountered on the market. On the
other hand, some models can be easily used both with solid and mercury electrodes
[34]. There have been some technical difficulties in constructing good dropping
mercury detectors, but fortunately, various significant improvements have appeared
such as introducing the horizontal mercury capillary and controlling the drop
expansion with a movable pin [35]. The works of Hanekamp et al. [36], Stulik &
Pacakova [37] are particularly interesting for evaluating the complementary aspect
of this type of detector in connection with solid electrode detectors.

For mercury electrode detectors, several equations have been derived [38-40]
according to whether the fluid flow is parallel (a), opposite (b), or perpendicular (c)
to the electrode. On the other hand, each equation can be simplified for high linear
fluid velocities. Under such conditions, the following equation can be proposed for
the limiting current:

$$i = knFCD^{1/2}U^{1/2} \, (mt)^{1/2} \tag{20}$$

where C, D, U have the same meaning as in the equations for solid electrodes (Table
3), m is the mercury mass flow, t the droptime, and k a constant depending on the
flowing model (a), (b), or (c). The influence of mobile phase flow (term U) and of cell
geometry (term k) can once again be noted.

To minimize peak broadening, a detector with a low-volume cell and a low time constant is required, as mentioned previously. An additional condition is introduced by the droptime t. The constraint $t < 0.2\sigma_c$ [41] can be respected. To optimize the signal-to-noise ratio, it is advantageous to keep a small electrode surface, that is a high drop rate, as has been verified by Hanekamp *et al.* [36,42].

2.7 Coulometric detectors

Early work on coulometric detectors was reviewed by Johnson & Larochelle [43]. Recent developments concerning both amperometric and coulometric detectors have been described by Stulik & Pacakova [37]. At the present time, the most widely-used cells for electrochemical monitoring of the HPLC effluent operate amperometrically. Nevertheless, the use of the coulometric mode where all of the electroactive solute is electrolyzed in the detection cell has certain advantages over amperometric detection. With the coulometric technique, if the number of electrons transferred per molecule is known, calibration is not necessary; the peak areas corresponding to amounts of transferred charge do not depend on the flow rate. A better reproducibility can therefore be expected, and, if the electrode area is large enough, the risk of the electrode response decreasing with time following adsorption phenomena or other surface modifications is not a real problem.

Along with these advantages, there are some limitations. Completion of the electrochemical conversion must be ensured under the conditions used. A large electrode area is necessary. This constraint has often been considered as equivalent to relatively large cell volumes. Therefore, a peak-broadening contribution incompatible with HPLC is encountered. In addition, noise can become excessive [44], and consequently low signal-to-noise ratios are observed. Such would be the case with a thin-layer cell if adapted for coulometric detection using a large electrode area.

But an interesting possibility for a coulometric cell has been proposed, that is, using a porous electrode. Various geometries and materials for constructing porous electrodes have been tested: stacks of metal gauze [45], sintered metals [46], packed beds of conducting particles [47], or reticulated vitreous carbon [48–50]. Kok [49] gives an interesting discussion on the electrochemical conversion efficiency of a porous electrode, by using a simplified approach which consists in regarding the porous electrode as a chromatographic column operating with an irreversibly retained solute (i.e. k' infinite). By considering that the HETP depends on particle diameter and flow velocity (u), Kok deduces the minimum length of a porous electrode necessary for at least 99% efficiency. The experimental results can be satisfactorily compared with these predictions and those calculated by considering the mass transfer by connective diffusion in the electrode pores and axial dispersion according to Newman & Tiedman [51]. Favourable area/volume ratios can be obtained under these conditions, which makes coulometric detectors very competitive with amperometric detectors available for HPLC [48–51].

This conclusion is interesting since coulometric detector cells for HPLC with porous electrodes can be obtained commercially. Nevertheless, it cannot be claimed that this cell gives a lower limit of detection than an amperometric cell. The main advantages of the coulometric cell are certainly that calibration is not necessary and that better reproducibility can be expected.

3   Main improvements in performance

Electrochemical detectors are inexpensive, sensitive, and highly specific detectors for HPLC. Their high level of sensitivity and the fact that their peak-broadening contribution is quite easily minimized make them attractive and efficient for micro-HPLC, as will be seen later. Their selectivity arises from the limited number of electroactive compounds and the ability to adjust the applied potential (see Appendix A2) so that the detector responds to some electroactive compounds and not to others. This feature can be minimized if, as in the case of oxidations for example, a high positive potential is required to detect the main compounds of interest. Selectivity can be further improved by using a differential pulse technique [41,52,53] in which the current is measured before and after the application of the potential pulse (usually from 5 to 100 mV in magnitude). Although this technique is interesting, it attenuates the original simplicity of electrochemical detection as well as its relatively low cost. The use of dual-electrode systems is more attractive. Finally, another interesting development for improving both sensitivity and selectivity is the possibility of adapting an adequate reaction detector.

3.1   Dual-electrode systems

A very simple approach for improving the selectivity of electrochemical detection involves the use of two working electrodes operating simultaneously at different potentials. Several studies have recently been developed for improvement of selectivity [54,59]. By using a thin-layer cell in the amperometric mode, it is relatively easy to adapt two working electrodes with appropriate electronic equipment as 'potentiostats' for controlling the potential of each working electrode. But it is also possible to use a coulometric–amperometric detector. In the amperometric mode, three types of adaptation of the configuration of these two electrodes with respect to the flow axis as shown in Fig. 2 can be distinguished:

- the parallel electrode configuration where the two working electrodes are positioned either next to each other according to model 'a' (parallel-adjacent) or on opposite sides of the channel — model 'c' (parallel-opposed model). In the latter configuration, the spacer which separates the two blocks of the cell should not be so thin that it allows products from one electrode to interfere with those of the other electrode;
- the series configuration where the two electrodes are placed along the flow stream on one side of the channel. The eluent flows successively over the upstream then over the downstream electrode (model 'b').

In each case, two different working potentials can be simultaneously provided, with electrodes of the same or different sizes and, if necessary, made of various materials. In the coulometric–amperometric mode, a special cell is required according to the series configuration where the upstream electrode of a dual amperometric detector is replaced by a coulometric cell [59].

   The parallel configuration used in amperometric mode permits in one chromatographic run, by using one electrode at a positive and one electrode at a negative

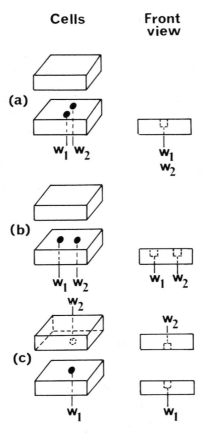

Fig. 2 — Dual electrochemical detector: various types in configuration. W_1, W_2 = working electrodes; (a) parallel–adjacent configuration; (b) series configuration; (c) parallel-opposed configuration.

potential, detection of oxidizable and reducible compounds, as reported for example by Lunte & Kissinger [57] for determination of pterins in biological samples.

More generally, by maintaining the two working electrodes at different applied potentials, parallel configuration is likely to allow better selectivity than when operating with dual wavelength in UV absorbance detection. Favourable cases such as detection of biogenic amines [58,60] or other compounds [56,61] are given in the literature.

Other information can be obtained with this mode to assist in peak identification [62] or the estimation of peak purity [56,63].

In the series configuration, the upstream electrode is generally used as a generator and the downstream electrode as a detector. In this case, the downstream electrode measures the products of the upstream electrode [64,65]. This behaviour can suggest a certain analogy with the fluorescence detector, the choice of working-electrode potentials corresponding to the choice of wavelengths in fluorescence detection. This series mode can improve selectivity when certain compounds overlap chromatographically but differ electrochemically [54].

The coulometric-amperometric configuration is efficient as reported by Haroon *et al.* [65] where the upstream electrode is porous, thus giving sufficiently a large surface. In this example, for vitamin K detection, the first electrode operating as a generator is at a sufficiently negative potential to reduce vitamin K_1 into hydroquinone derivative. At the downstream electrode, designated the 'detector electrode', a positive potential permits oxidation of the K_1 hydroquinone back to vitamin K_1 in the amperometric mode.

With such systems, it is also possible to consider completely and selectively oxidizing a species at the first electrode at a low potential and to make it undetectable by the second electrode. Such a system can separate both components of an interfering peak during a single chromatographic run [59].

3.2   Electrochemical detector miniaturization

Electrochemical detection is known to be a sensitive technique. It is possible to make detection cells with a very small volume. Also, the application of electrochemical detectors is attractive for micro-HPLC where the requirements for minimizing the peak-broadening contribution are very severe. Furthermore, it is important to note with this detection mode that only the volume encompassed by the working-electrode surface and the cell volume in front of this surface affect the elution band as extra-column variance. Another advantage with micro-HPLC comes from the low flow rates which reduce some of the noise produced by the detector. Finally, in the case of microcolumns, syringe pumps are used. Therefore, flow rate reproducibility is excellent and contributes favourably to the quality of the detection.

Modification of a conventional electrochemical cell with a much smaller spacer between the cell halves has been the first approach to using it with microbore columns [66]. Caliguri *et al.* [67] use a 10 μm thick spacer and obtain a limit of detection of 1 pg. Carlsson & Lundstrom [68] construct a more innovative cell, the geometry and materials being selected for low noise characteristics that are due to low differential capacity. These systems are suitable for use with microbore columns (inner diameter: 1.0 mm). There are also applications for electrochemical detection with capillary liquid chromatography, but the size of these detectors must be reduced even more. The works of Jorgenson *et al.* [69,70] are significant in this field. The working electrode is a 9 μm outer diameter graphite fibre which is positioned at the end of the capillary column.

In addition, dual electrochemical detectors having two working electrodes for HPLC are of interest for improving the selectivity of detection. This technique has also been extended by Goto *et al.* [71,72] to micro-HPLC with packed columns of bore size less than 1 mm. The amperometric detector is based on a thin-layer electrochemical cell with a small working electrode suitable for micro-HPLC. The cell has a volume of less than 1 μl, and it has been successfully used for the selective determination of catecholamines in human urine injected directly into a system consisting of a micro-HPLC with an alumina preconcentration micro-column. The same authors have also developed a coulometric parallel–opposed dual electrochemical detector for the selective and sensitive detection of catecholamines in human serum by micro-HPLC [73].

All these studies clearly demonstrate that the development of micro-HPLC is potentially interesting for future applications in bioanalysis.

3.3 New electrodes

It has previously been indicated that the vitreous carbon electrode is the one most widely used in electrochemical detection. Among the different research projects in this field, it is interesting to examine some new approaches to improving even further the performance of this detection mode: either new material for greater specificity and miniaturization, or new technology to amplify the amount of information and therefore to contribute to better selectivity and/or sensitivity.

As regards new materials which could be useful for the working electrode, the work of Kok *et al.* with a copper electrode is interesting [74,75]. After a study on the amperometric behaviour of a copper anode towards complexing agents [74], they successfully applied such an electrode to the HPLC-amperometric detection of amino acids with a flowthrough cell. This method has been shown to be selective, simple, and reliable compared to other frequently-used methods. The detection limits are of the same order of magnitude as with absorbance measurement after ninhydrin derivatization: analysis times are shortened and cost is reduced. Furthermore, this method has also been applied with microbore columns. There is a limitation due to the fact that only certain neutral or alkaline buffers can be used, while the chromatographic separations concerned are often performed with acid phases. This constraint can be overcome by post-column addition of an alkaline solution to an acid mobile phase. Other experiments indicate that this copper electrode can monitor other solutes which behave favourably as ligands.

It has been seen that making very small electrodes is attractive since, theoretically under the classical conditions, noise increases with the electrode area A, while the electrochemical reaction current depends on A exponent (c. $\frac{2}{3}$). Furthermore, the tendency is to use a very small detector cell. An alternative is a microarray electrode. The principle here is to obtain, on the surface of such an electrode, small 'islands' of the electrode material which are separated by an insulating material, although electrical contact exists. Several types of microarray electrodes have been proposed: stacks of carbon fibres glued between glass slides [76,77], graphite particles embedded in Kel-F [78], or reticulated vitreous carbon with pores filled with insulating epoxy [78].

In these cases, comparison of current densities with those at a normal glassy-carbon electrode shows that non-linear diffusion is the predominant mode of mass transport to the electrode surface [79]; the signal is therefore enhanced. Furthermore, a characteristic property of non-linear diffusion is that it is not influenced by convection in the solution. Therefore, flow fluctuations have less influence.

The depletion of the electroactive species near the electrode surface due to the electrochemical reaction, which limits the current with large area electrodes, is partially attenuated when the eluent flows over the insulating zones.

The analytical signal for those analyte molecules whose electrochemical reactions are limited by the mass transfer, can be enhanced relative to the interfering background signal by reducing the active area in order to minimize the response for electrochemical reactions that are limited by the rate of electron transfer (for example, solvent oxidation) or other surface processes.

Finally, in all cases, significant improvement of the signal-to-noise ratio is obtained.

Carbon fibre electrodes have also been used for electrochemical detection in the

case of open tubular liquid chromatography, but here the response is quasi coulometric [80].

Besides this present extension, it is interesting to note that the concept and
applications of multi-electrode coulometric detectors for liquid chromatography
have been introduced. This concept is similar to the introduction of a photodiode
array detector in liquid chromatography detection. A recent study [81] analysing
neurotransmitters as well as their precursor metabolites and associated compounds
in various tissues, is a very interesting example of this development. The above
mentioned study deals essentially with the three working electrode systems; however, results are also given for array cells with six to fifteen sensors and different
potential settings. Adopting this promising concept of an electrochemical cell with
numerous individual sensors, De Abreu & Purdy [82] report the construction of a 32
gold electrode array thin-layer flow cell with a new technology for electrode
manufacture; in addition, the connector is an etched copper circuit board that
provides electrical contact to individual electrodes of the array.

In conclusion, this technique deserves to be followed up in the near future.

3.4 Reaction detectors

In the search for more sensitive and selective detection in liquid chromatography,
post-column reaction systems offer interesting possibilities for improvements. In this
technique, a convenient reagent is generally introduced into the liquid chromatographic effluent at a constant rate, and this gives a more easily detectable product
after reaction with the analyte. Following this principle, much research has been
carried out in the last ten years. The technique is better known as post-column
chemical derivatisation in liquid chromatography. Frei & Lawrence [83,84] deal in
detail with its numerous possibilities. Two other recent reviews [85,86] provide more
briefly the main promising avenues of investigation. The initial aim was to try to
expand the range of applications of the main conventional liquid chromatography
detectors by adapting suitable derivatization techniques. Also many publications
report the introduction of good chromophore or fluorophore groups to the analytes
by adding an appropriate reagent and carrying out the reaction in an on-line post-
column reactor just before the detector. Progressively, this method has attracted a
great deal of attention, and it is now generally accepted and used with different
detection modes such as UV, fluorescence, or electrochemical detectors. The main
requirements — appropriate medium, reproducibility, sufficiently low band-
broadening contribution, excess reagent which does not interfere with the signal of
the new product — can be adequately controlled.

According to the time of the reaction, three different types of reactors can be
distinguished:

- helically-coiled open tubular reactor, convenient for fast reactions up to about
  30 s;
- bed reactor (or 'packed-bed reactor'), available for intermediate reaction times
  ranging from 0.5 to 4.0 min,
- helically-coiled tubular reactor with segmented streams, useful for slower reactions (up to 20 min).

For each type of reactor, the peak-broadening contribution has been evaluated in terms of the main reactor parameters and the residence time of the reacting species in the reactor [85,86]. This approach has certainly been very valuable in ensuring the present development of this technique, which has also been applied to micro-HPLC [87]. Nevertheless, there are still some shortcomings which can be minimized if not eliminated. Indeed, the need for post-column addition of reagents implies a reagent pump and possibly some mixing and dilution problems. In the case of coupling a reactor with an electrochemical detector, besides the more widely-applied addition of reagent produced off-line, a second version, which is the most interesting with this detection mode, involves producing the reagent on-line, that is, in the column effluent itself: the shortcomings mentioned previously are eliminated. The reagent is produced from a precursor of the reagent in the mobile phase, and after reaction with the analyte, the excess of reagent is monitored with an amperometric detector.

Takata & Muto [88] were the first to bring up the possibility of electrochemical generation of the reagent. Later, King & Kissinger [89] demonstrated that such a system can be used satisfactorily with electrochemically-generated bromine for monitoring some phenols and unsaturated compounds.

In the last four years, Frei *et al.* [86] have particularly developed this approach with pumpless reaction units involving electrochemical reagent production followed by detection of the derivatives or related products by electrochemical or other modes of detection. At present, the most typical successful examples are given by the on-line microcoulometric production of bromine or iodine from potassium bromide or iodide dissolved in the mobile phase. This technique leads to significant results with suitable groups in compounds such as unsaturated organics [89], phenolics [89], methoxysubstituted aromatics or phenolic ethers (important moieties in many compounds of pharmaceutical interest) [90], thioethers (also of pharmaceutical importance) [91], and thiol derivatives [92]. In the last example, that is, detection of N-acetyl cysteine with on-line-generated iodine [92], since iodine has a lower standard redox potential than bromine, improved selectivity compared to the bromine system is obtained.

King & Kissinger [89] used two commercial thin-layer cells, the first, after a slight modification, as a generator of bromine and the second as an amperometric detector of the bromine excess with a coiled tubing reactor between them. Kok *et al.* [90] adapted a home-made coulometric cell for on-line production of reagent. Indeed, the reagent should not diffuse into the counter-electrode compartment to prevent its reconversion or interference with other counter-electrode products. Also, the working electrode of the generator must be physically separated from the counter-electrode, by a membrane, for example. This concept of a pumpless reactor unit placed before the detector itself offers a very interesting potential for improving selectivity and sensitivity in liquid chromatography detection [86].

Indeed, a certain variety of pumpless reactor unit can be conceived as a solid-phase reactor where the solid is a support loaded with a chemical reagent or, ideally, a catalytic material generating an electrochemically detectable product.

There are several studies where the reactor unit is a simple photochemical reactor compatible with a very low peak-broadening contribution, changing the analyte into an electroactive species [93,96].

Another possibility is to obtain, with an electrochemical reactor unit, a species detectable under better conditions in a detection unit placed in series. This combination can be favourably illustrated by the recent work of Langenberg & Tjaden [97] and Kok *et al* [98]. The authors applied this direct electrochemical conversion of a compound to obtain more favourably fluorescent products in the case of vitamin K_1 and phenothiazines, respectively.

These recent examples illustrate the potential improvements which are possible with on-line post-column reaction detectors in HPLC. An electrochemical unit, associated with a pumpless reactor unit, is especially promising.

4 Applications

Since the adoption of electrochemical detectors for HPLC over fifteen years ago, their use has increased enormously. There are over one thousand reports on their use in the literature. Indeed, when the analyte is electroactive, this detection mode can be highly sensitive and selective, and it is now certain, in the case of small, readily-oxidized molecules, that electrochemical detection is often superior to the more generally applicable and reliable UV detectors. During its first decade the technique essentially developed from a specialized method for catecholamine analysis, while more recently it has proved to be a very useful technique in the liquid chromato-graphic determination of a variety of pharmaceuticals, their metabolites, and other compounds in biological samples. Nevertheless, the technique should continue to be developed. Only a few examples of applications are given here to illustrate its potential and to point out some of the difficulties of determining whether a given compound is electroactive or not in such a potential range. Generally, experiments are necessary to confirm or disprove the successful adaptation of liquid chromato-graphy/electrochemical detection.

A first approach to evaluating the possibilities of obtaining a satisfactory analysis with HPLC-EDTs can consist in examining the diagram of Fleet & Little [18] (Fig. 3). This diagram gives the approximate ranges of potential where various well-known types of compounds become electroactive; the oxidizable derivatives are on the right and the reducible ones on the left. The first group includes compounds such as phenolic and amine derivatives and heterocyclic molecules with nitrogen (e.g. quinoleine, indole, azine) and sulfur atoms (such as phenothiazine). The second series involves the molecules with nitro, diazo, and quinone groups which are easily reducible and have carbonyl functions which are more difficult to modify electro-chemically. Furthermore, if, by analogy, published potential values are chosen as guides for adjusting the working-electrode potential, it is necessary to consider the chosen reference electrode (standard calomel or Ag/AgCl electrode, for example) in order to interpret this value. Unfortunately, the authors do not always give this indication. The nature of the electrode and the mobile phase composition, pH included, can also influence the appropriate potential value to be used. Finally, the kinetic characteristics of the electrochemical reaction depend on the nature of the electrode, the analyte structure, and the medium (mobile phase). Consequently,

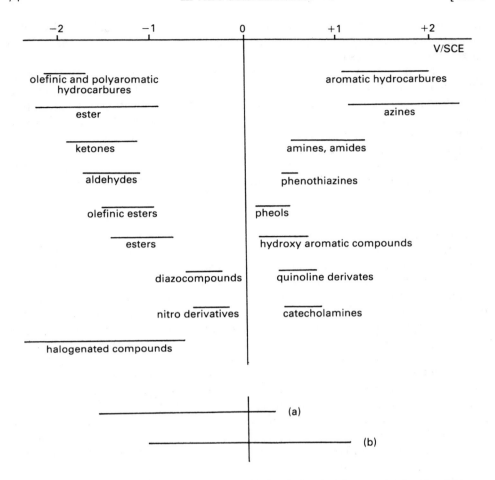

Fig. 3 — Appropriate ranges of potential where various compounds become electroactive. (a)= useful field of dropping mercury electrode; (b) = useful field of glassy carbon electrode.

some functional groups may be present in the analyte, but the necessary conditions for obtaining a Faradic current at the chosen potential according to the previous guide table, may be unfavourable.

Fig. 3 gives the approximate appropriate potential range for using the mercury electrode (a) or the glassy-carbon electrode (b). This diagram clearly indicates that the mercury electrode is particularly suitable for monitoring analytes by reduction electrochemical reaction. For the most widely-used oxidation detection mode, it is possible to reach higher potentials with glassy carbon than with mercury electrodes.

Now some applications in the main fields mentioned in Fig. 3 will be discussed.

4.1 Catecholamines and derivatives
Natural catecholamines are catechol compounds with the following basic structure:

HO⌇⟨⟩R   Dopamine: R = CH₂–CH₂–NH₂
HO⌇   Epinephrine: R = CHOH–CH₂–NHCH₃
          Norepinephrine: R = CHOH–CH₂–CH₂

They are easily oxidizable with a mechanism depending on the pH. In weak acid and neutral medium, the electroactivity is mainly caused by the oxidation of the two hydroxyl groups (3,4) followed by a particular mechanism leading to a ring closure of the amine side chain [99]. In the 3.4–6.0 pH range, by maintaining the working-electrode potential at $+0.8$ V vs SCE, all these catecholamines are satisfactorily monitored after liquid chromatographic separation. Many studies have been published in the last fifteen years: for instance, for simultaneous determinations of catecholamines and their metabolites [100], for their determination in various tissues [101], in brain parts [102,103], in plasma with the inherent purification problems [104], and in urine [105,106]. The determination of free derivatives before and after enzymatic deconjugation has been studied [107,108]. Some works have focused on the analysis of trace amounts [109]. A new metabolite of dopamine has been determined and reported by Ito *et al.* [110].

Improved selectivity through detection by dual electrodes in series has been described by several authors [57,60]. Determination of catecholamines has been performed by using small bore columns and an electrochemical detector with a single electrode [111] or a dual electrode system [73]. New electrodes such as pyrolytic carbon fibre electrodes [112] and others [81] have been tested with these compounds. The above survey presents only a small part of the various studies of the catecholamines and their metabolites. Their successful performance has certainly contributed to the present extent of EDTs in HPLC.

4.2   Tryptophan metabolites
As there are structural similarities between tryptophan metabolites and catecholamines, the suitability of electrochemical oxidation for the detection of the former compounds has been [113] and still is being studied by different authors [114]. Indeed, it was found [113] that by operating at 1.0 V it is possible to detect only those tryptophan metabolites with a 5-hydroxy group on the phenol ring. This group is the site of oxidation. This property allows the selective detection of metabolites in the tryptophan hydroxylase pathway. Under operation, at higher potentials, other tryptophan metabolites also become detectable; it is likely that these indole derivatives are oxidizable because of the presence of ring nitrogen. The work done on disturbance of serotonine metabolism is particularly interesting.

4.3   Isoquinolic and morphinic alkaloids
Among these compounds, the precursor of which is tyrosin, Peterson *et al.* [115] pointed out the structural similarity between the opiate alkaloids and catecholamines. He reported the possibility of electrochemical detection of alkaloids, such as morphine oxymorphone, maltrexone, maloxone, and morphine and reported that, under the working conditions adopted, the modification of catechol structure is not favourable for convenient detection. Furthermore, codein, where the phenol func-

tion is present as a methoxy group, and pentazoxine, which does not possess a furan group, are not detected. Following Peterson's work, other articles were published [116,118], showing that the HPLC-EDT offers interesting possibilities under various conditions. Moreover, by using post-column reactor detection, consisting mainly of bromine electrogeneration, this detection mode can satisfatorily extend to this class of compounds [90].

4.4   Common analgesics: ring-substituted phenols, substituted naphthols
Acetaminophen (I), phenacetin (II) and its metabolites, substituted phenols, as well as naporoxen can illustrate [119] the interest of an electrochemical detector for such derivatives. (I) and (II) are substituted aminophenols, which are detectable by oxidation, and it is normal to find that (I) is oxidized at lower potentials than (II) which shows ethylsubstitution of the phenol. Even though other ring-substituted phenols, such as salicylic derivatives, require higher anodic potentials in order to be monitored, they are nevertheless well within the analytically useful range.

Finally, as shown in the chart by Fleet & Little in Fig. 3, HPLC-EDT is appropriate for common analgesics analysis.

4.5   Phenothiazines
Quantitative determination of phenothiazine drugs in the plasma of patients under treatment is relatively difficult because of the low concentrations and the numerous metabolites present. Methods based on the HPLC-EDT association are relatively simple and easy to adapt to particular applications.

Several papers have confirmed the advantages of this method [120–122], although some derivatives are not detected. However, it is interesting to point out the greater scope of possible applications if a reaction detector is used as an electrogenerating reagent [49]. Besides the possibility of electrogenerating bromine and electrochemically quantifying excess bromine, another original extension is that used for the determination of thioridazine in plasma [62]. The method consists in using electrogenerated bromine which oxidizes the thioridazine into a highly fluorescent product. Finally, coupling of an electrochemical reactor with a fluorescence reactor gives a very satisfactory example of detection of this drug in plasma.

4.6   Characterization of quinonoid pterin derivatives
A method employing liquid chromatography/electrochemistry appears to be very interesting for the determination of a variety of pterin species and their oxidation states [57], these compounds having a tendency to be electroactive. Here, a significant example of HPLC-EDT can be presented with the simultaneous determination of the oxidized and reduced forms of biopterin in tissue samples.

Biopterin is, under its reduced form (Fig. 4), a cofactor along with the enzymes, both contributing to the catalysis of the rate-limiting reactions in the synthesis of both the catecholamines and serotonin.

By using a dual-electrode amperometric detector in the parallel–adjacent configuration and choosing appropriate potentials, Lunte & Kissinger [123] selectively detected at the first upstream electrode the reduced forms of biopterin, that is 7,8 dihydrobiopterin alone or both 7,8 dihydrobiopterin and 5,6,7,8 tetrahydrobiopterin, and the oxidized form at the downstream electrode.

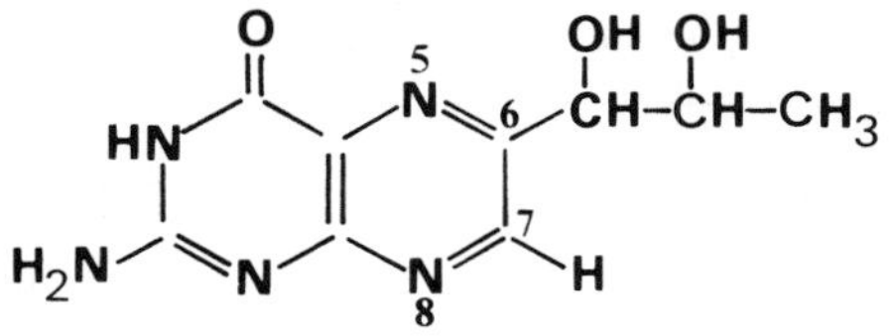

$$O \quad OH \quad OH$$

HN

H$_2$N

Fig. 4 — Biopterin: reduced form.

4.7 *Thiol and disulfide compounds*

There are several thiol and disulfide compounds such as glutathione, cysteine, and corresponding disulfides which are of biological or pharmaceutical interest. Their determination by liquid chromatography has received particular attention. This technique offers the possibility of performing the simultaneous quantitative determination of both oxidized and reduced thiols by using amperometric detection. Given the nature of the functional groups involved here, the use of a mercury electrode is better than that of a glassy-carbon electrode, which necessitates a higher potential. Besides various published studies with mercury electrodes or mercury film on gold surface electrodes [124,125], a very interesting development is offered with the use of a dual solid-electrode series configuration in the case of a thin-layer cell. Under these conditions, disulfides can be reduced at the upstream electrode at -1.0 V vs Ag/AgCl and the resulting thiols and any naturally present thiols are oxidized at the downstream electrode at $+0.15$ V vs Ag/AgCl [64,126].

While a number of specific detection techniques have been developed for the determination of thiols and disulfides, the problem is more difficult with organic compounds containing thioether groups which are generally detected by UV absorbance. Indeed, in this case, for electrochemical detection, the mercury electrode does not give a satisfactory response and, with other electrodes, oxidation potentials are high and electrode passivation and adsorption phenomena appear. However, as has been seen previously, it is interesting to use a reactor detection system. Kok *et al.* [91] have presented a very good application of this post-column detection mode. By electrogenerating bromine, thioethers are then oxidized (mostly into sulfoxides or sulfones) in the microreactor and the bromine excess is amperometrically detected. The authors show that in several cases the detection is more sensitive and more selective than UV absorbance detection. These same authors have recently published another application by electrogenerating iodine in order to determine *N*-acetylcysteine [92].

4.8 *Other applications*

In Table 5 are summarized various applications of electrochemical detection for illustrating the potential of this HPLC-detection mode along with previously detailed examples.

5 Conclusion

The association of HPLC and EDTs is at present too widely used for all aspects of its applications to be covered in this study.

Table 5 — Various applications of HPLC-EDTs in clinical and pharmaceutical analysis

Compounds	References	Comments
Catecholamines	[59,60,73,81,99–112]	see comments in the text
Tryptophan metabolites	[113,114]	id.
Alkaloids	[90,115,118]	id.
Analgesics	[119]	ring substituted phenols, substituted naphthols (see comments in the text)
Other derivatives	[127–129]	
Amine derivatives	[130]	aromatic amine carcinogen
Phenothiazines	[49,98,120–122]	see comments in the text
Quinonoid pterins	[57,123]	id.
Benzodiazepines	[131,132]	id.
Other pharmaceutical derivatives	[133–136]	nitrogen heterocyclic compounds and various substitutions
Uric acid	[137]	
Thyroid hormone	[138]	— tyrosine derivatives — dual coulometric-ampero-metric cell
Organometallic complexes	[139]	platinum-derived cancerchemotherapy agent
Bile acids	[140,141]	
Vitamin K_1	[65,97,142]	
Ascorbic acid	[143,144]	
Amino acids	[74,75]	copper electrode
Sulfur derivatives	[64,91,92, 124–126,145,146]	thiols, disulfides, thioethers (see comments in the text)

No mention has been made here, for example, of the use of a pre-column system for increasing the column life and which is necessary for obtaining a weak background when complex samples have to be analysed. Information about such systems is to be found easily from suppliers of HPLC. Clean-up of samples has not been discussed; the procedures to be followed with biological samples for various analytical techniques are well known [147]. Moreover, many applications mentioned above are published with various practical details for sample preparation. The use of a pre-column for trace enrichment [148] has also been voluntarily neglected.

Nevertheless, one of the objects of the present contribution is to demonstrate that often HPLC-EDTs can be easily used for biological sample analysis by staff who are not necessarily confirmed chromatographers and electrochemists. The development of this technique will continue to offer very interesting improvements in detectors as well as in chromatographic columns.

A1 Key words
This study is completed with key words and their references:

Thin-layer cell [17–19]
Wall-jet cell [15,20]
Mercury-electrode cell [34–40]
Current signal [10,11,15,27,33]
Noise [25,26]
Flow-rate influence [11,14]
Oxygen effect [42]
Elution gradient [27]
Coulometric detector [37,43–50]
Pulse technique [41,52,53]
Dual-electrode cell [54–60,63,64]
Array cell [81,82]
Reaction detector [49,84,85]
Post-column derivatisation [83,86,96]
Reagent electrogeneration [25,89–92]
Small bore columns [66–68,72]

A2 Choice of an appropriate potential for the working electrode
This potential depends on mobile phase composition, working-electrode material, and obviously on the solute. It is possible to obtain the chromatographic determination of the potential value by using amperometric detection, with a limited injected amount within the same sample volume, the working-electrode potential being increased step by step after each injection.

By starting from a sufficiently reductive potential with an oxidizable solute, a stable background line is first observed, then the signal is increased at each injection, and finally the signal magnitude remains constant (Fig. 5).

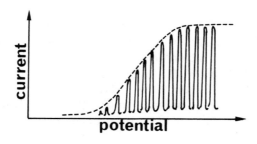

Fig. 5 — Choice of an appropriate potential for the working electrode (for each step, $\Delta E = 20\,\mathrm{mV}$).

During this last step, the current is limited by the solute diffusion. The potential at the beginning of the plate obtained can be chosen for the working electrode. It is to be noted that, if the detector must be operated on a reduction zone, the current–

potential curve is established by starting from a satisfactorily oxidative potential and changing the working electrode potential step by step as previously, but this time decreasing the potential.

References

[1] Van Deemter, J. J., Zuiderweg, F. J. & Klinkenberg, A. A. (1956) *Chem. Eng. Sci.* **5**, 271.

[2] Done, J. N., Kennedy, G. J. & Knox, J. H. (1973) In: Perry, S. G. (ed.) *Gas chromatography* 1972. London, Applied Science p. 145.

[3] Grushka, E., Snyder, L. R. & Knox, J. H. (1975) *J. Chromatogr. Sci.* **13**,25–37.

[4] Pryde, A. & Gilbert, M. T. (1979) In: *Application of high performance liquid chromatography*. London, Chapman & Hall.

[5] Snyder, L. R. (1968) *Principles of adsorption chromatography*. New York, Marcel Dekker.

[6] Novotny, M. V. & Ishii, D. (1985) *Microcolumn separations. J. Chromatogr. Library*. Vol. 30. New York, Elsevier.

[7] Kucera, P. (1984) *Microcolumn high performance liquid chromatography. J. Chromatogr. Library*. Vol. 28. New York, Elsevier.

[8] Ishii, D., Asai, K., Hibi, K. Jonokuchi, T. & Nagata, M. (1977) *J. Chromatogr.* **144**, 157–168.

[9] Ishii, D. & Takeuchi, T. (1980) *J. Chromatogr. Sci.* **24**, 462–472.

[10] Meschi, P. L. & Johnson, D. C. (1981) *Anal. Chim. Acta.* **124**, 303–314.

[11] Prabhu, S. & Anderson, J. L. (1987) *Anal. Chem.* **59**, 157–162.

[12] Stulik, K. & Pacakova, V. (1981) *J. Chromatogr.* **213**, 41–46.

[13] Morgan, D. M. & Weber, S. G. (1984) *Anal. Chem.* **56**, 2560–2567.

[14] Weber, S. G. (1983) *J. Electroanal. Chem. Interfacial Electrochem.* **145**, 1–7.

[15] Elbicki, J. M., Morgan, D. M. & Weber, S. G. (1984) *Anal. Chem.* **56**, 978–988.

[16] Moldoveanu, S. & Anderson, J. L. (1984) *J. Electroanal. Chem. Interfacial Electrochem.* **175**, 67–77.

[17] Kissinger, P. T., Refshauge, C., Dreiling, R. & Adams, R. N. (1973) *Anal. Lett.* **6**, 465–477.

[18] Fleet, B. & Little, C. J. (1974) *J. Chromatogr. Sci.* **12**, 747–752.

[19] Lieberman, S. H. & Zirino, A. (1974) *Anal. Chem.* **46**, 20–23.

[20] Dalhuysen, A. J., Van Der Meer, Th. H., Hoogerdoorn, C. J., Hoogvllet, J. C. & Van Bennekom, W. P. (1985) *J. Electroanal. Chem.* **182**, 295–313.

[21] Kissinger, P. T. (1977) *Anal. Chem.* **49**, 447A-456A.

[22] Stulik, K., Pacakova, V. & Starkova, B. (1981) *J. Chromatogr.* **213**, 41–46.

[23] Reuzig, F. R. & Frank, J. (1981) J. Chromatogr. **218**, 615–620.

[24] Gunasingham, H. & Fleet, B. (1982) *Analyst (London)* **10**, 896–902.

[25] Van Rooijen, H. W. & Poppe, H. (1981) *Anal. Chim. Acta* **130**, 9–22.

[26] Van Rooijen, H. W. & Poppe, H. (1983) *J. Liquid Chromatogr.* **6**, 2231–2254.

[27] Gunasingham, H., Tay, B. T., Ang, K. P. & Kon, L. L. (1984) *J. Chromatogr.* **285**, 103–114.

[28] Blaedel, W. J. & Klatt, L. N. (1966) *Anal. Chem.* **38**, 879–883.

[29] Matsuda, H. (1967) *J. Electroanal. Chem.* **15**, 325.

[30] Levich, V. G. (1962) *Physicochemical hydrodynamics*. 2nd ed. Englewood Cliffs N. J., Prentice Hall.

[31] Matsuda, H. (1967) *J. Electroanal. Chem.* **15**, 109.

[32] Yamada, J. & Matsuda, H. (1973) *J. Electroanal. Chem.* **44**, 189.

[33] Hanekamp, H. B. & Van Nieuwkerk, H. J. (1980) *Anal. Chim. Acta* **121**, 13.

[34] Beauchamp, M., Boinay, P., Fombon, J. J., Tacussel, J., Breant, M., Georges, J., Porthault, M. & Vittori, O. (1981) *J. Chromatogr.* **204**, 123–140.

[35] Michel, L. & Zatka, A. (1979) *Anal. Chim. Acta* **105**, 109–117.

[36] Hanekamp, H. B., Bos, P., Brinkman, U. A. Th. & Frei, R. W. (1979) *Fresenius Z. Anal. Chem.* **297**, 404–410.

[37] Stulik, K. & Pacakova, V. (1981) *J. Chromatogr.* **208**, 269–278.

[38] Kimla, A. & Strafelda, F. (1964) *Collect. Czech. Chem. Commun.* **29**, 2913.

[39] Strafelda, F. & Kimla, A. (1965) *Collect. Czech. Chem. Commun.* **30**, 3606.

[40] Okinaka, Y. & Kollthoff, I. M. (1957) *J. Electroanal. Chem.* **73**, 3326.

[41] Hanekamp, H. B. (1981) Thesis. University of Amsterdam, p. 14.

[42] Hanekamp, H. B., Voogt, W. H., Bos, P. & Frei, R. W. (1980) *Anal. Chim. Acta* **118**, 81–86.

[43] Johnson, D. C. & Larochelle, J. (1973) *Talenta*, **20**, 959–971.

[44] Roosendaal, E. M. M. & Poppe, H. (1984) *Anal. Chim. Acta* **158**, 323–333.

[45] Devynck, J., Pique, R. & Delarue, G. (1975) *Analusis* **3**, 417–423.

[46] Kennel, J. V. & Bard, A. J. (1984) *J. Electroanal. Chem.* **54**, 47.

[47 Davenport, R. J. & Johnson, C. D. (1974) *Anal. Chem.* **46**, 1971–1978.

[48] Curran, D. J. & Tougas, T. P. (1984) *Anal. Chem.* **56**, 672–678.

[49] Kok, W. Th. (1986) Thesis. University of Amsterdam, p. 22–24.

[50] Blaedel, W. J. & Wang, J. (1979) *Anal. Chem.* **51**, 799–802.

[51] Newman, J. S. & Tiedman, W. (1977) In: Tobias, C.N. & Gerischer, H. (eds) *Advance in electrochemistry and electrochemical engineering*. Vol. 11. New York, Wiley Interscience p. 355–430.

[52] Swartzfager, D. G. (1976) *Anal. Chem.* **48**, 2189–2192.

[53] Hanekamp, H. B., Voogt, W. H. & Bos, P. (1980) *Anal. Chim. Acta* **118**, 73–79.

[54] Blank, C. L. (1976) *J. Chromatogr.* **117**, 35–46.

[55] Fenn, R. J., Siggia, S. & Curran, D. J. (1978) *Anal. Chem.* **50**, 1067–1073.

[56] Roston, B. K., Shoup, R. E. & Kissinger, P. T. (1982) *Anal. Chem.* **54**, 1417A–1434A.

[57] Lunte, C. E. & Kissinger, P. T. (1983) *Anal. Chem.* **55**, 1458–1462.

[58] Mayer, G. S. & Shoup, R. E., (1983) *J. Chromatogr.* **225**, 533–544.

[59] Schieffer, G. W. (1980) *Anal. Chem.* **52**, 1994–1999.

[60] Elchisak, M. A. (1983) *J. Chromatogr.* **264**, 119–127.

[61] Radzik, D. M., Brodert, J. S. & Kissinger, P. T. (1984) *Anal. Chem.* **56**, 2927–2931.

[62] Roston, D. A. & Kissinger, P. T. (1982) *Anal. Chem.* **54**, 1798–1802.

[63] Lunte, L. E., Kissinger, P. T. & Shoup, R. E. (1985) *Anal. Chem.* **57**, 1541–1544.

[64] Allison, L. A. & Shoup, R. E. (1983) *Anal. Chem.* **55**, 8–12.

[65] Haroon, Y., Schubert, C. A. W. & Hanschka, P. V. (1984) *J. Chromatogr. Sci.* **22**, 89–93.

[66] Sagliano, N. & Hartwick, R. A. (1986) *J. Chromatogr. Sci.* **24**, 506–512.
[67] Caliguri, E. J., Lapella, P., Bottari, L. & Mefford, Z. N. (1985) *Anal. Chem.* **57**, 2423–2425.
[68] Carlsson, A. & Lundstrom, K. (1985) *J. Chromatogr.* **350**, 169–178.
[69] Saint Claire, R. L. & Jorgenson, J. W. (1985) *J. Chromatogr. Sci.* **23**, 186–191.
[70] White, J. G., Saint Claire, R. L. & Jorgenson, J. W. (1986) *Anal.Chem.* **58**, 293–298.
[71] Goto, M., Nakamura, T. & Ishii, D. (1981) *J. Chromatogr.* **226**, 33–42.
[72] Goto, M., Sakurai, E. & Ishii, D. (1982) *J. Chromatogr.* **238**, 357–366.
[73] Goto, M., Sakurai, E. & Ishii, D. (1983) *J. Liquid Chromatogr.* **6**, 1907–1925.
[74] Kok, W. Th., Hanekamp, H. B., Bos, F. & Frei, R. W. (1982) *Anal. Chim. Acta* **142**, 31–46.
[75] Kok, W. Th., Brinkman, U. A. Th. & Frei, R. W. (1983) *J. Chromatogr.* **256**, 17–26.
[76] Caudill, W. L., Howell, J. O. & Wightman, R. M. (1982) *Anal. Chem.* **54**, 2532–2535.
[77] Wightman, R. M. (1981) *Anal. Chem.* **53**, 1125A-1134A.
[78] Weisshaar, D. E., Tallman, D. E. & Anderson, J. L. (1978) *Anal. Chem.* **50**, 1051-1056.
[79] Sleszynski, N., Osteryoung, J. & Carter, M. (1984) *Anal. Chem.* **56**, 130–135.
[80] Knecht, L. A., Guthrie, E. J. & Jorgenson, J. W. (1984) *Anal. Chem.* **56**, 479–482.
[81] Matson, W. R., Langlais, P., Volicer, L., Gamache, P. H., Bird, E. & Mark, A. K. (1984) *Clin. Med.* **30**, 1477–1488.
[82] De Abreu, M. & Purdy, W. C. (1987) *Anal. Chem.* **59**, 204–206.
[83] Lawrence, J. F. & Frei, R. W., (1976) *Chemical derivatization in liquid chromatography*. Amsterdam, Elsevier.
[84] Frei, R. W. & Lawrence, J. F. (1981) In: *Reaction in analytical chemistry*. Vol. 1. New York, Plenum Press.
[85] Frei, R. W. (1982) *Chromatographia* **15**, 161–166.
[86] Frei, R. W., Jansen, H. & Brinkman, U. A. Th. (1985) *Anal. Chem.* **57**, 1529A-1539A.
[87] Jansen, H., Brinkman, U. A. Th. & Frei, R. W. (1985) *J. Chromatogr. Sci.* **23**, 279–284.
[88] Takata, Y. & Muto, G. (1973) *Anal. Chem.* **45**, 1864–1868.
[89] King, W. P. & Kissinger, P. T. (1980) *Clin. Chem.* **26**, 1484–1491.
[90] Kok, W. Th., Brinkman, U. A. Th. & Frei, R. W. (1984) *Anal. Chim. Acta* **162**, 19–32.
[91] Kok, W. Th., Halvax, J. J., Voogt, W. H., Brinkman, U. A. Th. & Frei, R. W.(1987) *Anal. Chem.* (in press).
[92] Kok, W. Th., Halvax, J. J. & Frei, R. W. (1987) *J. Chromatogr.* (in press).
[93] Lefevere, M. F., Frei, R. W., Scholten, A. H. M. T. & Brinkman, U. A. Th. (1982) *Chromatographia* **15**, 459–467.
[94] Danyalian, A. (1987) Thesis. University of Lyon p. 24–26 and p. 95–101.
[95] Danyalian, A. (1987) Thesis. University of Lyon p. 49–77.
[96] Magnet, C. (1982) Thesis. University of Lyon.
[97] Langenberg, J. P. & Tjaden, U. R. (1984) *J. Chromatogr.* **289**, 377–385.

[98] Kok, W. Th., Voogt, W. H., Brinkman, U. A. Th. & Frei, R. W. (1987) *J. Chromatogr.* (in press).

[99] Fenn, R. J., Siggia, S. & Curran, D. J. (1978) *Anal. Chem.* **50**, 1067–1073.

[100] Magnusson, O., Nilsson, L. B. & Winsterlund, D. (1980) *J. Chromatogr.* **221**, 237–247.

[101] Wagner, J., Palferyman, M. & Zraika, M. (1979) *J. Chromatogr.* **164**, 41.

[102] Keller, R., Oke, A., Mefford, I. & Adams, R. N. (1976) *Life Sci.* **19**, 995–1004.

[103] Felice, L. J., Felice, J. D. & Kissinger, P. T. (1978) *J. Neurochem.* **31**, 1461.

[104] Ong, H., Capet-Antonini, F., Yamaguchi, N. & Lamontagne, D. (1982) *J. Chromatogr.* **223**, 97–105.

[105] Felice, L. F. & Kissinger, P. T. (1976) *Anal. Chem.* **48**, 794–796.

[106] Riggin, R. M. & Kissinger, P. T. (1976) *Anal. Chem.* **49**, 2109–2111.

[107] Santagostino, G., Frattini, P., Schinelli, S., Cucchi, M. L. & Corona, G.L. (1982) *J. Chromatogr.* **233**, 89–95.

[108] El Ckisak, M. A. & Carlson, J. H. (1982) *J. Chromatogr.* **233**, 79–88.

[109] Westerink, B. H. C. (1983) *J. Liquid Chromatogr.* **6**, 2337–2351.

[110] Ito, S., Kato, T. & Fujita, K. (1986) *Chromatographia* **21**, 645–647.

[111] Church, W. H. & Justice, J. B. (1987) *Anal. Chem.* **59**, 712–716.

[112] Gonon, G., Fombarlet, G. M., Buda, M. J. & Pujol, J. F., (1981) *Anal. Chem.* **53**, 1386–1389.

[113] Richards, D. A. (1979) *J. Chromatogr.* **175**, 293–299.

[114] Trouvin, J. H., Gardier, A., El Gemayel, G. & Jacquot, C. (1987) *J. Liquid Chromatogr.* **10**, 121–136.

[115] Peterson, R. G., Rumack, B. H., Sulliman, J. B. & Makowsky, A. (1980) *J. Chromatogr.* **188**, 420–425.

[116] Lake, L. L., Di Fazio, C. A. & Duckworth, E. N. (1982) *J. Chromatogr.* **233**, 410–416.

[117] Wilson, T. D. (1984) *J. Chromatogr.* **301** , 39–46.

[118] Janicot, J. L., Caude, M. & Rosset, R. (1986) *Analusis* **14**, 441–455.

[119] Meinsma, D. A., Radzik, D. M. & Kissinger, P. T. (1983) *J. Liquid Chromatogr.* **6**, 2311–2335.

[120] Mc Kay, G. Cooper, J. K., Midha, K. K. Hall, K. & Hawes, E. M. (1982) *J. Chromatogr.* **233**, 417–422.

[121] Curry, S. H., Brown, E. A., Hu, O. Y. P. & Perrin, J. H. (1982) *J.Chromatogr.* **231**, 361–376.

[122] Stoll, A. L., Baldessarini, R. J., Cohen, B. M. & Finnklestein, S. P. (1984) *J. Chromatogr.* **307**, 457.

[123] Lunte, C. E.& Kissinger, P. T. (1983) *J. Liquid Chromatogr.* **6**, 1863–1872.

[124] Rabenstein, D. L. & Saetre, R. (1977) *Anal. Chem.* **49**, 1036–1039.

[125] Bergstrom, R. F., Kay, D.R. & Wagner, J. G. (1981) *J. Chromatogr.* **222**, 445–452.

[126] Allison, L. A., Keddington, J., & Shoup, R. E. (1983) *J. Liquid Chromatogr.* **6**, 1785–1798.

[127] Riggin, R. M., Rau, L. D., Alcorn, R. L. & Kissinger, P. T. (1974) *Anal. Chem.* **7**, 791.

[128] Wehmeyer, K. R., Doyle, M. J., Wright, D. S., Eggers, H. M., Halsall, H. B.

& Heineman, W. R. (1983) *J. Liquid Chromatogr.* **6**, 2141–2156.

[129] Tesarova, E., Pacakova, V. & Stulik, K. (1987) *Chromatographia* **23**,102–108.

[130] Mefford, I., Keller, R. W., Adams, R. N., Sternson, L. A., & Yllo, M. S.(1977) *Anal. Chem.* **49**, 683.

[131] Hanekamp, H. B., Voogt, W. H., Bos, P. & Frei, R. W. (1980) *J. Liquid chromatogr.* **3**, 1205.

[132] Hanekamp, H. B., Voogt, W. H., Frei, R. W. & Boos, P. (1981) *Anal. Chem.* **53**, 1362–1365.

✓ [133] Miner, D. J., Skibic, M. J. & Bopp, R. J., (1983) *J. Liquid Chromatogr.* **6**, 2209–2230.

✓[134] Nordholm, L. & Dalgrard, L. (1982) *J. Chromatogr.* **233**, 427–431.

✓ [135] Suckow, R. F. (1983) *J. Liquid Chromatogr.* **6**, 2195–2208.

✓[136] Magic, S. E. (1976) *J. Chromatogr.* **129**, 73.

[137] Riggin, R. M., Rau, L., Alcorn, R. L. & Kissinger, P. T. (1974) *Anal. Lett.* **7**, 791.

[138] Hepler, B. R. & Purdy, W. C. (1983) *J. Liquid Chromatogr.* **6**, 2275–2310.

[139] Ding, X. D. & Krull, T. S., (1983) *J. Liquid Chromatogr.* **6**, 2173–2193.

[140] Kemula, W. & Kutner, W. (1981) *J. Chromatogr.* **204**, 131–134.

[141] Kutner, W., Debowsky, J. & Kemula, W. (1980) *J. Chromatogr.* **191**, 47–60.

[142] Brunt, K., Bruins, C. H. P. & Doornbos, D. A. (1981) *Anal. Chim. Acta* **125**, 85.

[143] Pachla, L. A. & Kissinger, P. T. (1976) *Anal. Chem.* **48**, 364–367.

[144] Strohl, A. N. & Curran, D. J. (1979) *Anal. Chem.* **51**, 1045–1049.

[145] Inoue, K., Otake, T. & Kioguku, K. (1982) *Yakagaku Zasshi* **102**, 1041–1048.

[146] Iriyama, K., Masahiko, Y. & Iwamoto, T. (1986) *J. Liquid Chromatogr.* **9**, 2955–2968.

[147] Farinotti, R. & Dauphin, A. (1982) *J. Pharm. Clin.* **1**, 257–286.

[148] Nielen, M. W. F., Koordes, R. C. A., Frei, R. W. & Brinkman, U. A. Th. (1985) *J. Chromatogr.* **330**, 113–119.

1.3 ION-SELECTIVE ELECTRODES IN CLINICAL CHEMISTRY

P. Nabet

The use of ion-selective electrodes in clinical chemistry, either on single or on pluriparametric apparatus, is developing rapidly, and more and more frequently these systems are being adopted instead of atomic absorption or emission photometry (the flame photometry of clinical chemists). The reasons for this have often been discussed:

- While flame photometry, with improvement of the apparatus, has become a very reliable technique, with intra- and inter-assay variation coefficients of less than 0.4%, it needs high-quality inflammable gases (such as propane or acetylene) and compressed air. This sets real problems, and, for instance, in our hospital laboratory, the security staff has required that gases should not be located in the vicinity of the patients' area.
- Flame photometry also needs very precise and frequent controls, since it is well known that small variations in the sample atomization flow can lead to errors (variations in sample viscosity, in gas pressures, in the diameter of pipes, and the distribution system . . .).
- The functioning cost of this apparatus is rather high.
- Finally, the sample is destroyed, which could be a problem during some pathological states or for pediatric samples.

Apparently, the use of selective electrodes does not give rise to the same problems:

- The apparatus needs only ordinary electric current, can be moved anywhere, and placed at the patient's bedside or in the doctor's consulting room.
- The apparatus can be set up on standby position and used sample by sample or in series.
- Theoretically, the functioning cost is low.
- The sample can remain intact after the measurement, and can be used for other determinations.
- The technology allows continuous measurements in total blood, with the possibility of monitoring the studied parameter.

At the present time, only a few electrolytes, i.e. Na^+, K^+, Ca^{2+}, Cl^-, and HCO_3^-, have been subjected to routine clinical analyses and ISE-based instrumental determination. Some proposals are being examined for other parameters such as Li^+, F^-, NH_3^+, or Mg^{2+}, but they have not yet reached routine applications.

However, although the use of selective electrodes seems simple, it raises a number of questions (some of which are already resolved, or on the way to being resolved) which show that this technology is more complicated than it seems, as Kissel points out [1]:

> When one draws up a long list of possible ISE error sources, the simplicity and utility of the technology can sometimes appear diminished. The summary point simply is that one must pay attention to the analytical situation even with

ISEs. Too often in the author's experience those unfamiliar with ISE technology assume a simple 'dip and read' of the sensor into any unknown matrix will produce satisfactory results.

1. Definitions and nomenclature

1.1 Principles of ISEs[†]

The principle of ISEs belongs to the general technique of potentiometry. A potential difference is measured between one half-cell formed by the selective electrode itself and a second half-cell formed by a reference electrode. As a result:

(i) There are several interfaces between these two half-cells. Very often these interfaces may be the location of disturbances during the measurements and the cause of errors, especially with the complex biological solutions. Among the different interfaces, those situated between the external reference electrode and the sample have caught the attention of the apparatus designers. The use of salt bridges of equitransferent ions at high concentrations (e.g. KCl 3 M) leads to local precipitations of proteins, which in time will disturb the measurements. Several types of geometry for these junctions have been used on the commercially available analysers [2], which could have an effect on the recorded values [3]. To minimize these effects, the constructors have often chosen a flow-free junction with permanent circulation of a calibration sample. The assembly proposed by the SMAC apparatus from TECHNICON (Tarrytow, USA) seems to be of interest; it allows rinsing and regeneration of the reference electrode (Fig. 1).

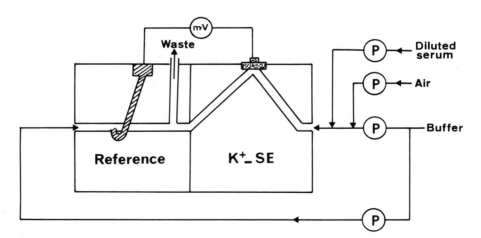

Fig. 1 — Flowthrough electrode proposed by TECHNICON. According to Andersen [4]
P = pump, mV = millivoltmeter.

The general problem of reference electrodes and their assembly has been discussed by Brezinski [5].

(ii) A current is present in the circuit. It is always very small, and mostly zero-current

† Additional information on the principles and set-up of ISEs can be found in Chapter 2(§2.2) and Chapter 3(§3.1).

conditions are observed, so high-quality measuring instruments are required to avoid errors during the measurements. In particular, the input impedance must be very high, i.e. c. 10^{12}–10^{15} Ω. Consequently, the problems of insulation and screening are important.

So the routine use of ISEs depends not only on the progress in the field of electrodes but also on the development of electronics (screening, integrated electronics) and computing (frequent and even permanent standardization, complex calculations).

1.2   Ion activity
Consider a classical potentiometric measuring cell. The complete measuring unit is generally as follows:

External reference electrode (e.g. Hg/Hg$_2$Cl$_2$/KCl)	Sample under test	Ion-selective membrane	Internal reference electrode (e.g. HCl/AgCl/Ag)

Selective electrode unit

This is the classical assembly for a glass electrode (H$^+$- and Na$^+$-selective) with liquid junctions.
Direct contact simplifies this assembly:

External reference electrode	Sample under test	Ion-selective membrane	Conducting metal

Selective electrode unit

The potential of these electrodes is related to the activity of ions examined by the Nernst relationship:

$$E = E_o + \frac{2.3\,RT}{z_i F}\log(a_i) \tag{1}$$

where E_o is equal to the sum of all junction potentials, considered to be constant, and z_i and a_i are, respectively, the charge number and the activity of the ion i for which the electrode is selective.

However, the electrode selectivity for one ion is never perfect, and there are always interfering ions. Thus, the Nernst equation has been completed by Nicolsky and Eisenman:

$$E = E_o + \frac{2.3\,RT}{z_i F}\log\left/\left[a_i + \sum_{j=1}^{n}(K_{i,j}^{Pot}\, a_j^z\, z_i/z_j) \right]\right. \tag{2}$$

where z_i and a_i are respectively the charge and the activity of the primary ion to be measured, z_j and a_j are, respectively, the charge and the activity of the interfering ion, and $K_{i,j}^{Pot}$ is the selectivity coefficient of the electrode for the primary ion i vs interfering ion j.

The lower the selectivity coefficient, the higher the selectivity of the electrode.

The activity of an ion is related to its concentration by the following equation:

$$a_i = C_i \gamma_i \tag{3}$$

in which C_i is the concentration of ion i and γ_i is its coefficient of ionic activity.

The coefficient of ionic activity depends on the temperature and the ionic strength of the medium. Thus it is necessary to maintain the temperature as constant as possible and to try to work at a stable ionic strength for both samples and standards. These conditions are fairly well respected for determinations in serums or blood since the composition of these fluids is relatively stable (ionic strength ranging between 0.15 and 0.16 M). However, this is less true in urines (ionic strength between 0.1 to 1 M), which have to be diluted by a non-interfering ionic-strength buffer.

The most recent discrete clinical instruments based on ISEs allow Na^+ and K^+ to be determined in total blood as well as in serums or urines. Measurements in urines can be performed simply by switching the instrument to the position 'urines', in which case the urine sample is diluted by the ionic-strength buffer. In Figs 2 and 3, determinations of Na^+ and K^+ in patients' serums and urines by two commercial analysers are compared to those obtained with a flame photometry device. The dispersion of the experimental values is in fact less for urines than for serums. This is probably due to the mode of standardization, the dilution of urine samples by the buffer, and perhaps an electronic correction.

Variations in ionic strength can have a greater or lesser influence on the measurements according to the nature of the measured ion and to the required precision. For the measurement of blood pH, for instance, the standardization is usually made with aqueous solutions whose ionic strength is c. 0.1 M; the ionic strength of blood is 0.16 M, which leads to a discrepancy of 0.01 to 0.02 pH unit, when the instrument itself is capable of a precision of 0.0025 pH unit. Although the true value of blood pH is unobtainable, pH determinations in blood samples are in practice considered as clinically significant since measurements are always performed under the same conditions. Obviously, the problem is quite different for the analyst who needs accuracy and precision.

1.3   Notions of direct potentiometry and indirect potentiometry

The requirement for sample dilution, previously reported for urines, leads to the widely discussed notions of direct or indirect potentiometry.

In agreement with Kissel [1, p. 76–78], direct potentiometry may be defined in electrochemistry as the analysis method whereby both standard and test samples are processed under the same conditions, the test samples being either neat, diluted with saline, or neutral buffers, or supplemented with a solution decomplexing the cations to be determined; measurements are generally performed in the linear-response range of the electrode, each sample giving only one electrical signal. In indirect

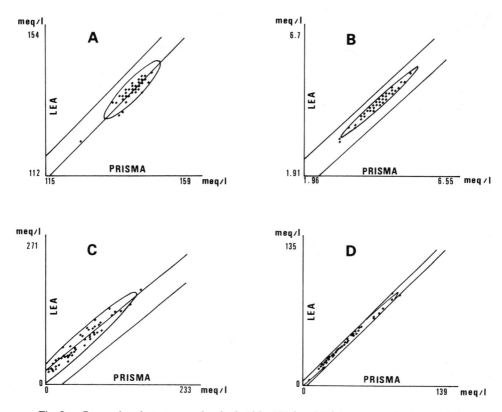

Fig. 2 — Comparison between results obtained for NA$^+$ and K$^+$ in serum and urines with the LANGE analyser (ISE) and the PRISMA analyser (flame photometry).
(coefficient $r = 0.912$)
A: Na$^+$ in serum. Correlation line: $Y = 1.01\,X - 5.3$ (correlation coefficient $r = 0.912$)
B: K$^+$ in serum. Correlation line: $Y = 0.98\,X + 0.17$ ($r = 0.99$)
C: Na$^+$ in urine. Correlation line: $Y = 1.1\,X + 19.47$ ($r = 0.975$)
D: K$^+$ in urine. Correlation line: $Y = 0.97\,X + 1.38$ ($r = 0.997$)

potentiometry, the change in the electrode potential is measured after addition to the tested sample of one or several reagents. This indirect method permits the determination of an ion activity by using an electrode which is not specific for this ion (one can say that indirect potentiometry is to electrochemistry what complexometry is to absorptiometry).

These two definitions are clear and unambiguous. In clinical chemistry, unfortunately, a widespread troublesome practice consists in characterizing direct potentiometry by the use of electrodes in undiluted biological fluids (blood, serum, or urine) and indirect potentiometry by the use of diluted samples. This common practice is unsound, even if the dilution is able to unmask ions, as in the case of calcium ions. Clear definitions are most important since techniques and clinical chemistry apparatus are now available which make use of the technology of indirect potentiometry to measure ions for which sufficiently selective electrodes do not yet exist: determi-

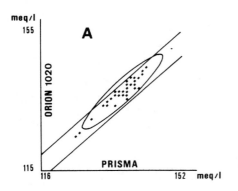

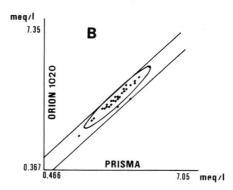

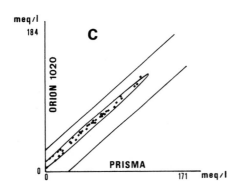

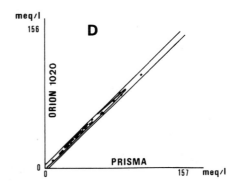

Fig. 3 — Comparison between results obtained for NA$^+$ and K$^+$ in serum and urines with the
ORION 1020 analyser (ISE) and the PRISMA analyser (flame photometry).
A: Na$^+$ in serum. Correlation line: $Y = 1.07\,X - 9.09$ ($r = 0.938$)
B: K$^+$ in serum. Correlation line: $Y = 1.02\,X + 0.05$ ($r = 0.969$)
C: Na$^+$ in urine. Correlation line: $Y = 0.98\,X + 7.61$ ($r = 0.996$)
D: K$^+$ in urine. Correlation line: $Y = 0.99\,X + 0.83$ ($r = 0.999$)

nation of fluorine ion [6], ISE-based assays of non-ionic parameters such as
haemoglobin, haematocrit, and haemolysis [7]. This technology will probably be
more and more widely applied, and ORION (Cambridge, USA) have recently
presented their Model 960, which could allow the use of indirect potentiometry
through multiple incremental techniques (KAP^{TM}) or GRAN function techniques
(GAP^{TM}) [8].

1.4 Plasmatic or seric water
It is worth recalling that ISEs are sensitive to ion activities, the ions to be determined
being free in an aqueous phase in most of the cases encountered in clinical chemistry.
To evaluate the concentration of a given ion, several parameters must be taken into
account, such as the ionic strength of the sample, the speciation of ions due to pH
effects or to natural ligands (e.g. proteins), or the presence (or absence) of

substances occupying an inert volume, for example lipids or proteins found in serum, plasma, or total blood. This last case is particularly important in clinical chemistry for measurements involving serum or plasma samples: between 40 to 120 g/l of proteins and 4 to 200 g/l of lipids can be found in serum or plasma.

Clinical chemistry and physiopathology have for a long time taken their arguments from the data obtained by emission photometry (flame photometry) or atomic absorption photometry (in particular for NA^+ and K^+). These techniques refer the values to the total volume of the sample, i.e. water, proteins, and lipids; the sample is, moreover, largely diluted before analysis (1/100 to 1/200), which minimizes the effects of inert substances in the total volume of the sample, and also the effects of electric junctions. When an ion activity (in an aqueous phase) is determined with an ISE in a greatly diluted sample, the effects of inert substances are minimized, and the values obtained will obviously be very close to those obtained by flame photometry for the same diluted sample. On the contrary, the comparison of the results obtained in undiluted normal serum by potentiometry and by flame photometry shows that the electrochemical technique gives, on average, values 7% higher than those displayed by flame photometry. This difference could be due to the fact that in normal serum, water and inert substances occupy 93% and 7% of the volume, respectively. However, this explanation seems to be too simple since clinical studies on multiple-myeloma cases [9] and measurements of true plasmatic water volumes show that only part of this 7% discrepancy is attributable to the relative volumes of water and of inert substances [10].

Clinical chemistry remains quite divided on the choice of the best method to be used: diluted or undiluted samples.

Two arguments plead in favour of the diluted-sample procedure:

• The dilution allows the ionic strength of the medium as well as the junction phenomena to be fully controlled, ensuring better standard conditions. This is particularly true when aqueous primary solutions of the ionic species to be determined are used for electrode calibration.
• Up to now, physicians have had to consider data obtained by flame photometry, which are very similar to those obtained by ISE measurements in diluted samples.

Two arguments can also be retained in favour of the undiluted-sample technique:

• The first lies in the simplicity of the technique itself (direct procedure avoiding sample handling, allowing measurements in whole blood and at the patient's bedside).
• The second is of medical interest, as illustrated by the following report. In 1980, Frier *et al.* [10] described four cases of children presenting severe acidocetosis due to a non-equilibrated diabetes mellitus and consequently a very important hyperlipaemia. For one of the cases, flame photometry gave a Na^+ value of 86 meq/l and a K^+ value of 2.7 meq/l. Despite the discovery of a discrepancy between the value of osmolality (415 mmol/kg) and the apparent values of plasma electrolytes, no connection was made at that time between hyperlipaemia (> 210 g/l) and the low values of electrolytes, and a therapy by intravenous 9%-NaCl infusion was started. Twelve hours after the child's admission, the

apparent values of electrolytes were 113 meq/l for Na^+ and 2.7 meq/l for K^+. The real values in the aqueous phase were at least 224 meq/l for Na^+ and 4.8 meq/l for K^+, the osmolality being 486 mmol/kg: an intracellular dehydration probably followed, and death occurred around the 33rd hour.

It is evident that ISE measurements in undiluted serum samples from this child would not have displayed an apparent hyponatremia which led to the therapeutic error. It is also certain that such a case of acute hyperlipaemia is rarely encountered in pathology. However, a number of papers and reviews have been devoted to this problem of plasma water deplacement by lipids or proteins [2,9,14–16]: it appears that the effects of hyperproteinaemia differ from those of hyperlipidaemia since proteins interact with water whereas ions and lipids do not.

The differences (5 to 7%) between values given by ISEs in undiluted samples and values obtained by flame photometry or potentiometry in diluted samples can be minimized for normal or nearly normal serums. For this, it is possible to make use of the correction proposed by Waugh [16], which can be introduced directly into the computer section of the commercial analyser, or else the analyser can be standardized with serums of human origin which behave like patients' samples.

For pathological samples (hyperlipaemia, hyperproteinaemia), however, the solutions are not so simple: it is a dangerous illusion to make systematic corrections, in particular when the electrodes (assays on diluted samples) are part of a pluriparametric device where protides and lipids are determined at the same time as ions.

Some authors recommend the use of potentiometry in diluted samples to fit in with the results given by flame photometry and, in the case of hyponatremia, they also recommend verification that the serum is not opalescent (which is the sign of hyperlipaemia) or does not present a hyperproteinaemia. In addition to the problems related to the multiplication of assays and to possible errors of the technical staff, one must remember that there are high-level hyperlipaemias with absolutely clear serums.

For other authors, it is better to measure the real activity of the ion in the aqueous phase of the blood or serum, and to base medical argumentation on these data which are closer to the physiological reality.

While this discussion is possible for ions such as Na^+, K^+, Mg^+, or Li^+, the determination of Ca^{2+} requires the absolute integrity of the biological sample.

We advocate the use of potentiometry in undiluted samples, even in whole blood if possible, and as near as possible to the patient's bedside.

2 ISE characteristics

2.1 Selectivity coefficient

Principles for determination of the selectivity coefficient of an electrode for an ion i vs an interfering ion j, that is, the term $K_{i,j}^{Pot}$ in the Nicolsky–Eisenman equation (2), are reviewed in a recent publication [12].

This determination can be performed by several methods, the two most frequently used being the separate-solution and the mixed-solution methods. The latter is recommended by the IUPAC. The curve relating E to $\log(a_i)$ is set out by using a series of solutions having a constant activity of the interfering ion (a_j) and increasing activities of the primary ion (a_i). The curve obtained shows two asymptotes, one of which is parallel to the X axis, that is, $E = f(a_j)$, the other representing the Nernstian theoretical response of the ISE, that is, $E = f(a_i)$ (Fig. 4). The intersection of the two

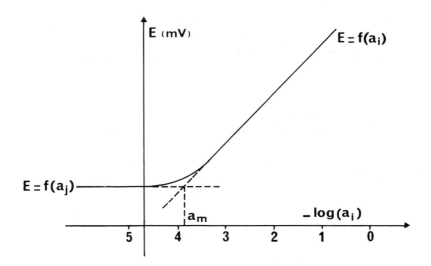

Fig. 4 — Measurement of selectivity coefficient by the mixed-solution method.

asymptotes gives a_m, the practical detection limit of the electrode, from which the selectivity coefficient is calculated: $K_{i,j}^{Pot} = a_m/a_j$.

2.2 Precision
From the Nernst equation for an ISE (1), it appears that the change in activity a_i of a primary ion i is approximately equal to $(4z)\%$ per mV at 25°C, and this independently of the selectivity or the activity range, so if a precision of 1% is needed, the e.m.f. must be determined to 0.25 mV for a monovalent ion (0.125 mV for a divalent ion). At present, this is the limit attainable by ISEs in direct potentiometry.

2.3 Time constants of ISEs
Three time constants can be defined:

- the response time of the electrode, that is, the time after which the electrode attains a new equilibrium potential when a change in the measured ion activity has taken place (from some milliseconds to several minutes);
- the slow shift of the e.m.f. for a solution of stable activity (some minutes to several hours);
- the lifetime of the electrode itself, that is, the period during which the electrode can be used satisfactorily (some months to several years).

The electrode response time is mainly determined by diffusion phenomena into the solution and into the selective membrane. As illustrated by Fig. 5, a sharp injection into the electrode and a stop-flow measurement improve the response time when using flowthrough electrodes.

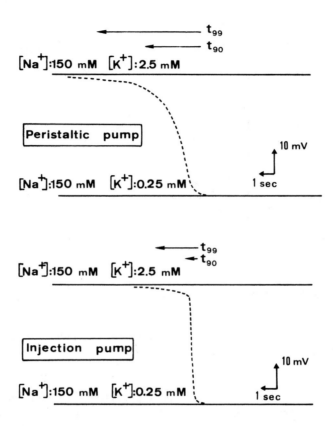

Fig. 5 — Response time of a flow-through K$^+$-selective electrode as a function of liquid injection.
t_{90} and t_{99} = times required for the electrode potential to reach respectively 90% and 99% of its final value

The e.m.f. drift could be the consequence of several factors such as the solubilization of the carrier included in the membrane or modifications in the membrane hydration: thus, direct-contact Na$^+$- and K$^+$-selective electrodes involving a neutral carrier included in silicone recover their original slope, specificity, and response time characteristics if they are kept for a few days under vacuum, in a hot, anhydrous atmosphere (Table 1).

2.4 Accuracy and/or fidelity

Besides the role, emphasized above, played by the ionic strength, the pH and the presence of proteins or lipids in the accuracy and the fidelity of ISE measurements, it

Table 1 — Action of the hydratation state on the functioning of an Na^+ selective electrode (direct contact electrode. ETH 227 in PVC + NPOE)

Days after fabrication	$\dfrac{1}{\text{selectivity}}$ $1/K^{Pot}_{Na^+,K^+}$	Response time(s)	Slope in mixed solutions $[K^+]=5\times10^{-3}$ M $[Na^+]$variable		
			$10^{-1}\to1\times$ 10^{-3} M	$1\to10\times$ 10^{-3} M	$10\to100\times$ 10^{-3} M
4	64	10	36	45	48
9	15	20	32	42	49
10	6	30	36	45	48
15	7	30	36	47	46
16†	87	< 10	40	48	40

† After several hours in anhydrous atmosphere at 37°C under partial vacuum.

must also be remembered that ISEs are very sensitive to interferences of species adsorbing on the electrode surface (for example surfactants or proteins), to ions of opposite charge which can hinder the permselectivity (for example hydrophobic anions for electrodes with neutral-carrier membranes). Finally, the important role of the reference electrode must not be forgotten. These are phenomena that each ISE-user must be aware of and always keep in mind.

3 Main types of ISEs used in clinical chemistry
The ISEs mainly used in clinical chemistry are currently those measuring H^+ (pH electrodes), Na^+, K^+, Ca^{2+}, and Cl^- ions, and they are incorporated into commercial instruments. They are either glass electrodes (H^+, Na^+) or electrodes with liquid or semiliquid membranes (Na^+, K^+, Ca^{2+}, Cl^-).

A number of reviews have been published on the use of these ISEs in clinical chemistry [18–26], to which the reader is referred for more detailed information.

3.1 Glass electrodes
The best known is obviously the pH electrode. Its principle dates from 1909, and this H^+-selective electrode is currently the most often used ISE in clinical chemistry. Commercial instruments for the control of acid–base equilibria in biological fluids, in particular whole blood, are based on this electrode. A capillary hydrogen-ion selective glass microelectrode, thermostatted and connected to the reference electrode by a liquid junction, is able today to appreciate 0.0025 pH unit. When used in an electrolyte solution which contains hydrogen carbonate ions in equilibrium, through a thin gas-permeable membrane, with carbon dioxide dissolved in a biological fluid (namely blood), it allows the CO_2 partial pressure to be determined. If the pH at which the measurement is performed is lowered sufficiently, the CO_2 of hydrogen carbonates is released and the measurement of total CO_2 is possible.

As early as 1924, it was shown that specific membranes for alkali cations such as

Na^+, K^+, Li^+ could be made with some special types of glass. Na^+-specific electrodes have been proposed, using glass containing 11% Na_2O, 18% Al_2O_3, 71% SiO_2, and trivalent metallic oxides [27]. Most commercially available devices for clinical chemistry use such electrodes. There are today other special types of glass sensitive to selenium and lead, but they are not yet in clinical use. Research is very active in this area.

3.2   Liquid or semiliquid membrane electrodes

These electrodes, in particular those for Ca^{2+} and K^+ and some Na^+- and Li^+-selective electrodes, offer at the present time the greatest possibilities in the field of human clinical biology. Generally, the carriers or exchangers, i.e. salts, insoluble complexes (e.g. sodium dodecylphosphate), or ionophores (valinomycin, ETH 227, etc . . .) are included in a polymer such as polyvinyl chloride or silicone.

Various works have investigated the influence exerted on the electrode performances by the nature of the polymer and of the solvent used for inclusion of the carrier in the polymer, by the presence or absence of an insoluble salt such as tetraphenylborate, and by the lipophilicity of the membrane.

All K^+-selective electrodes of the currently available commercial devices use polymer-included valinomycin. Besides glass electrodes, ETH 227 or ETH 237, sometimes monensin, or else a crown-ether [bis(1,2 crown-4)] have been most frequently used for Na^+ determination. Such Na^+-selective membrane electrodes containing an ionophore have been recently included in commercial analysers, for example the HITACHI pluriparametric analysers (ETH-type-based electrode) or the Microlyte from KONE (monensin-based electrode). For Li^+, ETH 149, ETH 1810 [28], or N,N'-diheptyl-5,5-dimethyl-N,N'-bis(3-oxapentyl)-3,7-dioxanonade diamide [8] have been proposed, and for Ca^{2+}, mixtures such as (ETH 1001 + 2-nitrophenyl-octylether + sodium tritiophenylborate), (calcium di-*n*-decylphosphate + calcium di-*n*-octylphenylphosphate + polyvinyl chloride), or else cyclic amide polyether derivatives [29].

Novel ionophores of potential utility in clinical chemistry are frequently presented in the literature, which indicates that, except valinomycin for K^+, the other ionophores or carriers already proposed for Na^+, Li^+, or Ca^{2+} are not fully suitable.

4   Clinical chemistry devices based on ISEs

Tables 2 and 3 give a list of ISE-based devices used in clinical chemistry. The electrodes are incorporated in either pluriparametric analysers able to measure other parameters than ions (generally by spectrophotometry) or in instruments designed for the determination of ions only.

Most of these commercial instruments (if not all) use flowthrough-type measuring cells which allow a permanent circulation of a reference solution. Some devices accept only serum or only urines; others accept both serum and urines. At present, however, more and more instruments apply to all biological fluids, including whole blood.

While all pluriparametric analysers possess a sampler, some of the instruments which specifically measure ions remain manual, and others are equipped with a sampler. Sampling rates range from 20 to 150 samples per hour.

Table 2 — Pluriparametric analysers for clinical chemistry using ion-selective
electrodes

Spectrum (ABBOTT) Chicago, IL 664, USA	Na^+ (glass), K^+ (valinomycin), Cl^- Undiluted sample
Astra 4 Astra 8 (BECKMAN) Beckman Inst. Inc., Fullertone, CA, USA	Na^+ (glass), K^+ (valinomycin), pCO_2 Diluted sample Na^+, K^+, Cl, PCO_2
Ektachem 400 (EASTMAN KODAK)	
Ektachem 700 Eastman Kodak CO, Rochester, NY, USA	Disposable electrodes Undiluted sample
Hitachi 704 Hitachi 705 (HITACHI) Hitachi 737 Boehringer, Mannheim, W Germany	Na^+ (ETH 157), K^+ (valinomycin), Cl^- Diluted sample Na^+ (glass). K^+ (valinomycin), Cl^-
SMAC I and II Chem I (TECHNICON) RA 1000 Tarrytow, NY, USA	CO_2 (pH electrode) Diluted sample
MONARCH GENESIS (IL) Instrumentation Laboratory Ltd, Warrington, England, UK	Na^+ (glass), K^+ (valinomycin), Cl^- Diluted sample
PARAMAX (DADE) Irvine, California, USA	Na^+ (glass), K^+ (valinomycin) Undiluted sample
PROGRESS (KONE) Espoo, Finland	Na^+, K^+ Diluted sample
ACA III ACA V DUPONT de Dimension NEUMOURS Wilmington, DE, USA	Na^+, K^+ Undiluted sample
Easy ERIS MERCK Darmstadt, W Germany	NA^+, K^+ Diluted sample

Table 3 — Single analysers for clinical chemistry using ion-selective electrodes

IL 501, 502, 508 (IL)	NA^+, K^+	Na^+ (glass) K^+ (valinomycin)
IL 446	pCO_2	pCO_2: pH electrode
Instrumentation Laboratory Ltd, Warrington, England, UK		
E4A (BECKMANN)	Na^+, K^+ Na^+, K^+, Cl^-,	Na^+ (glass), K^+ (valinomycin) Cl^- (Ag/AgCl),
ELIV		
	CO_2	CO_2: pH electrode
Fullertone, CA, USA		
CORNING 178 (CORNING)	pH, pCO_2	Na^+ (glass), K^+ (valinomycin)
CORNING 614 (CORNING)	Na^+, K^+	Ca^{2+}: neutral ionophore
CORNING 634 (CORNING)	pH, Ca^{2+}	
Medsield, MA, USA		
CHEM-LITE (KONE)	Na^+, K^+, Cl^-	Na^+ (monensine), K^+ (valinomycin),
Microlyte	N^+, K^+	Cl^- (Ag/AgCl)
Espoo, Finland		
NOVA 1 (NOVA BIOMEDICAL)	Na^+, K^+	
NOVA 2 (NOVA BIOMEDICAL)	Ca^{2+}	
NOVA 3 (NOVA BIOMEDICAL)	Cl^-, pCO_2	
NOVA 5 (NOVA BIOMEDICAL)	Na^+, K^+, Cl^-	Na^+ (glass)
NOVA 6 (NOVA BIOMEDICAL)	Na^+, K^+, Ca^{2+}	K^+ (valinomycin)
NOVA 7 (NOVA BIOMEDICAL)	Ca^{2+}, Total	Cl^- (ion exchanger)
	Ca, pH	CO_2: pH electrode
NOVA 8 (NOVA BIOMEDICAL)	Ca^{2+}, pH	Ca^{2+}: neutral ionophore
NOVA 9 (NOVA BIOMEDICAL)	Na^+, K^+,	
	Total Ca	
	(In the urines)	
NOVA 10 (NOVA BIOMEDICAL)	Na^+, K^+	
	Total Ca, Cl^-	
NOVA 11 (NOVA BIOMEDICAL)	Na^+, K^+, Li^+	
STAT PROFILE (NOVA)	pH, pCO_2, pO_2, Ca^{2+}, Na^+, K^+	
Waltham, MA, USA		
AVL 980	Na^+, K^+, Ca^{2+}	Na^+ (glass), K^+ valinomycin)
AVL 982 AVL (Suisse)	Na^+, K^+	Cl^- (Ag/AgCl), Ca^{2+}: ion
AVL 983	Na^+, K^+, Cl^-	exchanger
Shaffhouse, Switzerland		
ANA-LYTE + 1 (BAKER)	Na^+, K^+	
Phillipsburg, NJ, USA		
	Na^+, K^+	Na^+ (glass)
ORION 1020		
SS20 (ORION RESEARCH)	Ca^{2+}	K^+ (valinomycin)
Cambridge, MA, USA		
KNA 1	Na^+, K^+	Na^+ (glass), K^+ (valinomycin) Ca^{2+}: ion exchanger
ICA 1 (RADIOMETER)	pH, CA^{2+}	pH (glass)
Copenhagen, Denmark		
STAT-ION (TECHNICON)	Na^+, K^+, Cl^-	
Tarrytow, NY, USA		

Although ISEs are increasingly applied in clinical chemistry, their technology is not yet recognized everywhere, and a number of firms and laboratories remain faithful supporters of flame photometry. For example, some analysts are investigating new absorptiometric techniques for ion assays [30]. Why, in these conditions, are ISE-based devices so numerous on the clinical chemistry market? Some general reasons and some specific reasons for each ion can be put forward:

General reasons
Clinical chemists are faced with a number of assays on samples collected from varied individuals, healthy and sick, adults and children. The biological samples themselves are of different types, for example, serum, plasma, cerebrospinal fluid, urines, whole blood, fluids from artificial kidneys, etc. So, clinician chemists need simple, fully automated techniques which can be used by skilled but polyvalent staff. ISEs, whose performances and ease of utilization have been recalled above, seem to fit these requirements. For instance, in our clinical chemistry laboratory at Nancy (France), between 380 and 450 ISE-based Na^+ and K^+ assays are performed every day, 365 days per year, not counting other determinations such as pH, pO_2, pCO_2, Ca^{2+}, and so on.

Specific reasons
ISEs give access to prominent physiological data such as the actual activity of some cations, which is quite important for calcium, in particular. It has been well known for a long time that most cellular events (for example, excitability of neuronal and muscular cells, hormone actions) are influenced by the level of ionized calcium in biological fluid water. In human plasma, this ionized calcium, which represents 40 to 45% of the total calcium, could previously only be satisfactorily evaluated by means of tedious and expensive biological methods. The appearance of ISEs has allowed this important parameter to be easily determined while highlighting the role of some regulating factors and the requirements for precise sampling conditions (anaerobiosis, whole blood, calcium heparinate, . . .) [31–35]. Moreover, ISEs have allowed the role of inert substances (proteins, lipids) in the false hyponatremias and hypokaliemias to be defined precisely, and ISE-based measurements have revealed the possible binding of Na^+ to some proteins, to bicarbonates, and to heparin. Finally, it must be remembered that the exploration of the acid–base equilibrium in blood is possible only by measuring the pH, the pO_2, and pCO_2: three determinations based on the sole use of ISEs.

5 Specific problems in the use of these clinical chemistry instruments
Besides the previously described problems linked to electrodes themselves, such as the electrode specificity, junction phenomena, the role of the reference electrode and of the measuring-cell flow, an important feature which has appeared with the clinical use of ISEs is quality control. To deliver reliable results, clinical chemistry laboratories must have their own quality control system, which must be as efficient as possible. To fulfil this requirement, techniques and instruments are tested regularly (every ten samples for some analysers), using standard solutions. All authors insist on the importance of both the nature of these standards and the standardization procedures [1,2,36–39]. Standard solutions must be as similar as possible to patient

samples, and control serums are generally used. Under the patronage of the National Center for Hospital Equipment (Centre National des Equipements Hospitaliers, Saint-Quentin en Yvelines, France), the French Society for Clinical Biology (Société Française de Biologie Clinique) has proposed a procedure for evaluating ISE-based analysers for Na^+ and K^+ [39]. Series of commercially available bovine serums (BIO-FRANCE REACTIFS, Vandoeuvre, France), with varied levels of Na^+ and K^+ (Table 4), overloaded with lipids (between 1 and 20 g/l) and containing proteins

Table 4 — Quality-control serums for ion-selective electrodes (study of the sensitivity and of interferences)

Serum no.	1	2	3	4	5	6	7	8
$[Na^+]$ (mM)	50	80	100	120	140	160	180	200
$[K^+]$ (mM)	1	15	2	3	4	6	8	6

(50–200 g/l), ammonium (0–70 mM), and lithium ions (0–8 mM), allow the determination of the electrode slope and of the Na^+ (or K^+) electrode selectivity towards the main interfering ions (Li^+, NH_4^+, K^+, or Na^+) and also allow the influence of inert substances and, if necessary, the contamination rate from one serum to another to be evaluated. Such control solutions, periodically checked in the same way as patient serums, can provide a good quality control of ISE-based analysers.

Acknowledgements
I thank very much Messrs Mécrin, Herbenval, Crémel, and Bost for their help.

References
[1] Kissel, T. R. (1986) In: Rossiter, B. W. & Hamilton, J. F. (eds) *Physical methods of chemistry. Vol. 2: Electrochemical methods*. New York, John Wiley & Sons p. 175.
[2] Cowell, D. C., Browning, D. M., Clarke, S., Kilshaw, D., Randell, J., & Singer, R. (1985) *Med. Lab. Sci.* **42**, 252–261.
[3] Bijster, P., Vader, H. L., & Vink, C. L. J. (1983) *Ann. Clin. Biochem.* **20**, 116–120.
[4] Andersen, I. B. (1986) *Clin. Chem.* **32**, 231.
[5] Brezinski, D. P. (1983) *Analyst* **108**, 425–442.
[6] Kissa, E. (1978) *Clin. Chem.* **33**, 253–255.
[7] Kobos, R. K., Abbott, S. D., Levin, H. W., Kilkson, H., Peterson, D. R., & Dickinson, J. W. (1987) *Clin. Chem.* **33**, 153–158.

[8] Comer, J. & Avdeef, A. (1987) *Int. Labmate* p. 11–14.
[9] Ladenson, J. H., Apple, F. S., Aguanno, J. J., & Koch, D. D. (1982) *Clin. Chem.* **28**, 2383–2386.
[10] Frier, B. M., Steer, C. R., Baird, J. B., & Bloomfield, S. (1980) *Arch. Dis. Childhood* **55**, 771–775.
[11] Decker, U. P. & Geyer, R. (1984) *Acta Hydrochim. Hydrobiol.* **12**, 19.
[12] Marcca, C. & Cakrt, M. (1983) *Anal. Chim. Acta* **154**, 51–60.
[13] Bell, J. A., Hilton, P. J., & Walker, G. (1972) *Brit. Med. J.* **4**, 709–710.
[14] Albrink, M. J., Haid, P. M., Man, E. B., & Peters, J. P. (1955) *J. Clin. Invest.* **34**, 1483–1488.
[15] Langhoff, E. & Ladefoged, J. (1985) *Clin. Chem.* **31**, 1811–1814.
[16] Waugh, W. H. (1969) *Metabolism* **18**, 706–712.
[17] Meier, P. C., Ammann, D., Morf, W. E., & Simon, W. (1980) In: Koryta, J. (ed.) *Medical and biological applications of electrochemical devices*. New York, John Wiley & Sons p. 13–91.
[18] Ammann, D., Morf, W. E., Anker, P., Meier, P. C., Pretsch, E., & Simon, W. (1983) *Ion-Selective Electrode Rev.* **5**, 3–92.
[19] Ladenson, J. H. (1983) *Anal. Proc.* **20**, 554–556.
[20] Van Lente, F. (1984) *Lab. Med.* **15**, 165–170.
[21] Oesch, U., Anker, P., Ammann, D., & Simon, W. (1985) In: Pungor, E., and Buzas, I. (eds.) *Ion-selective electrodes*. Vol. 4. Budapest, Akademiai Kiado p. 81–101.
[22] Czaban, J. D. (1985) *Anal. Chem.* **57**, 345A–356.
[23] Oesch, U., Ammann, D., & Simon, W. (1986) *Clin. Chem.* **32**, 1448–1459.
[24] Martin, M. J., & Roufe, P. (1986) *Anal. Proc.* **23**, 303–304.
[25] Pivert, G. (1987) *Eur. Rev. Biomed. Technol.* **8**, 309–314.
[26] Eisenman, G. (ed.) (1967) *Glass electrodes for hydrogen and other cations: principles and practice*. New York, Marcel Dekker.
[27] Metzger, E., Ammann, D., Asper, R., & Simon, W. (1986) *Anal. Chem.* **58**, 132–135.
[28] Kimura, K., Kumani, K., Kitazawa, S., & Shono, T. (1984) *Anal. Chem.* **56**, 2369–2372.
[29] Wong, T. S., Spoo, J., Kerst, K. C., & Spring, T. G. (1985) *Clin. Chem.* **31**, 1464–1467.
[30] Sachs, C. (1984) *Rev. Fr. Lab.* **133**, 6–13.
[31] Khalil, S. A. H., Moody, G. J., de Oliveira Neto, G., & Thomas, J. D. R. (1985) *Anal. Proc.* **22**, 10–11.
[32] Wandrup, V., & Kvetny, J. (1985) *Clin. Chem.* **31**, 856–860.
[33] Bowers, G. N., Brassard, C., & Sena, S. F. (1986) *Clin. Chem.* **32**, 1437–1447.
[34] Sachs, C. (1987) *Information Scientifique du Biologiste* **13**, 101–108.
[35] Toffaletti, J., Bird, C., Berg, C., & Abrams, B. (1986) *Clin. Chem.* **32**, 1548–1550.
[36] Fyffe, J. A. (1982) *J. Autom. Chem.* **4**, 79–81.
[37] Fyffe, J. A., & Toffaletti, J. (1986) *Clin. Chem.* **32**, 1790–1792.
[38] Approved IFCC methods: *Reference method* (1986) *for pH measurement in blood*. Scientific Committee, Analytical Section, Expert panel on pH and blood gases. *Lab. Medica*, Dec. 1986–Jan. 1987, p. 33–37.

[39] Société Française de Biologie Clinique (1982) *Evaluation coordonnée des analyseurs de Na⁺ et K⁺: photométries de flamme et électrodes sélectives. Cahiers du CNEH* **23**, 1–75.

1.4 ELECTROCHEMICAL BIOSENSORS IN CLINICAL ANALYSIS

J. L. Romette

1 Historical background

Bioprobes or biosensors represent nowadays a very attractive category of sensors. Many research groups throughout the world are working to improve them and are trying to find ways of commercializing the technology. Biosensors are not a recent discovery: their principle was demonstrated during the sixties by Clark & Lyons [1] and then by Updike & Hicks [2]. However, they have only been developed commercially over the last ten years. Table 1 shows the historical evolution of the technology involved in biosensors.

Table 1 — Historical and technological evolution of biosensors

ION- AND GAS-SELECTIVE ELECTRODES	ISFETs MODIFIED ELECTRODES	OPTOELECTRONIC PROBES
		ENFET
	Immobilized whole cells	Optical sensor — enzymatic — immunoenzymatic
Immobilized enzymes	Thermistor — immunoenzymatic — enzymatic	Thermistor — immunoenzymatic — enzymatic
	Tissue slice or cellular organelle sensor	Tissue slice or cellular organelle sensor
	Microbial sensor	Microbial sensor
	Immunological sensor	Immunological sensor
Enzyme electrodes	Enzyme sensor	Enzyme sensor
1960	1970	1980

From the start, research moved towards formal description of new products instead of developing those already described. This meant that some potential customers reading the publications thought they had found a solution to their analytical problem, but were unable to obtain the product on the market. This situation produced a lack of credibility concerning the value of the technique.

For instance, the first works on glucose determination date from 1962 [1], but the first glucose electrode was marketed only in 1978. Meanwhile, different sensors for sugars, amino acids, etc., have been described, but the experiments have been limited to the research laboratory.

Only a few sensors are commercially available at present, which does not mean that there are no potential customers. In fact, because of the strategic nature of some analyses in the control of biotechnological processes, particularly in the food-processing industries, important industrial companies are themselves developing analytical tools for their own use. Whenever research has been finalized, giving a marketed product, it can be stated that no exclusive user possessing the know-how had a vested interest in limiting the diffusion of the analytical technique involved.

2   The concept of biosensors

A biosensor is composed of two main elements:

- a biological element which represents the active part of the biosensor. This can be any biocatalytic activity retained on/in a support in close contact with the second element, i.e. the transducer;
- an electronic device or transducer. Specific for the physicochemical changes occurring inside the biological element, its role is to transduce the biocatalytic effect as an electric signal.

The association of several electrochemical probes with enzymes, micro-organisms, tissue slices, and compounds with immunological properties has been extensively described. The most significant examples of these realizations are mentioned in the next section. Noticeable progress has been made in both immobilization techniques to retain the biocatalytic activity and in the transducer improvement to get better analytical characteristics.

2.1   Immobilization techniques

A number of reports dealing with the techniques for biocatalyst immobilization have been published: more than 500 references can be easily found in the literature. However, the greatest improvements of these methods for increasing the biocatalyst activity and stability are the fruit of recent studies. Fig. 1 summarizes the most commonly used immobilization techniques (see also Chapter 2 ($2.3)). Two main families can be distinguished:

Physical techniques

These exploit the physical properties of the support: polarity, porosity, permeability. In essence, the work consists in making a kind of network around the biocatalyst to keep it in close contact with the associated transducer. Many methods have been described, among which are adsorption, microencapsulation, and inclusion inside a gel. The active compound is retained inside the insoluble matrix owing to steric hindrance or on the support by electrostatic interactions. Different materials can be used as a support, for example glass fibres, ion-exchange resins, derivatized celluloses, acrylic polymers, and polysaccharide gels. The main advantage of these physical techniques is their mildness, ensuring a high yield of activity retention: no chemical denaturation is produced by the immobilization process. However, a major drawback arises from this mildness, i.e. a weak mechanical strength which may allow a leakage of the biocatalyst. Moreover, the support does not stabilize the biocatalyst activity.

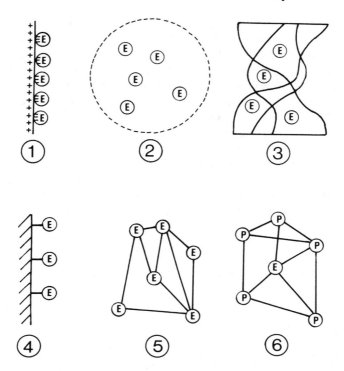

Fig. 1 — Enzyme (E) immobilization procedures. 1, absorption; 2, encapsulation; 3, entrapment; 4, covalent binding; 5, cross-linking; 6, co-polymerization (P: inactive compound).

Chemical techniques

Reactive groups of the biocatalyst react with the support to form stable links such as covalent bonds. Crosslinking agents are frequently used to create polymeric structures in which the biocatalyst is incorporated. Among these techniques, the most frequently used are the covalent immobilization of the biocatalyst on the surface of an insoluble support, the polymerization of the active molecule using a crosslinking agent, or its copolymerization with inactive chemicals. The support can be silica, cellulose, polyaspartic or polyglutamic acid, nylon, or proteins. The usual crosslinking agents include azide, carbodiimide, diazonium salts, isothiocyanate derivatives, ethyl chloroformate, and glutaraldehyde. The chemical methods of immobilization generally produce a deactivation effect on the biocatalyst mainly because some bonds with the support can occur at the active site level. Such a deactivation can be significantly reduced by protecting the active site of the biocatalyst molecule before the action of the crosslinking agent. For instance, an enzyme active site can be protected by using a structural analog of the substrate, which is not transformed by the enzyme, but for which the enzyme has an affinity. The immobilization yield of these techniques is usually lower than that of physical methods. Nevertheless, the support may confer a greater stability to the biocatalyst, mainly when copolymerization is used: a good protection against deactivation effects of the bulk solution, due, for example, to ionic strength, extreme pH, or heat, can be obtained. In the case of an enzyme immobilization inside a protein matrix, the support can be used to maintain typical concentrations of the substrate or product around the enzyme by exploiting the existing diffusional constraints (i.e. oxygen for oxidases). The enzyme behaviour when immobilized is different from that of the enzyme in solution since substrates and products are submitted to diffusion. In fact, the coupling of diffusion and reaction often results in a potentialization of the enzyme reaction through local concentrations in effectors inside the active support.

The active support can be membrane-shaped or granular. It is also possible to produce active surfaces by a direct coating of the sensitive part of transducers (e.g. glass membrane of pH electrodes or electrode glassy-carbon surfaces).

The confinement of the biocatalyst activity is the first step towards the construction of a biosensor. The second step is to obtain close contact of this active element with the transducer.

2.2   Main electrochemical transducer-biocatalyst combinations

2.2.1   pH and ion-selective electrodes
2.2.1.1   pH electrodes. Many enzymatic reactions produce ionic species, in particular protons which may change the pH of a weakly buffered bulk solution.

In 1974, Bucher [3] described the 'pH stat' titration procedure, which consists in estimating the amount of base or acid which must be added to the bulk solution to maintain the pH at a chosen set end value. This technique may be easily applied to evaluate the catalytic activity of an enzyme such as lipase in human serum, following the reaction:

$$\text{triglyceride} + H_2O \xrightarrow[\text{(EC 3.1.1.3)}]{\text{Lipase}} \text{diglyceride} + \text{fatty acid}$$

Usually, olive oil is used as a substrate. Olive oil and the sample solution containing lipase (serum) are mixed together and titrated with sodium hydroxide 10^{-2} M at constant pH. The consumption of NaOH is recorded vs time; the slope of the curve gives an evaluation of the enzyme activity.

A pH electrode is used in this assay. The principle and characteristics of hydrogen ion-selective electrodes have been plentifully described in a number of publications devoted to basic and applied electrochemistry: they will not be recalled here.

The use of pH electrodes as electrochemical transducers in biosensors is limited by the great variability of sample pH values. Moreover, the conditions for using a pH electrode suppose a low buffered sample and a high biocatalytic activity. A low buffering power of the reacting medium can be obtained only by large sample dilution, thus increasing the risk of experimental errors. The use of a high level of biocatalytic activity penalizes the technique from an economic viewpoint.

Nevertheless, some applications of the pH electrode/biocatalyst biosensor combination can be found in the literature. A typical example is reported in the present book by Tran Minh & El Yamani (Part 3, Chapter 1.3): determination of pesticide residues with a butyrylcholinesterase electrode.

2.2.1.2   Other ion-selective electrodes. While ion-selective electrodes are widely used in clinical analysis (see, in this book, Chapter 1 (§1.3) by Nabet and Chapter 2 (§2.2) by Michel & Blond), they are not frequently used in biosensor production.

Nevertheless, some works are interesting and could give rise to new developments in the future, e.g. tissue slices associated with an iodide electrode [4].

The iodide electrode has been used by Boitieux *et al.* [5] in an immunosensor for the determination of hepatitis B surface antigen (HB$_s$Ag).The technique is summarized in Fig. 2. The immunological procedure is based on a kind of 'Elisa test' adapted to an electrochemical detection of the labelling enzyme activity. The conjugate assembly is performed in a 'sandwich' mode (Fig. 2A). Specific HB$_s$Ag antibodies are immobilized on a gelatin matrix which covers the crystal of the iodide electrode. A test cell has been designed to use this bioprobe as shown in Fig. 2B. The test cell is filled with a diluted solution of antigen which binds to part of the immobilized antibodies. Then the electrode surface is washed with a solution of peroxidase-labelled antibodies. The peroxidase activity is determined by introducing the enzyme substrate and iodide in the test cell according to:

$$H_2O_2 + 2I^- + 2H^+ \xrightarrow{\text{peroxidase}} I_2 + 2H_2O$$

The signal recorded by the iodide electrode can be easily related to the amount of enzyme retained on the gelatin support and therefore to the amount of HB$_s$Ag in the sample (Fig. 2C).

The sensitivity of this technique is of the same order as that obtained with radio-immunological assays.

2.2.1.3   Field effect transistors. A new type of sensor sensitive to ions, the ion sensitive field effect transistor (ISFET), has been described by Janata & Huber in 1979 [6]. This device is produced by microelectronic techniques in an integrated circuit form (Fig. 3). The substrate of the transistor is usually silicon in which doping atoms have been diffused, producing an extrinsic semiconductor.

In an electric field-free region of this extrinsic semiconductor containing only one type of doping atom, there are equal densities of mobile charges (excess electrons in n-type areas, and holes in p-type areas where electrons are lacking) and ionized doping atoms (positively charged for n-type areas and negatively charged for p-type areas).

If an electric field is applied to the surface of the semiconductor, the density of the mobile charge carriers will be either enhanced or depleted, depending on the field. When the field increases the concentration of holes, the surface is said to be accumulated and behaves much like a metal. When, on the contrary, the field forces the mobile holes away from the surface, a charge region consisting of the ionized acceptor atoms forms in the semiconductor far from the surface. If the surface potential deviates sufficiently far from the bulk potential, the surface will 'invert'. Further increase in the normal surface field will not significantly change the surface potential but will only change the density of electrons and consequently the electrical

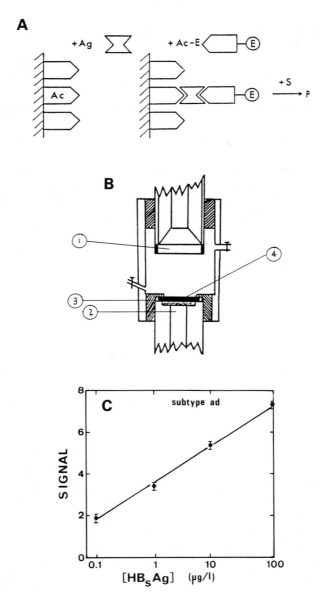

Fig. 2 — Determination of HB$_s$Ag with an enzymo-immunological sensor. A: 'sandwich' method for the assembly of conjugates. B: diagrams of the immunosensor and of the measurement cell. 1, reference electrode; 2, iodide electrode; 3, magnetic device fastening the protein membrane to the iodide sensor; 4, gelatin matrix on which the immuno-enzymatic reaction occurs. From Boitieux *et al.* [5]. C: calibration curve for HB$_s$Ag subtype ad. From [5].

conductivity of this layer. The physical effect which is measured is the change in electric current carried by the surface inversion layer, the drain current.

The device must be operated under conditions that cause the surface inversion

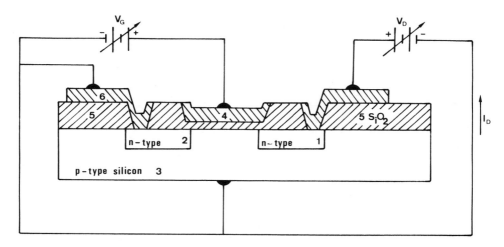

Fig. 3 — Schematic diagram of an insulated gate FET (IGFET). 1, drain; 2, source; 3, substrate; 4, gate; 5, insulator; 6, metal connexions. From Janata & Huber [6].

layer to form. The device (Fig. 3) resembles a parallel plate capacitor. The semiconductor substrate forms one plate of the capacitor and the metal gate the other. In normal operating mode, the gate voltage V_G is applied between these two plates. The polarity and magnitude of V_G are such that the semiconductor field effect gives rise to a p-to-n inversion layer at the substrate-gate interface. This n-type inversion layer forms a conducting channel between the source and drain regions. If this inversion layer is not present, application to the drain of a positive voltage V_D with respect to the source and substrate results in no noticeable current into the drain because the drain to substrate p–n junction diode is reverse biased. However, if the n-type inversion layer exists along the surface of the semiconductor between the source and drain, a continuous path for the current flow I_D exists. The intensity of the drain current I_D will be determined by the electrical resistance of the surface inversion layer and the voltage differences V_D between the source and drain. So it becomes possible to deduce V_G from I_D by imposing the V_D value. Fig. 4 gives the typical response curves obtained by varying the different values of V_G, V_D, and I_D. These curves show an insaturation part in which I_D is a linear function of V_G.

The gate voltage V_G can be considered as a sum of potentials from different sources, i.e. electrical, chemical, electrochemical. For instance, if an ion-sensitive membrane is applied on the surface of the FET, it becomes possible to measure its potential through I_D variations in the presence of ion activity in an external sample medium (Fig. 5).

The first applications of these devices as a part of a biosensor are recent and are for the moment limited to the research laboratory. However, the results obtained so far promise a brilliant industrial development in the near future.

Caras & Janata [7] have described a penicillin-sensitive ENFET (enzyme field effect transistor) (Fig. 6A). The enzyme reaction is catalysed by penicillinase (EC 3.5.2.6) according to:

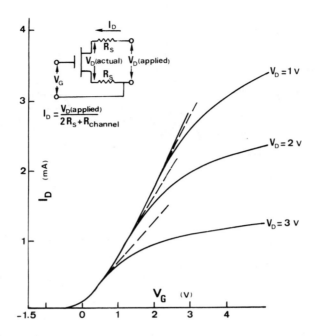

Fig. 4 — Typical response curves of an IGFET obtained for different values of V_D, I_D, and V_G. From [6].

penicillin $\rightarrow$ penicilloic acid.

The product is a strong acid that can be easily detected by a pH electrode [8,9]. In this case, the ISFET replaces the pH electrode. Fig. 6B shows the experimental set. Note that the FET gate is a reference electrode (Ag/AgCl) located outside the device itself, and that the measurement is of a differential type. The enzyme is immobilized in a protein matrix crosslinked by glutaraldehyde. Fig. 6C shows the signal obtained when the biosensor is in contact with a solution of penicillin. This type of assay can be extended to other determinations where pH shifts are detected. However, the first necessary step in this development is to improve the production of reliable FETs. Most present problems concern shifting of the signal, short lifetime, and poor reproducibility in manufacture.

2.2.2 Gas electrodes

2.2.2.1 pO_2 (Clark-type) electrode/oxidase activity. In 1956 Clark [10] described a compact oxygen probe which has found a wide range of analytical applications in both applied and fundamental biosciences.

In essence, a Clark-type oxygen electrode consists of a platinum, gold, or rhodium cathode and a silver anode, which are connected electrically by a film of electrolyte solution (potassium chloride), and between which a polarization potential of c. 0.6 V is applied. A hydrophobic gas-selective membrane insulates the

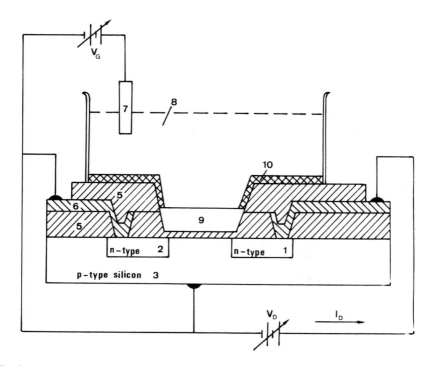

Fig. 5 — Schematic diagram of an ISFET. 1, drain; 2, source; 3, substrate; 5, insulator; 6, metal lead; 7, reference electrode; 8, sample solution; 9, ion-sensitive membrane; 10, encapsulant. Note that 7, 8, 9 replace the metal gate 4 of the IGFET (Fig. 3. From [6].

electrochemical compartment from the bulk solution. Oxygen in the sample diffuses through this membrane and is reduced at the cathode level. The electrochemical reaction produces a current which is proportional to the partial pressure of oxygen in the sample since oxygen diffusion through the membrane is the limiting step in the reduction process.

A typical example of this transducer as a part of a biosensor has been described by Romette *et al.* [11] (Fig. 7). An enzyme electrode for the determination of glucose in blood is obtained by copolymerization of glucose oxidase (EC 1.1.3.4) and gelatin, using glutaraldehyde. The active matrix is fixed on to a modified pO_2 sensor with fast response characteristics. The enzyme reaction is the following:

$$glucose + O_2 \rightarrow gluconic\ acid + H_2O_2\ .$$

Oxygen consumption is detected by the transducer, the signal is easily related to the glucose concentration of the sample.

This technique has been successfully developed through a commercially available autoanalyser (Enzymat, Seres Inc., France). Fig. 8 shows the calibration curve obtained with this biosensor.

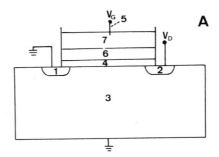

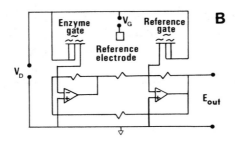

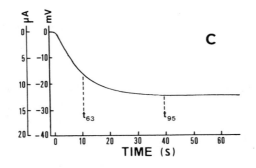

Fig. 6 — The penicillin-sensitive ENFET as described by Caras & Janata [7]. A: schematic diagram of the ENFET: 1, drain; 2, source; 3, substrate; 4, insulator; 5, reference electrode acting as a gate; 6, albumin support containing the enzyme; 7, enzyme substrate solution. B: the differential system of analysis. C: differential signal obtained with the penicillin-sensitive ENFET for a pulse change (0–10 mM) in penicillin concentration (pH 7.2, 25°C).

2.2.2.2 pCO$_2$ electrode/decarboxylase activity. The first pCO$_2$ electrode was described by Stow *et al.* in 1957 [12], but it was developed by Severinghaus & Bradley [13].

The electrode principle is to follow the acido-basic changes occurring in the electrolyte solution when the sensor is exposed to CO$_2$, either dissolved or in gaseous form. Since CO$_2$ may react with water to produce carbonic acid, an easy way to

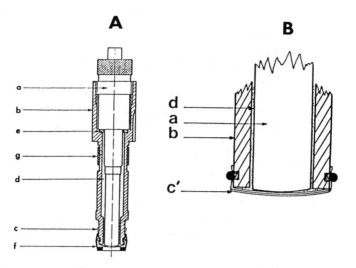

Fig. 7 — Glucose sensor using imobilized glucose oxidase and a pO_2 electrode. From Romette *et al.* [11]. A: pO_2 sensor: a, electrode support; b, jacket; c, gas-selective membrane; d, electrolyte; e, O-ring; f, measuring cell gasket; vent joint. B: glucose electrode: a, electrode support; b, jacket; c', active gas-selective membrane covered with the enzyme layer; d, electrode.

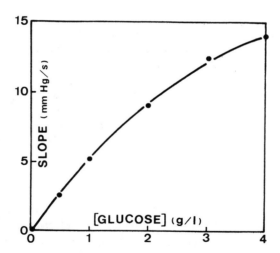

Fig. 8 — Calibration curve obtained with the glucose sensor described in Fig. 7 (solution of glucose in phosphate buffer 0.05 M, pH 7.5, at 25°C. From [11].

detect its presence is to monitor the pH of the electrolyte. Practically, pH is a linear function of the logarithm of CO_2 partial pressure.

Like all gas sensors, the pCO_2 electrode has a hydrophobic gas-selective membrane separating the electrochemical compartment from the bulk solution. The internal compartment is composed of a glass pH electrode and a reference electrode. The nature of the gas-selective membrane (e.g. silicone, polyethylene, Teflon, Mylar) defines the sensor characteristics. Because the chemical reaction occurs in the electrolyte layer, its thickness at the electrode tip is very important. Thus a nylon net is inserted between the pH glass electrode and the internal side of the gas-selective membrane to maintain a fixed electrolyte volume. Different equipments are commercially available, for clinical analysis (blood) or for bioprocess control. Usually, these transducers are coupled with decarboxylase activities to produce biosensors.

The applications of this type of biosensor are very limited because most of the time the CO_2 content of the sample is as variable as is the enzyme substrate. Different methods have been tried to stabilize the CO_2 level before estimating the enzyme substrate concentration.

An example of application has been described by Tran *et al.* [14] for L-lysine determination using L-lysine decarboxylase (EC 4.1.1.18). The enzyme was immobilized directly on the CO_2 gas-selective membrane by copolymerization with gelatin as an inert support, using glutaraldehyde as a crosslinking agent. This L-lysine enzyme electrode is described in Fig. 9. L-Lysine decarboxylase catalyses the following reaction:

$$\text{L-lysine} \rightarrow \text{cadaverine} + CO_2 \ ,$$

where the forward reaction proceeds to completion. The CO_2 produced was estimated in real time with the pCO_2 sensor. Some pyridoxal 5'-phosphate, which is a cofactor for L-lysine decarboxylase, had to be added to the enzyme at the moment of immobilization.

Electrode responses to different L-lysine concentrations are shown in Fig. 10. The calibration curve obtained by plotting $d(pCO_2)/dt$ at the inflexion point vs lysine concentration is given in Fig. 11. To develop the method into a fully automatic procedure, a fluid diagram has been established as described in Fig. 12.

2.2.2.3 NH$_3$-sensitive electrode/dehydrogenase activity. The principle of the NH_3-sensitive electrode is derived from the pCO_2 sensor. The gas-selective membrane is usually Teflon. Ammonia gas dissolved in the sample diffuses through this membrane in gaseous form, then is dissolved in the electrolyte solution basically composed of ammonium chloride O.1 M. Because of the electrolyte pH value, ammonia reacts with water to produce NH_4^+ and OH^-, which changes the acid-base equilibrium of the electrolyte. An internal pH electrode follows these changes to produce a signal proportional to the NH_3 content of the sample. Different enzyme reactions can be coupled with the transducer, involving, for example, urease, deanimase, or dehydrogenase activities. Fig. 13 shows a biosensor specific for L-alanine and using L-alanine dehydrogenase. When this bioprobe is in contact with a sample containing the amino acid and the cofactor (NAD^+), ammonium ions are

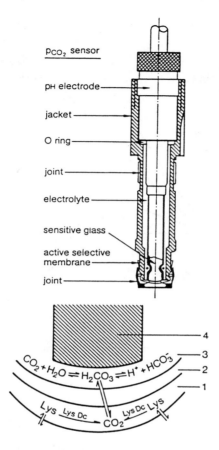

$\mathrm{p_{CO_2}}$ sensor

pH electrode

jacket

O ring

joint

electrolyte

sensitive glass

active selective membrane

joint

$$CO_2 + H_2O \rightleftharpoons H_2CO_3 \rightleftharpoons H^+ + HCO_3^-$$

$$Lys \xrightarrow{Lys\ Dc} \quad \xleftarrow{Lys\ Dc\ Lys} \atop CO_2$$

Fig. 9 — Schematic representation of the L-lysine electrode: 1, gelatin membrane containing immobilized lysine decarboxylase; 2, gas-selective membrane; 3, bicarbonate electrolyte solution; 4, glass membrane electrode. From Tran *et al.* [14].

produced and transformed into ammonia gas. The signal due to bulk pH evolution can be readily related to the amino acid content of the sample. This sensor could be easily marketed if the cost of analysis was not so high because of the large amount of cofactor used; this cost significantly diminishes its interest when compared with the more classical colorimetric methods.

Rechnitz *et al.* [15] have coupled an ammonia probe with tissue slices to measure glutamine concentration (Fig. 14). The tissue slices (50 μm thick) are extracted from pig kidney, which exhibits a very high glutaminase activity mainly located in mitochondria [16]. The enzyme reaction produces ammonium and glutamic acid. It seems possible to obtain a good selectivity for glutamine and a long lifetime of the sensor, as shown by Fig. 15.

2.2.3 Metal or carbon electrodes
Imposed potential amperometry is the most commonly used technique in this type of

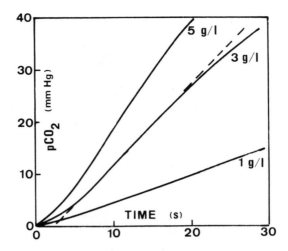

Fig. 10 — Typical response of the L-lysine electrode as pCO_2 with time (phosphate buffer 0.02 M, pH 6.0, at 25°C). From [14].

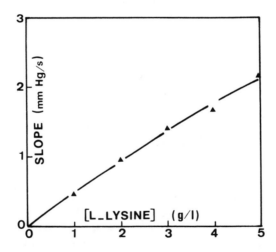

Fig. 11 — Calibration curve of the L-lysine electrode relating the slope $d(pCO_2)/dt$ to the amino acid concentration. From [14].

biosensor arrangement. The chosen potential, specific for the analyte, is imposed between two electrodes immersed directly in the sample.

Unlike gas electrodes, the electrochemical compartment is not insulated from the bulk solution, and interferences may occur since other species which are electrochemically active at the selected potential may be present in the sample together with the

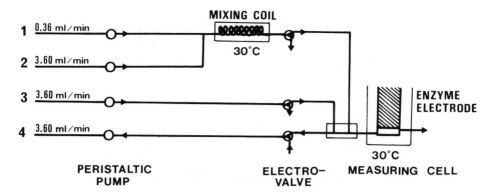

Fig. 12 — Flow system for automatic determination of L-lysine: 1, sample; 2, 0.2 M phosphate buffer, pH 5.8; 3, 0.2 M pH 6.5; 4, air. From [14].

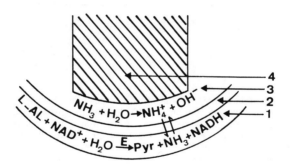

Fig. 13 — Schematic representation of an alanine electrode: 1, L-alanine dehydrogenase membrane; 2, gas-selective membrane; 3, ammonium chloride electrolyte solution; 4, glass membrane. L-AL = L-alanine; E = L-alanine dehydrogenase; Pyr = pyruvate.

substance to be detected. The current resulting from the oxido-reduction of the analyte is related to the analyte concentration. Different applications of this technique to biosensors have been described for glucose determination in different media. For instance, glucose concentration can be evaluated through the oxidation of ferrocyanide into ferricyanide:

$$\text{glucose} + O_2 \xrightarrow{\text{glucose oxidase}} \text{gluconic acid} + H_2O_2$$

$$H_2O_2 + \text{Ferrocyanide} \longrightarrow \text{Ferricyanide} + e^- + H_2O \ .$$

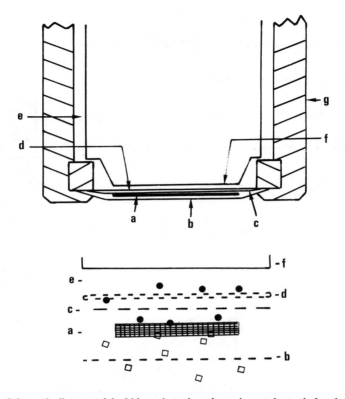

Fig. 14 — Schematic diagram of the kidney tissue-based membrane electrode for glutamine: a, slice of porcine kidney tissue; b, monofilament Nylon mesh; c, protective dialysis membrane; d, gas-permeable membrane; e, internal electrolyte solution; f, pH-sensing glass membrane; g, plastic electrode body (components d–g represent the Orien Model 95–10 ammonia gas-sensing electrode). Bottom, expanded view of the electrode sensing tip showing components and phases: □, glutamine; ●, ammonia (identifying letters as above). From Rechnitz *et al.* [15].

The overall current is proportional to the ratio of ferrocyanide to ferricyanide concentration and is calibrated vs glucose concentration using glucose standards.

The same system can be applied to galactose determination by substituting potassium iodide to ferrocyanide.

An interesting glucose sensor has been proposed by Cass *et al.* [17]. These authors have developed an assay specific for β-D-glucose using glucose oxidase and ferrocene instead of oxygen as an electron acceptor. The biosensor is composed of a carbon electrode to which the enzyme is covalently linked at the same time as 1,1'-dimethylferrocene. The catalysed reaction is as follows:

$$\text{glucose} + 2 \text{ ferricinium}^+ + H_2O \rightarrow \text{gluconic acid} + \text{ferrocene} \ .$$

The ferricinium$^+$ ion is regenerated at the carbon anode (at a potential of $+ 160$ mV) according to:

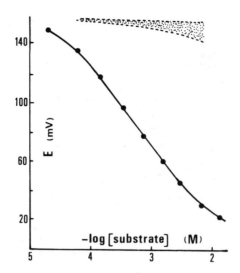

Fig. 15 — Calibration curve obtained for different glutamine concentrations (phosphate buffer 0.1 M, pH 7.8, at 26°C). The grey area represents the interference signal level. From [15].

$$2 \text{ ferrocene} \rightarrow 2 \text{ ferricinium}^+ + 2\,e^- \ .$$

The regeneration current is the biosensor signal, and it can be calibrated as a function of the glucose concentration. The interest of this sensor lies in its ability to determine glucose concentrations even when there is no oxygen inside the sample.

A number of papers report tentative construction of biosensors based on these amperometric transducers. One of the marketed applications of the technique is the analyser proposed by Yellow Springs Instrument since 1979. The YSI Model 23 A, designed for the determination of glucose in blood and urine, uses the glucose oxidase reaction.

The measurement cell of this analyser is a kind of reservoir containing a few millilitres of the diluted sample (Fig. 16). On one side of the cell there is a vibrating Teflon membrane wall to mix and oxygenate the sample during the enzymatic reaction that occurs inside the gel layer. The gel containing the enzyme is immobilized inside a dialysis bag fixed with an O-ring in close contact with the transducer. The transducer is specific for hydrogen peroxide produced by the enzyme reaction. Its selectivity for H_2O_2 is improved by the dialysis bag containing the enzymatic gel. The walls of this bag are microporous materials allowing only the small molecular weight molecules to diffuse. Strong interferences may exist, however, diminishing the interest of this biosensor for bioprocess control. The rinsing step needs catalase addition to remove residual H_2O_2. The signal level may be dependent on the oxygen content of the sample if the latter is far from the pO_2 obtained by air bubbling.

To conclude this section, the present development of other transducer–biocatalyst combinations in which the detection is not of an electrochemical nature must be

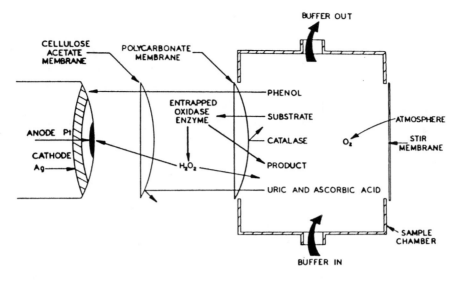

Fig. 16 — YSI Model 23A glucose analyser: detail of the measurement cell. From Yellow
Spring Instrument Co., User manual of the analyser.

underlined: optical fibres for pH detection, thermistors for the enthalpimetric
measurement of enzyme reactions, optrodes coupled with fluorophores, for
example.

3   Examples of electrochemical biosensor combinations for clinical analysis
Tables 2 to 5 give illustrative examples of the different electrochemical transducer/
biocatalyst combinations which have been applied to the *in vitro* determination of
various substances of biological or clinical interest.

4   Economic data
4.1   *Enzymatic analysis*
Biosensors may be considered as a new analytical technique among the enzymatic
methods of analysis. The enzymatic analysis market comprises mainly the medical
market, either public (hospital) or private (clinical analysis laboratory). This market
is quite extensive and is divided into two parts:

Equipment
This part of the market is composed of multiparametric auto-analysers, small
analysers for one parameter determination, colorimeters using analytical kits . . .
This segment has a market value of $585 M. New advances involve the increasing use
of small automatic systems estimating one or two parameters simultaneously in less
than 2 minutes. Customers for these are either specialized hospital departments
(surgery, obstetrical, emergency) or private laboratories. The analysis is not a
routine type involving a large number of samples (i.e. 150 samples/h) such as for

Table 2 — Examples of biosensors based on electrochemical transducers and applied to *in vitro* determination of biological compounds — enzyme electrodes

Parameter	Enzyme	Transducer	References
		pH and ion-selective electrodes	
Acetylcholine	acetylcholine esterase	pH	[18]
Glucose	glucose OX	pH	[8]
		pH-FET	[19,20]
		SoS-FET	[21]
Lactate	lactate DH		[22]
L-phenylalanine	L-amino acid OX + HPO	I^-	[23]
Penicillin	penicillinase	pH	[8]
		pH (Sb, el.)	[24]
Urea	urease	pH	[8]
		pH (iridium dioxide el.)	[25]
		pH (Sb el.)	[26]
		pH-FET	[20,27,28]
		SoS-FET	[21]
		metal or carbon electrode (amperometry)	
Cholesterol (free)	cholesterol OX	H_2O_2	[29]
Creatine kinase	creatine N-P transferase + hexokinase + glucose OX	ferrocene	[30]
D-galactose	galactose OX	ferrocene	[31]
Glucose	glucose OX	H_2O_2	[32]
		Pt-FAD†	[33]
		ferrocene	[34]
		H_2O_2-IsFET	[35]
		benzoquinone	[36]
	glucose DH	quinoprotein	[37–39]
L-glutamate	glutamate DH + glutamate-pyruvate transaminase	Pt-NADH	[40]
Glutamate-oxaloacetate and glutamate-pyruvate transaminase	oxaloacetate DC + pyruvate OX	H_2O_2	[41]
L-lactate	lactate DH	ferrocyanide	[42,43]
		NADH	[38,44]
		microcarbon-NADH	[45]
NAD	glucose DH	Pt-NADH	[46]
Oxalate	oxalate OX	H_2O_2	[47]
L-phenylalanine	L-amino acid OX	H_2O_2	[48]
Thioglycosides	thioglycoside glucohydrolase + glucose OX	H_2O_2 H_2O_2	[49] [49]
Uric acid, urate	urate OX	H_2O_2	[50,51]

Table 2 Continued

Parameter	Enzyme	Transducer	References
		gas electrode	
N-acetyl-L-methionine	amino acylase + amino acid OX	pNH_3	[52]
Asparagine	asparaginase	pNH_3	[53]
Catecol	catechol 1–2 OX	pO_2	[54]
Cholesterol (free)	cholesterol OX	pO_2	[55]
Cholesterol (free and total)	cholesterol esterase + cholesterol OX	pO_2	[56]
Choline	choline OX	pO_2	[57,58]
Creatinine	creatinine deaminase	pNH_3	[59]
	creatinine iminohydrolase	iridium-metal oxide pNH_3 semi-conductor	[60]
Gluconate	gluconate kinase + 6 PD gluconate DH	pCO_2	[61]
Glucose	glucose OX	pO_2	[11]
Glutamine	glutaminase + glutamate OX	pO_2	[62]
L-lactate	lactate DH + HPO	pO_2	[63]
	lactate OX + lactate DH	pO_2	[64]
	lactate OX	pO_2	[65]
Leucine	L-amino acid OX	pNH_3	[66]
L-lysine	L-lysine DC	pCO_2	[14]
	L-lysine OX	pO_2	[67]
L-lysine, L-arginine	L-amino acid DC + diamine OX	pO_2	[68]
L-lysine, tyrosine, L-phenylalanine	specific DC	pCO_2	[69
Oxalate	oxalate OX	pCO_2	[47]
Oxaloacetate	artificial enzyme	pCO_2	[70]
Salicylate	salicylate hydrolase	pO_2	[71]
Thioglycosides	thioglycoside glucohydrolase + glucose OX	pCO_2	[49] [49]
Total proteins	pepsin + tyrosinase	pO_2	[72]
Urea	urease	pCO_2	[73]
		pNH_3	[74,75]
		pNH_3 semi-conductor	[76]

† spectroelectrochemical detection
DC = decarboxylase, DH = dehydrogenase, el. = electrode, FET = field-effect transistor, HPO = horse radish peroxidase, IsFET = ion-selective field-effect transistor, OX = oxidase, SoS = silicon on sapphire.

continuous flow auto-analysers, but an individual type, with a fast response time (i.e. 50 to 60 samples per day).

Worldwide Leaders are: Technicon Co., Beckman, Du Pont, Abbott, Union Carbide (USA), and Hitachi (Japan). They represent 64% of the equipment market.

Table 3 — Examples of biosensors based on electrochemical transducers and applied to *in vitro* determination of biological compounds — microbial electrodes

Parameter	Microbial species	Transducer	References
		pH and ion-selective electrodes	
Cephalosporin	*Citrobacter freundii*	pH	[77]
Nicotinic acid	*Lactobacillus arabinosus*	pH	[78]
		gas electrode	
Aminolgycoside antibiotics	*Escherichia coli*	pO_2	[79]
Arginine	*Streptococcus faecium*	pNH_3	[80]
L-ascorbic acid	*Enterobacter agglomerans*	pO_2	[81]
Asparagine	*Serratia marcescens*	pNH_3	[82]
Aspartame	*Bacillus subtilis*	pO_2	[83]
L-aspartate	*bacterium cadaveris*	pNH_3	[84]
Cholesterol	*Nocardia erythropolis*	pO_2	[85]
Glucose	*Pseudomonas fluorescens*	pO_2	[86,87]
Glutamic acid	*E. coli*	pCO_2	[88]
Glutamine	*sarcina flava*	pNH_3	[89]
L-histidine	*Pseudomonas* sp.	pNH_3	[90]
Lactate, pyruvate, succinate	*E. coli*	pO_2	[91]
Nicotonamide	*E. coli* or *Bacillus pumilus*	pNH_3	[92]
Nystatin	*Saccharomyces cerevisiae*	pO_2	[93]
Phenylalanine	*Leuconostoc mesenteroides*	pO_2†	[94]
Serine	*Clostridium acidiurici*	pNH_3	[95]
Thiamin	*S. cerevisiae*	pO_2	[96]
Tyrosine	*Aeromonas phelogenes*	pNH_3	[97]
Urea	nitrifying bacteria	pO_2	[98]
	Proteus mirabilis	pNH_3	[99]
		metal or carbon electrode (amperometry)	
Steroid hormones (*3-keto-4-ene steroids*)	*Nocardia opaca*	*rotating glassy carbon electrode*	[100]
Vitamin B_1	*Lactobacillus fermentum*	Pt anode-Ag_2O cathode	[78]

† Lactate electrode (lactate oxidase + pO_2 electrode)

Reagent

This part of the market represents $830 M. Leaders are: Technicon Co., Du Pont (USA), Boerhinger (FRG), Harleco (USA), and Merck (FRG). They represent 30% of the worldwide market. This market is tightly linked to the equipment market. Most of the companies competing in the reagent market use a common marketing strategy. It is the reagent, rather than the equipment, that generates the greater part of the profits. In some cases, the equipment is given to the customer, at no extra charge.

It is possible to extract from this economic overview some large profitable segments, which are listed in Table 6.

Table 4 — Examples of biosensors based on electrochemical transducers and applied to *in vitro* determination of biological compounds — immunoelectrodes

Parameter	Enzyme/antibody	Electrode	References
Cyclic AMP	urease-labelled antibody	pNH_3	[101]
Hepatitis B surface antigen (HB_sAg)	peroxidase-labelled anti-HB_5Ag	I^-	[5,102]
α-Foetoprotein (AFP) antigen	catalase-labelled AFP	pO_2	[103]
Human chorionic gonadotropin (hCG)	catalase-labelled anti-hCG	pO_2	[104,105]
Immunoglobulin G (IgG)	catalase-labelled anti-IgG	pO_2	[106,107]
Insulin	glucose oxidase- or catalase-labelled insulin antibody	pO_2	[108,109]
Serum albumin (SA)	catalase-labelled SA antibody	pO_2	[108]
Bovin serum albumin	unrease-labelled anti-IgG	pNH_3	[110]
Theophilline	catalase-labelled theophilline antibody	pO_2	[111]

Table 5 — Examples of biosensors based on electrochemical transducers and applied to *in vitro* determination of biological compounds — tissue electrodes

Parameter	Tissue	Electrode	References
L-ascorbate	gourd or cucumber (ascorbate oxidase)	pO_2	[81,112]
AMP	rabbit muscle (AMP deaminase)	pNH_3	[113]
Dopamine	banana pulp tissue slices (polyphenol oxidase)	pO_2	[114]
Glutamine	porcin kidney (glutaminase)	NH_4^+	[115]

There is a strong probability of future development in two segments:

- a narrow segment with high growth: ion-selective electrodes and biosensors, auto-analysers for microbiology and immunology;
- a new segment representing a strong competitive technology: monoclonal antibodies and dry chemistry with their specific equipments.

The dry chemistry concerns the specialized market but also the general public through home analysis (pregnancy tests, glycemia tests for diabetics).

Industry is also concerned with enzymatic methods of analysis and biosensors. The industrial market is less well defined, since some industrial groups develop in-house analytical procedures. Usually enzymatic methods of analysis are used at the laboratory level and rarely on the production line. Samples are pretreated before measurement (dilution, dialysis, centrifugation, bleaching, . . .) just as biological fluids can be treated in clinical analysis. However, a glucose analyser developed for

Table 6 — Market segments of biosensors in Europe. From Sorade's estimation in *Biosensors market in western Europe*, Frost & Sullivan Reports, Frost & Sullivan Ltd, London (1986)

Segments	Annual growth (%)	1986 Volume ($M)
Reagents — Immunology†	+ 15.3	93
— Enzymology	+ 5.7	47
— Microbiology	+ 9.9	31
— Coagulation	+ 12.3	22
Equipment — Monoparametric analyser	+ 10	27
— hematology analyser	+ 6.5	10
— HPLC	+ 7.5	4
— Gamma counter	+ 7.6	3.6

† serology, blood bank, research excluded.

medical analysis can easily be used for industrial purposes. This is true also for analytical methods using biosensors.

In the past, technical innovations in enzymatic analysis always originated from medical applications and were then transferred to industrial application. However, current industrial needs for real-time analysis to control a process cannot be satisfied by existing methods of analysis, and they induce research in this field. Potentially, developments in biosensors can open up a new industrial market because they are able to deliver a fast response on a crude sample.

On the other hand, developments in molecular biology have stimulated new opportunities in both microbial and mammalian cell culture. The development of these production methods results in a greater need for control systems to operate the processes with maximum efficiency. On-line analytical techniques, able to deliver data in real time, are needed by biotechnological companies and have motivated the development of biosensors.

4.2 Biosensor market

The commercially developed probes (Table 7) use purified oxidase enzymes, mostly coupled with oxygen or hydrogen electrochemical transducers. The major reason for this choice is the storage stability and the specificity of oxidases, and also because the enzymatic reaction does not require any cofactors.

The enzyme reaction catalysed by such oxidases is represented by the equilibrium:

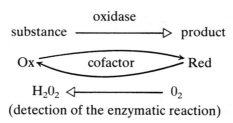

(detection of the enzymatic reaction)

Table 7 — Commercially available devices based on electrochemical biosensors

Company	Instrument	Parameter	Enzyme	Concentration range (mM)	Sample rate per hour	Stability (no. of samples)
Japan						
Fugi	Gluco 20A	glucose	GOD	0–30	80	500
	UA-30A	maltose	maltese + GOD			
		urate	uricase	7–65	50	
Daiichi Kagaku	Autostat GA 1110	glucose	GOD	1–40	35	
A.I.C.	Glucoroder-E	glucose	GOD	0–40	80	6000
Toyobo Omron	HER 100	lactate	lactate OX	0–8.3	20	10 days
USA						
Y.S.I.	Model 23A	glucose	GOD	1–45	40	300
		alcohol	alcohol OX	0–65	20	7 days
	Model 27	lactose	galactose OX + GOD	0–55	60	10 days
		galactose	id.	0–55	40	10 days
Europe						
Z.W.G. (DDR)	Gkm 01	glucose	GOD	1–50	60	1000
		urate	uricase	0.2–1	40	
Aplama (Hung.)	Enzalyst	glucose	GOD	0.5–30	20	

Setric GI (France)		lactate	lactate dehydrogenase	0.05–8	60	
Solea (France)	Glucoprocess	glucose	GOD	0–1	40	500
		maltose	maltase + GOD	0–0.5	20	
Seres (France)	Enzymat	glucose	GOD	0.2–20	40	3000
		lactose	lactase + GOD	1–20	40	1000
		sucrose	invertase +			
			GOD	0.3–10	40	1000
		lysine	lysine OX	0.1–2	40	1600
		glutamate	glutamate OX	0.1–2	40	1600
		alcool	alcool OX	1–20	40	400
		lactate	E. coli	1–50	40	1000
		pyruvate	E. coli	1–50	40	1000
		succinate	E. coli	1–50	40	1000
		glutamin	glutaminase + glutamate	0.1–3	40	1000

GOD = glucose oxidase, OX = oxidase.

In the USA the first analyser using an enzyme electrode was the LACTATE 640 from the Roche company (marketed in 1976). This semi-continuous flow analyser was for the clinical analysis of L-lactate in blood. It used cytochrome b_2, retained by a microporous membrane, on a platinum electrode surface [116]. The electrode signal corresponds to the electrochemical reoxidation of hexacyanoferrate III.

The analyser was taken off the market a few years ago. In 1979, following the work of Clark & Duggan [117], the Yellow Spring Instruments company commercialized the analyser Model 23 A. This glucose analyser was also aimed at the medical market. From 1980 to 1983, 800 of these analysers were sold in the USA. The measurement principle is based on the electrochemical determination of hydrogen peroxide production by the glucose oxidase reaction. The enzyme is entrapped inside a gel contained in a kind of dialysis bag, in close contact with the anode (see Chapter 2 (§2.3). Recently, the same company commercialized the Model 27 which allows lactose, sucrose, starch, and ethanol determinations.

In Japan, the Fuji Electric company is about to market two analysers, the manual GLUCO 20A, and the fully automatic UA300, for glucose determination. The biosensor detects the H_2O_2 production of the GOD reaction. The enzyme is immobilized inside a photopolymerized collagen support. A macroporous H_2O_2-selective membrane has been studied by Osawa *et al.* [118] to avoid the high risk of interference encountered by YSI analysers. The same approach is applied to estimate L-lactate concentrations using a lactate oxidase with the LACTATE HER100 made by the Toyobo–Omron company following the studies of Tsuchida & Yoda [119]. The glucose analyser AUTO STAT GA1110 made by the Daiichi Kagaku company is a continuous flow type analyser. Fully automatized and computerized, its target is the medical market. The H_2O_2 produced by the GOD reaction is measured.

The Glucoroder E is a glucose analyser made by the AIC company (a subsidiary company of Asahi). Initially planned under three versions — M100 manual, AS200 semi-automatic, AD300 fully automatic — only the last model is commercialized today. The measurement principle is an oxygen consumption detected by a pO_2 sensor. The glucose oxidase is immobilized inside an undescribed support.

In Europe, following the studies of Kulis *et al.* [26], the Aplama company markets the GO4 ENZALYST specifically for glucose determination. The equipment is manual. The H_2O_2 produced by the GOD reaction is estimated by using peroxidase activity and hexacyanoferrate II as an electron exchanger. The enzymes are maintained in close contact with a glassy-carbon electrode polarized vs a silver/silver chloride reference electrode. The signal is constituted by a current used to reduce the hexacyanoferrate II at the cathode.

The ZWG company commercializes a glucose analyser GLUCOMETER 01 based on work by Pfeiffer *et al.* [120]. A pO_2 electrode is covered by a piece of silk tissue impregnated with gelatin polymerized by glutaraldehyde which contains the enzyme. The same equipment is under development for urate determination [121]. Betermann *et al.* [122], with the same company, are working on the incorporation of a series of sensors inside a continuous flow analyser. The final equipment will be able to measure 120 samples per hour.

In France, the Sétric Génie Industriel company, following the studies of Durliat *et al.* [42], is selling an analyser for L-lactate determination, for both the industrial and medical markets. The enzyme, lactate dehydrogenase, is free, separated from the sample by a microporous dialysis membrane. Ferricyanide is used as an electron exchanger. The signal results from the electrochemical reoxidation current of the electron exchanger.

The GLUCOPROCESSEUR is a semi-automatic glucose analyser made by the Solea Tacussel company following the studies of Coulet & Bertrand [123]. Glucose oxidase is immobilized on the surface of a collagen support fixed onto an anode measuring the H_2O_2 production of the enzyme reaction. A second electrode without enzyme allows a differential measurement. New measurements such as maltose and cholesterol are in development.

The Seres company, using the method described by Romette *et al.* [11], sells a fully automated system, ENZYMAT, using different oxidase enzymes immobilized inside a gelatin matrix by glutaraldehyde. The active support is fixed on a pO_2 electrode estimating the O_2 consumption of the enzymatic reaction. Ten parameters can be measured by this equipment. The measurement time is around 2 minutes per sample.

A glutamine sensor can be discussed as an example of the kind of biosensor used in such an analyser. The active element is a gelatin support reticulated by glutaraldehyde containing two enzymes catalysing the sequential reactions:

$$\text{glutamine} + H_2O \xrightarrow{\text{glutaminase}} \text{L-glutamate} + \text{ammonia}$$

$$\text{L-glutamate} + O_2 \xrightarrow[\text{oxidase}]{\text{glutamate}} \alpha\text{-ketoglutarate} + H_2O_2 \ .$$

The support is coupled with a pO_2 probe (Clark-type electrode). The L-glutamine concentration range is from 0.1 to 3 mM. Compared to the classical methods, as described in Table 8, this new method is simpler and faster, allowing real-time and on-line measurements.

The major shortcoming of these analytical systems is the limited operational stability of the sensors. Oxidase enzymes typically denaturate through the biocatalysis itself rather than by thermodenaturation. Bourdillon *et al.* [130] demonstrated that oxidase enzymes have a catalytic power limited to a fixed number of substrate molecules. This number is specific to the enzyme. The immediate consequence of such behaviour is a fixed number of measurements available for an average sample substrate concentration, after which the biosensor must be replaced.

The phenomenon may be reduced by using another electron exchanger instead of oxygen. Recent studies describe the electrochemical recycling of oxidase cofactor directly at the electrode level. These interesting developments are not obviously compatible with easy-to-use analytical techniques. Another disadvantage of the use of oxidase is the sample O_2 dependence of the reaction rate. Oxidase enzymes catalyse a two-substrate reaction (the reaction rate equation is 'ping pong' and not Michaelis type). The reaction rate depends not only on the analysed substrate concentration such as glucose, lysine, or lactate, but also on the sample O_2 content.

Table 8 — Methods for L-glutamine analysis

Enzyme	Source	Detection	Concentration range	References
Glutaminase	mitochondria	NH_3 electrode	0.1–5 mM	[4]
Glutamine synthetase	E. coli	+ Fe(III) chromogen	0.5–10 mM	[124]
Glutaminase + glutamine DH	pig kidney	NADH		[125]
Glutaminase + glutamine DH	E. coli	NAD(P)		[125]
Glutaminase	E. coli	Nessler's reagent	0.05–4 mM	[126]
Glutaminase	pig kidney tissue	NH_3 electrode	0.02–6 mM	[127]
Glutaminase	Sarcina flava	NH_3 electrode	0.06–10 mM	[128]
Glutaminase	E. coli	NH_4^+ electrode		[129]
Glutamine DC	E. coli	CO_2 electrode		[125]

DC = decarboxylase, DH = dehydrogenase

The sample O_2 content may vary widely from one sample to another in blood analysis (from 18 to 60 mm Hg) or in fermentation broth analysis (from 0 to 60 mm Hg). Different procedures are used to stabilize the O_2 level during the measurement step. For example, the YSI company includes in the measurement process a strong sample dilution with an air-saturated buffer and introduces air directly inside the measurement cell through an air-permeable vibrating membrane. Seres, in the ENZYMAT, uses the enzyme support properties to supply the enzyme reaction with enough O_2 to make the reaction rate dependent only on the substrate concentration of the analysed sample.

All these systems are attractive because of the complement they provide to more classical analytical methods such HPLC or colorimetric determinations. They have major advantages: lower cost, fast response, and simplicity of procedure.

5   Conclusion — new trends
Because of the intense commercial interest in biosensor development, much of the more sophisticated and exciting new biosensor technology is being kept secret or is in the process of being patented. However, some potential developments can be mentioned.

At the transducer level, relatively large platinum and gold electrodes in many of the amperometric biosensors are being replaced by carbon and graphite substrates, and the search for new electrode coatings and material continues. Much research is being carried out on conducting organic polymers such as polypyrrole, polythiophene, and polyacetylene. Attempts to overcome early problems of instability and relatively poor mechanical properties of these polymers have been partly resolved by copolymerization with compounds such as polystyrene.

Large-scale development will come not only with inexpensive electrode materials, but also with miniaturization. Many workers have attempted to approach this by starting with the available microelectronics technology, or optical techniques using optical fibres. For instance, IsFETs are adapted for biosensor use. For potentiometric biosensors this will open up the possibilities of multi-head, multiparameter devices. The problem of 'pin-hole' generation on exposure of the silicon oxide layer of FET-based biosensors to aqueous samples can now be overcome to a large extent by complete encapsulation in multiple layers of photosetting polymers; sapphire can replace silicon as a substrate.

At the current rate of development, the silicon-processing industries will reach the theoretical maximum in packing densities in ten years' time. The inevitable 'electronic devices at a molecular level' will emerge some time in the future. The concept of the biochip springs from two basic intriguing aspects of molecular biology: biopolymers can self-assemble and molecules such as DNA can store, replicate, and transmit information. Such developments are expected to lead to an integrated-circuit biosensor and perhaps to an intelligent bioprobe.

Optoelectronic and optical fibre biosensors offer several advantages over the electrical methods discussed so far, not only in terms of safety, cost, and mechanical flexibility, but also in suitability for disposable construction. These devices have smaller temperature coefficients and require no reference component, and thus have an inherently simple design.

From the biochemical standpoint, the research is focused on three types of bioactivity:

- biocatalytic sensors, using immobilized enzymes, cells, plant or animal tissue slices;
- immunoreaction incorporating antibody and antigen conjugates for selective measurement of drugs, proteins, hormones, . . . ,
- receptor-based sensors using animal sensory structures.

The purification of new biocatalytic active substances is still a very attractive field. The production of an artificial enzyme is on the point of opening up the range of the parameters determined with biosensors. The first-generation biosensors resulting from this research are the active elements of the analysers commercially available today.

The second generation of biosensors will use immunological reactions. Antibody–antigen interactions, or the more complicated complement-mediated processes, can be extremely selective and sensitive. In attempts to construct this type of biosensor, bioluminescence, chemiluminescence, and fluorescence principles are being explored. These would have enormous potential because of the sensitivity which can be achieved. One of the major problems to be solved in the design of an efficient immunobiosensor is the speed of measurement. In the near future the cost of monoclonal antibodies should decline to a point where their use in biosensors becomes economically attractive.

Perhaps the most fascinating research in biosensors involves the development of chemoreceptor structures as the active element. The sensitivity and the selectivity may be enormously improved. Response and recovery times at least 100 times more rapid than those of biocatalytic biosensors appear to be feasible. The potential in analytical chemistry of such chemoresponse phenomena has not yet been fully understood. Thus, this research represents the future for biosensor development.

Even if the development of laboratory analysers remains the most important application of biosensors, the most challenging area of their application is that of internal devices for diagnosis and management.

Maintenance of a blood-compatible sensing surface for extended periods has proved extremely difficult. Approaches based on interposing biological buffer layers (heparin, antiplatelet agents) between the sensing surface and the blood have met with limited success. Promising results have been obtained with a sensor coating based on the patient's own serum albumin.

Biosensor technology has yet to mature sufficiently to acquire commercial respectability, and is currently in danger of being pushed too far and too fast.

References

[1] Clark, L. C., & Lyons, C.H. (1962) *Ann. N.Y. Acad. Sci.* **102**, 29–45.

[2] Updike, S. J., & Hicks, G. P. (1967) *Nature* **214**, 986–988.

[3] Bucher, T. H. (1974) In: Bergmeyer, H.U. (ed.) *Methods of enzymatic analysis*. 2nd ed., vol.1. New York, Academic Press p. 254.

[4] Arnold, M. A., & Rechnitz, G. A. (1980) *Anal. Chem.* **52**, 1170–1174.

[5] Boitieux, J. L., Desmet, G., & Thomas, D. (1978) *Clin. Chim. Acta* **88**, 329–336.

[6] Janata, J., & Huber, R. J. (1979) *Ion-selective Electrode Rev.* **1**, 31–79.

[7] Caras, S., & Janata, J. (1980) *Anal. Chem.* **52**, 1935–1937.

[8] Nilsson, H., Akerlund, A. C., & Mosbach, K. (1973) *Biochim. Biophys. Acta* **320**, 529–534.

[9] Enfors, S. O., & Nilsson, H. (1979) *Enzyme Microb. Technol.* **1**, 260–274.

[10] Clark, L. C. (1956) *Trans. Am. Soc. Artif. Intern. Organs* **2**, 41–47.

[11] Romette, J. L., Froment, B., & Thomas, D. (1979) *Clin. Chim. Acta* **95**, 249–256.

[12] Stow, R. W., Baer, R. F., & Randall, B. F. (1957) *Arch. Phys. Med.* **38**, 646–650.

[13] Severinghaus, J. W., & Bradley, A. F. (1958) *J. Appl. Physiol.* **13**, 515–520.

[14] Tran, N. D., Romette, J. L., & Thomas, D. (1983) *Biotechnol. Bioeng.* **25**, 329–340.

[15] Rechnitz, G. A., Arnold, M. A., & Meyerhoff, M.E. (1979) *Nature* **278**, 466–467.

[16] Schoolwerth, A. C., Nazar, B. L., & Lanoue, K. F. (1978) *J. Biol. Chem.* **253**, 6177–6183.

[17] Cass, A. E. G., Davis, G., Francis, G. D., Hill, H. A. O., Aston, W. J., Higgins, I. J., Plotkin, E. V., Scott, L. D. L., & Turner, A. P. F. (1984) *Anal.Chem.* **56**, 667–671.

[18] Durand, P., David, A., & Thomas, D. (1978) *Biochim. Biophys. Acta* **527**, 277–282.

[19] Caras, S. D., Petelenz, D., & Janata, J. (1985) *Anal. Chem.* **57**, 1920–1923.

[20] Hanazano, Y., Nakako, M., & Shiono, S. (1986) *IEEE Trans. Electron. Devices* **33**, 47–51.

[21] Kimura, J., Kuriyama, T., & Kawana, Y. (1986) *Sensors Actuators* **9**, 373–387.

[22] Chen, A. K., & Liu, C. C. (1979) *Process Biochem.* **14**, 12–22.

[23] Guilbault, G. G., & Nagy, G. (1973) *Anal. Chem.* **45**, 417–419.

[24] Flanagan, M. T., & Carroll, N. J. (1986) *Biotechnol. Bioeng.* **28**, 1093–1099.

[25] Ianniello, R. M., & Yacynych, A. M. (1983) *Anal. Chim. Acta* **146**, 249–253.
[26] Kulys, J. J., Laurinavicius, V., Pesliakiene, M., & Gureviciene, V. (1983) *Anal. Chim. Acta* **148**, 13–18.
[27] Miyahara, Y., Moriizumi, T., & Ichimura, K. (1985) *Sensors Actuators* **7**, 1–10.
[28] Anzai, J., Tezuka, S., Osa, T., Nakagima, H., & Matsuo, T. (1986) *Chem. Pharm. Bull.* **34**, 4373–4376.
[29] Bertrand, C., Coulet, P. R., & Gautheron, D. C. (1981) *Anal. Chim. Acta* **126**, 23–34.
[30] Davis, G., Green, M. J., & Hill, H. A. O. (1986) *Enzyme Microb. Technol.* **8**, 349–352.
[31] Dicks, J. M., Aston, W. J., Davis, G., & Turner, A. P. F. (1986) *Anal. Chim. Acta* **182**, 103–112.
[32] Thevenot, D. R., Sternberg, R., Coulet, P. R., Laurent, J., & Gautheron, D. (1979) *Anal. Chem.* **51**, 96–100.
[33] Durliat, H., & Comtat, M. (1984) *Anal. Chem.* **56**, 148–152.
[34] Hill, H. A. O. (1985) *Anal. Proc.* **22**, 201–202.
[35] Murakami, T., Nakamoto, S., Kimura, J., Kuriyama, T., & Karube, I. (1986) *Anal. Lett.* **19**, 1973–1986.
[36] Ikeda, T., Hamada, H., Miki, K., & Senda, M. (1985) *Agric. Biol. Chem.* **49**, 541–543.
[37] D'Costa, E. J., Higgins, I. J., & Turner, A. P. F. (1986) *Biosensors* **2**, 71–87.
[38] Mullen, W. H., Churchouse, S. J., Keedy, F. H., & Vadgama, P. M. (1986) *Clin. Chim. Acta* **157**, 191–198.
[39] Mullen, W. H., Churchouse, S. J., Keedy, F. H., & Vadgama, P. M. (1986) *Anal. Proc.* **23**, 145–146.
[40] Shubert, F., Kirstein, D., Appelqvist, R., Gorton, L., & Johansson, G. (1986) *Anal. Lett.* **19**, 1273–1288.
[41] Kihara, K., Yasukawa, E., & Hirose, S. (1984) *Anal. Chem.* **56**, 1876–1880.
[42] Durliat, H., Comtat, M., Mahenc, J., & Baudras, A. (1976) *J. Electroanal. Chem.* **66**, 73–76.
[43] Durliat, H., Comtat, M., & Mahenc, J. (1979) *Anal. Chim. Acta* **106**, 131–135.
[44] Mizutani, F., Sasaki, K., & Shimura, Y. (1983) *Anal. Chem.* **55**, 35–38.
[45] Suaud-Chagny, M. F., & Gonon, F. G. (1986) *Anal. Chem.* **58**, 412–415.
[46] Palleschi, G. (1986) *Anal. Lett.* **19**, 1501–1510.

[47] Bradley, C. R., & Rechnitz, G. A. (1986) *Anal. Lett.* **19**, 151–162.
[48] Lubrano, G. J., & Guilbault, G. G. (1978) *Anal. Chim. Acta* **97**, 229–236.
[49] Kuan, S. S., Ngeh-Ngwainbi, J., & Guilbault, G. G. (1986) *Anal. Lett.* **19**, 887–899.
[50] Narijo, M., & Guilbault, G. G. (1974) *Anal. Chim. Acta* **75**, 169–180.
[51] Jaenchen, M., Gruening, G., & Betermann, K. (1985) *Anal. Lett.* **18**, 1799–1820.
[52] Nikolelis, D. P., & Hadjiioannou, T. P. (1983) *Anal. Lett.* **16**, 401–416.
[53] Nikolelis, D. P. (1984) *Anal. Chim. Acta* **161**, 343–348.
[54] Neujarhr, H. Y. (1980) *Biotechnol. Bioeng.* **22**, 913–918.
[55] Satoh, I., Karube, I., & Suzuki, S. (1977) *Biotechnol. Bioeng.* **19**, 1095–1099.

[56] Mascini, M., Tomassetti, M., & Iannello, M. (1983) *Clin. Chim. Acta* **132**, 7–15.

[57] Matsumoto, K., Seijo, H., Karube, I., & Suzuki, S. (1980) *Biotechnol.Bioeng.* **22**, 1071–1086.

[58] Campanella, L., Mascini, M., Palleschi, G., & Tomassetti, M. (1986) *Clin. Chim. Acta* **151**, 71–83.

[59] Guilbault, G. G., & Coulet, P. R. (1983) *Anal. Chim. Acta* **152**, 223–228.

[60] Winquist, F., Lundstrom, I., & Danielsson, B. (1986) *Anal. Chem.* **58**, 145–148.

[61] Jensen, M. A., & Rechnitz, G. A. (1977) *J. Membr. Sci.* **5**, 117–127.

[62] Romette, J. L., & Cooney, C. L. (1987) *Anal. Lett.* **20**, 1069–1081.

[63] Cheng, F. S., & Christian, G. D. (1979) *Clin. Chim. Acta* **91**, 295–301.

[64] Mizutani, F., Yamanaka, T., Tanabe, Y., & Tsuda, K. (1985) *Anal. Chim. Acta* **172**, 153–166.

[65] Weaver, M. R., & Vadgama, P. M. (1986) *Clin. Chim. Acta* **155**, 295–308.

[66] Johansson, G., Edstrom, K., & Ogren, L. (1976) *Anal. Chim. Acta* **85**, 53–57.

[67] Romette, J. L., Yang, J. S., Kusukabe, H., & Thomas, D. (1983) *Biotechnol. Bioeng.*, **25**, 2557–2566.

[68] Macholan, L. (1978) *Coll. Czechosl. Chem. Com.* **43**, 1811–1817.

[69] Berjonneau, A. M., Thomas, D., & Broun, G. (1974) *Path. Biol.* **22**, 197–202.

[70] Ho, M. Y. K., & Rechnitz, G. A. (1987) *Anal. Chem.* **59**, 536–537.

[71] Rahni, M. A., Guilbault, G. G., & De Oliveira, G. N. (1986) *Anal. Chim. Acta* **81**, 219–225.

[72] Toyota, T., Kuan, S. S., & Guilbault, G. G. (1985) *Anal. Chem.* **57**, 1925–1928.

[73] Tran Minh, C., & Broun, G. (1973) *C.R. Acad. Sci. (Paris)* **276**, 2215–2217.

[74] Johansson, G., & Ogren, L. (1976) *Anal. Chim. Acta* **84**, 23–29.

[75] Mascini, M., & Guilbault, G. G. (1977) *Anal. Chem.* **49**, 795–798.

[76] Winquist, F., Spetz, A., Lundstrom, I., & Danielsson, B. (1984) *Anal.Chim. Acta* **163**, 143–149.

[77] Matsumoto, K., Seijo, H., Watanabe, T., Karube, I., Satoh, I., & Suzuki, S. (1979) *Anal. Chim. Acta* **105**, 429–432.

[78] Matsunaga, T., Karube, I., & Suzuki, S. (1978) *Anal. Chim. Acta* **99**, 233–240.

[79] Kingdon, C. F. M. (1985) *Appl. Microbiol. Biotechnol.* **22**, 165–168.

[80] Rechnitz, G. A., Kobos, R. K., Riechel, S. J., & Gebauer, C. R. (1977) *Anal. Chim. Acta* **94**, 357–365.

[81] Vincke, B. J., Devleeschouwer, M. J., & Patriarche, G. J. (1985) *Anal. Lett.* **18**, 1593–1606.

[82] Vincke, B. J., Devleeschouwer, M. J., & Patriarche, G. J. (1983) *J. Pharm. Belg.* **38**, 225–229.

[83] Rennenberg, R., Riedel, K., & Scheller, F. (1985) *Appl. Microbiol. Biotechnol.* **21**, 180–181.

[84] Kobos, R. K., & Rechnitz, G. A. (1977) *Anal. Lett.* **10**, 751–758.

[85] Wollenberger, U., Scheller, F., & Atrat, P. (1980) *Anal. Lett.* **13**, 825–836.

[86] Karube, I., Mitsuda, S., & Suzuki, S. (1979) *Eur. J. Appl. Microbiol.Biotechnol.* **7**, 343–350.

[87] Vais, H., Oancea, F., Faghi, A.M., Delcea, C., & Margineanu, D.G. (1985)

Rev. Roum. Biochim. **22**, 57–62.

[88] Hikuma, M., Ooana, H., Yasuda, T., Karube, I., & Suzuki, S. (1980) *Anal. Chim. Acta* **116**, 61–67.

[89] Rechnitz, G. A., Riechel, T. L., Kobos, R. K., & Meyerhoff, M. E. (1978) *Science* **199**, 440–441.

[90] Walters, R. R., Moriarty, B. E., & Buck, R. P. (1980) *Anal. Chem.* **52**, 1680–1684.

[91] Burstein, C., Adamowicz, E., Bouchent, K., & Romette, J. L. (1986) *Appl. Biochem. Biotechnol.* **12**, 1–15.

[92] Vincke, B. J., Devleeschouwer, M. J., Dony, J., & Patriarche, G. J. (1984) *Int. J. Pharm.* **21**, 265–275.

[93] Karube, I., Matsunaga, T., & Suzuki, S. (1979) *Anal. Chim. Acta* **109**, 39–44.

[94] Matsunaga, T., Karube, I., Terakoa, N., & Suzuki, S. (1981) *Anal. Chim. Acta* **127**, 245–249.

[95] Di Paolantonio, C. L., Arnold, M. A., & Rechnitz, G. A. (1981) *Anal. Chim. Acta* **128**, 121–127.

[96] Mattiasson, B., Larsson, P. O., Lindahl, L., & Sahlin, P. (1982) *Enzyme Microb. Technol.* **4**, 153–157.

[97] Di Paolantonio, C. L., & Rechnitz, G. A. (1982) *Anal. Chim. Acta* **141**, 1–13.

[98] Okada, T., Karube, I., & Suzuki, S. (1982) *Eur. J. Appl. Microbiol. Biotechnol.* **11**, 149–154.

[99] Vincke, B. J., Devleeschouwer, M. J., & Patriarche, G. J. (1983) *Anal. Lett.* **16**, 673–684.

[100] Wollenberger, U., Scheller, F., & Atrat, P. (1980) *Anal. Lett.* **13**, 1201–1210.

[101] Boitieux, J. L., Lemay, C., Desmet, G., & Thomas, D. (1981) *Clin. Chim. Acta* **113**, 175–182.

[102] Lowe, C. R., & Goldfinch, M. J. (1983) *Biochem. Soc. Trans.* **11**, 448–451.

[103] Aizawa, M., Morioka, A., & Suzuki, S. (1980) *Anal. Chim. Acta* **115**, 61–67.

[104] Yamamoto, N., Nagasawa, Y., Shuto, S., Sawai, M., Suto, T., & Tsubomura, H. (1978) *Chem. Lett.* **3**, 245–246.

[105] Wingard, L. B. (1982) *J. Electroanal. Chem.* **141**, 307–312.

[106] Aizawa, M., Wada, M., & Suzuki, S. (1979) *J. Solid-phase Biochem.* **4**, 251–268.

[107] Yamamoto, N., Nagasawa, Y., Shuto, S., Tsubomura, H., Sawai, M., & Okumura, H. (1980) *Clin. Chem.* **26**, 1569–1572.

[108] Mattiasson, B., Mosbach, K., & Svenson, A. (1977) *Biotechnol. Bioeng.* **19**, 1643–1651.

[109] Kobos, R. K., & Ramsey, T. A. (1980) *Anal. Chim. Acta* **121**, 111–118.

[110] Meyerhoff, M. E., & Rechnitz, G. A. (1979) *Anal. Biochem.* **95**, 483–493.

[111] Itagaki, H., Hadoka, Y., Suzuki, Y., & Haga, M. (1983) Chem. Pharm. Bull. **31**, 1283–1288.

[112] Macholan, L., & Chemlikova, B. (1986) *Anal. Chim. Acta* **185**, 187–193.

[113] Fiocchi, J. A., & Arnold, M. A. (1984) *Anal. Lett.* **17**, 2091–2109.

[114] Sidwell, J. S., & Rechnitz, G. A. (1985) *Biotechnol. Lett.* **7**, 419–422.

[115] Arnold, M. A. (1985) *Anal. Chem.* **57**, 565–566.

[116] Racine, P., Engelhardt, R., Higelin, J. C., & Mindt, W. (1975) *Med. Instrum.* **9**, 11–14.

[117] Clark, L. C., & Duggan, C. A. (1982) *Diabetes Care* **3**, 174–180.
[118] Osawa, H., Akiyama, S., & Hamada, T. (1983) In: Seiyama, T., Fueki, K., Shiokama, J., & Susuki, S. (eds.), *Proceedings of the First International Meeting on Chemical Sensors*, *Fukuoka*, 19–22 *September* 1983. Amsterdam, Elsevier p. 163–168.
[119] Tsuchida, T., & Yoda, K. (1984) *Lactate Analyser HER* 100, Techn. Inf. Toyobo Co., Japan.
[120] Pfeiffer, D., Scheller, F., Janchen, M., & Weise, H. (1980) *Anal. Lett.* **13**, 1179–1185.
[121] Scheller, F. W. (1985) *Biosensors* **1**, 135–160.
[122] Bertermann, K., Lutter, J., Lucius, I., Elze, P., & Scheller, F. (1983) *Z. Med. Lab. Diagn.* **24**, 315–322.
[123] Coulet, P. R., & Bertrand, C. (1979) *Anal. Lett.* **12**, 581–590.
[124] Meister, A. (1985) In: Colowick, S. P., & Kaplan, N. O. (eds.) *Methods in enzymology*. Vol. 2. New York, Academic Press, p. 380–385.
[125] Guilbault, G. G. (1976) *Handbook of enzymatic methods of analysis*. New York, Marcel Dekker.
[126] Kikuchi, M. (1974) *Anal. Biochem.* **57**, 83–90.
[127] Rechnitz, G. A., Arnold, M. A., & Meyerhoff, M. E. (1979) *Nature* **28**, 466–467.
[128] Rechnitz, G. A., Reichel, T. L., Kobos, R. K., & Meyerhoff, M. E. (1978) *Science* **199**, 440–441.
[129] Guilbault, G. G., & Shu, F. R. (1971) *Anal. Chim. Acta* **56**, 333–338.
[130] Bourdillon, C., Vaughan, T., & Thomas, D. (1982) *Enzyme Microb. Technol.* **4**, 175–188.

1.5   ELECTROCHEMICAL DETECTION TECHNIQUES IN CLINICAL MICROBIOLOGY

G.A. Junter

1   Introduction

The rapid detection of bacterial infections and the identification of the organisms responsible, together with the short-term establishment of chemotherapeutic guidelines suitable for controlling the infectious disease, are essential priorities in clinical microbiology. The latter is most commonly based on *in vitro* tests of bacterial susceptibility to antibiotics (i.e. antibiograms), even though the antibacterial activity of antibiotics *in vitro* is not always accurately reflected at the site of infection. Therefore, the last twenty years have seen the development of various instrumental techniques or procedures aimed at detecting bacteria in clinical samples and evaluating the *in vitro* efficacy of antimicrobial agents.

The techniques used for detecting microorganisms and monitoring their growth, for example during their exposure to an inhibitor, are usually classified as direct, when cell number or cell mass are measured, or indirect, when the measured parameter reflects the cellular activity [1].

Direct counting techniques, which may or may not distinguish between living and dead organisms, are suitable for enumerating individual bacterial cells: optical microscopic observations or electronic particle counting (mainly Coulter Counter from Coulter Electronics Ltd, Luton, England [2]) [3]. Bacterial growth can also be followed by conventional enumeration procedures: plate counts providing the number of colony-forming units (individual viable cells, aggregates or chains), or the most probable number method [3]. These techniques are essentially manual, giving rise to delayed results, low frequency and high cost of analyses. Some automated systems ensure serial dilutions of samples, distribution and plating of aliquots onto nutrient agar plates, incubation and reading of the plates; though these systems do offer substantial help in routine counts, there is still a long delay in providing results. In the area of microscopic techniques, it is worth mentioning the recent appearance in France of a fully automated counting unit, associating an epifluorescence microscope with a video camera and allowing both the enumeration and the identification of microorganisms (Bioconcept System, Nice, France). Suitable more particularly for highly contaminated samples, this device should allow the quasi-continuous and real-time counting of organisms by increasing the frequency of observations.

The spectrophotometric measurement of light scattering by bacterial suspensions is a prevalent instrumental technique for monitoring bacterial growth. Here, bacterial growth is reflected by the biomass increase. Narrow-beam and multichannel high-performance photometers have been developed for continuously measuring the opacity of bacterial cultures and studying the kinetics of bacterial growth in the presence of antibiotics. Several semi-automated or automated industrial devices based on these photometric measurements have been proposed for enumerating bacteria in clinical samples (mainly urines) and for testing the susceptibility of microorganisms to antibiotics: Abbott MS-2 (Abbott Diagnostic Division, Dallas, USA), Autobac (Pfizer Diagnostics, New York, USA), Abac (Biotrol, Paris, France), API System (Analytab Products, and Plainview, USA), for example. A number of evaluations by comparison with conventional procedures have promoted the implantation of these devices in laboratories (see e.g. [4] and references therein). Optical techniques, however, show some significant drawbacks when applied in clinical microbiology. They cannot distinguish between viable and dead cells and are unsuitable for initially turbid media, e.g. blood. Moreover, biophotometric growth curves of bacterial cultures exposed to antibiotics may be difficult to interpret, in particular when the antibacterial agent induces important changes in the morphology of organisms.

The determination of intracellular components of microorganisms, such as proteins, nucleic acids, adenosine triphosphate, or flavin nucleotides, is another direct means to follow the growth of a microbial culture. For example, the bioluminescence ATP assay method has been applied with success to the detection and enumeration of microorganisms in clinical samples, more particularly in urines; several automated bioluminescence-measuring instruments designed for laboratory use exist at the present time (Lumac Biocounter, 3M, St. Paul, USA) [5–9].

In addition to these various techniques designed for the direct determination of cell populations, significant work has been devoted to the practical development of indirect procedures for detecting and monitoring microbial activity. While microbial metabolic activity can be reflected by cell proliferation (which is a consequence of that activity), it can also be monitored by measuring substrate input, metabolite, or end-product output, heat production, or else the reduction of electron acceptors allowing aerobic (i.e. oxygen) or anaerobic (e.g. NO_3^-, SO_4^{2-}) respiration of organisms. One of the diverse techniques reported in the literature is the gas or gas–liquid chromatographic determination of gaseous or volatile metabolic end products, which is very useful for the identification of bacteria [10–13]. The measurement of carbon dioxide evolved by bacteria as a consequence of carbon-substrate oxidation has become more widely used with the development of radio-metric techniques: the appearance of $^{14}CO_2$ during the bacterial metabolism of ^{14}C-labelled glucose is followed by the use of a liquid scintillation counter or a gas counting chamber connected to an electrometer. This radiorespirometric procedure has been applied more particularly to the detection of bacterial infections in blood samples. A semi-automated device, the Bactec (Johnston Laboratories, Cockeys-ville, USA), has been subjected to a number of laboratory evaluations [14–25]. Microcalorimetry is another indirect procedure widely used in clinical microbiology [26–30]: power–time curves have been applied to the detection and monitoring of bacterial growth in biological media, to the identification of bacteria, and more particularly to the evaluation and characterization of the antibacterial effects of antibiotics or other drugs (susceptibility tests and bioassays [31–35], investigation of the kinetics of action [36–45]).

Several physical or chemical parameters reflecting either the overall activity of cultures or a precise metabolic pathway of organisms can also be monitored by electrochemical procedures. In particular, the measurements of electrical conducti-vity, redox potential, hydrogen-ion (pH) and dissolved oxygen content of the culture broth have been used for a long time in basic and applied microbiology. Neverthe-less, recent improvements have led to renewed interest in them, particularly in the area of clinical microbiology. This chapter presents both the historical background and the current and potential applications of these electrochemical measurements in clinical microbiology.

2 Impedance measurements

As early as 1899, Steward [46] observed that the electrical conductivity of culture media containing living bacteria changed during the growth of organisms. Following this pioneer work, some attempts were made during the first part of the twentieth century to apply electrical conductivity measurements to the study of bacterial metabolism [47–53]. However, it is only during the last twenty years that electrical impedance measurements in microbial cultures have been increasingly developed, with subsequent applications in the field of clinical microbiology.

2.1 Electrical impedance: basic notions

The impedance of an electrolytic conductor, such as a culture medium, measured between two metal electrodes immersed in the solution and submitted to an alternative potential difference, is a complex entity with both a resistive and a reactive component. On the one hand, the conductor offers an ohmic resistance to the movement of electric charges under the influence of the potential gradient; on the other hand, the temporary displacement of charges gives rise to reactive (i.e. capacitive) effects with no energy loss. In particular, the double layer of electric charges, which appears at the metal/liquid interface as a consequence of electrode polarization, behaves like a capacitor.

A number of workers concerned with the application of impedimetric measurements to biological materials, for example Schwan [54,55], have investigated the impedance characteristics of electrodes, namely of those used practically in clinical electrographic recording (electrocardiograms, encephalograms, etc.); they have proposed various equivalent electrical circuits for these impedances, consisting of different combinations of resistances and capacitances (see e.g. [56]).

Whereas the overall electrode impedance is both resistive and capacitive and is frequency-dependent, the bulk impedance of the electrolyte is mainly resistive and frequency-independent over a wide range of frequencies including those of interest in biological applications [56].

Within the framework of the present chapter, i.e. application of impedimetric techniques to bacterial culture broths, the reader can refer to a very simple equivalent representation of the measured impedance, consisting of a capacitance C (equivalent capacitance of the electric double layers at the electrode/medium interfaces) in series with a resistance R (electrode polarization resistance added to the bulk resistance of the culture broth) (Fig. 1). The impedance of this schematic equivalent circuit is $(R^2 + 1/C^2\omega^2)^{\frac{1}{2}}$, where ω is the angular frequency of the sinusoidal signal applied between the electrodes.

2.2 Equipment for impedance measurements

The essential equipment for impedance measurements in electroanalytical chemistry includes an alternative voltage generator capable of working at frequencies ranging from 1 to 300 kHz, a measuring-bridge circuit of the Wheatstone type in which the impedance cell is inserted (Fig. 2), and an electronic detector amplifying the output signal.

Ur was one of the first to suggest that microbial activity might be detected by monitoring changes in impedance of the culture medium [59,60]. He presented [59] a bridge circuit in which two measuring cells (namely a reference and a sample cell)

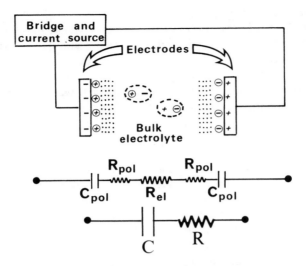

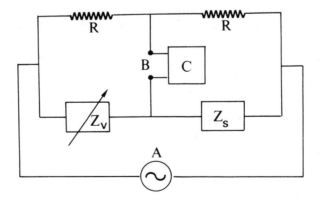

Fig. 1 — Schematic representation and equivalent circuit of an impedance-measuring cell. C_{pol} and R_{pol} are the polarization capacitance and resistance of each electrode, R_{el} is the resistance of the electrolyte, and C and R are the overall capacitance and resistance, respectively. From Eden & Eden [57,58].

Fig. 2 — Classical impedance-measuring bridge circuit. R, resistor; A, sinusoidal oscillator; B, bridge output; C, detector; Z_s, measured impedance; Z_v, variable impedance (series or parallel combination of variable capacitors and resistors). Z_v is adjusted to balance Z_s, i.e. the output signal is nulled, and the capacitive and resistive parts of Z_s are deduced from those of Z_v.

form opposing arms: the impedance of the test cell was continuously compared to that of the control cell. Ur made use of a commercial apparatus based on this impedance-measuring circuit ('Strattometer', Stratton and Co. Medical Ltd, Hatfield, UK) to study different events involved in blood coagulation [59–61]. With Brown, he later applied this device to impedance measurements in culture media [62–65]. A detailed description of the Strattometer instrument is given in reference [63]. Comparable impedance bridges, with both a reference and a sample cell, were also described later by Tsuchiya & Terashima [66] (Fig. 3).

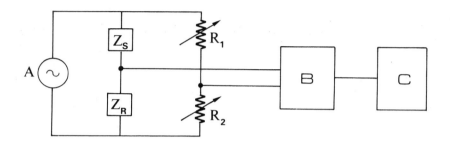

Fig. 3 — Bridge network with two measuring cells for comparative impedance measurements (from Tsuchiya & Terashima [66]). A, oscillator; R_1, R_2, variable resistors; Z_s, impedance of the sample cell; Z_r, impedance of the reference cell; B, amplifier; C, recorder. The sample cell receives the microbial inoculum, and the reference cell, identical with the test cell and containing the same quantity of the same nutrient broth, remains sterile during the measurements. The bridge is initially balanced by acting on R_1 and R_2. Then the offset voltage of the bridge is continuously recorded during bacterial growth. This output potential is proportional to
$$[R_2/(R_1 + R_2) - Z_r/(Z_r + Z_s)].$$

Most of the recent developments of impedance and conductance measurements in applied microbiology, however, are indebted to the existence of commercially available automated systems, namely the Bactometer Microbial Monitoring System (Bactomatic, Inc., Princeton, USA) and the Malthus Microbiological Growth Analyser (Malthus Instruments, Stoke-on-Trent, UK): it is worth giving some brief technical details of these two systems before approaching their applications in clinical and environmental microbiology (this chapter and Chapter 2 of Part 3, respectively).

2.2.1 Bactometer devices

The first commercial impedance-measuring device to be extensively employed in quantitative microbiology is the Bactometer 32 model. This instrument is based on an electrical circuit [67] similar to that described by Tsuchiya & Terashima [66] (Fig. 3). Once the sample cell has been inoculated, however, the bridge is automatically balanced and the impedance ratio $Z_r/(Z_r + Z_s)$ is recorded.

Impedance measurements are performed (generally at a fixed frequency of 2 kHz) in disposable multichambered units composed of couples of reference and sample chambers. The basic module [68,69] consists of eight pairs of measuring chambers (8 reference and 8 sample chambers) made of electrically nonconductive material and attached to a plastic support (Fig. 4, I). A stainless-steel lead frame is imbedded in the support, and the terminal portions of the leads project in spaced pairs into each chamber to form electrode couples (Fig. 4, II). The multichambered modules (up to four, i.e. 32 sample/reference pairs) are provided with sterile or contaminated broth and loaded into the incubator section of the system; during incubation, the instrument ensures cyclic determinations of the impedance ratio output for each pair of chambers. The duration of the complete scanning cycle for 32 pairs of chambers is 96 s.

Following the Bactometer 32 model, more recent sophisticated versions have been commercialized (Bactometer Model 120B, 123), which increase the number of samples that can be monitored, decrease the interval between two successive readings, and improve data handling and display. For example, the 120B model is capable of monitoring up to 128 samples in about 6 minutes; the results are automatically displayed on a video screen as digital data or impedance change curves. These versions, essentially designed for the screening of great numbers of samples, mostly use the same disposable modules as the earlier model (Fig. 4, I). The measuring circuit, however, is no longer a bridge arrangement, but a single cell electronic circuit which allows continuous readings of either impedance, conductance or capacitance [58].

Detailed descriptions of the different Bactometer devices can be found in the literature, e.g. Models 32 [70,71], 120B [72], or 123 [58] (Fig. 4, III).

2.2.2 *Malthus instruments*
The principle and electronics of the Malthus system have been presented by Richards *et al.* [73]. These authors have detailed different circuits, including one or two measuring cells and a detector with a phase-sensitive demodulator, designed for selectively measuring changes in conductance of the sample cell. They have concluded that two-cell bridge circuits, giving a higher resolution, are preferable for simultaneous measurements when only a few samples are examined (multichannel system), whereas unbalanced circuits are more particularly suitable for the analysis of a large number of samples (single measuring system ensuring cyclic determinations). Richards *et al.* have underlined that the temperature in the single measuring cell of an unbalanced circuit must be precisely controlled and regulated during the conductance determinations, since the conductance is temperature-dependent: they have stated that a convenient way to gain such constant temperature is to incubate the cell in a water bath [73].

The commercial Malthus devices (Malthus Microbiological Growth Analysers) make use of unbalanced electrical circuits, as do the more recent Bactometer instruments and for the same reasons (i.e. the number of analyses that most laboratories have to carry out daily). As in the competitive Bactomatic production, a range of increasingly sophisticated models is available, with from 8 up to 256 channels under microcomputer control. Individual measuring cells consist of reusable glass tubes (10 cm^3) or glass bottles (100 cm^3) fitted with the electrode assembly consisting of a pair of platinum electrodes fused to a ceramic strip (Fig. 5, I). In large-capacity models, individual conductance tubes are arranged in racks in

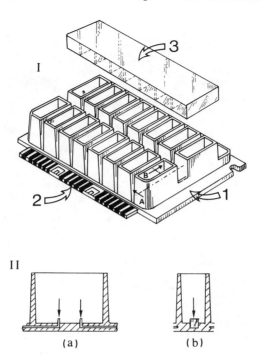

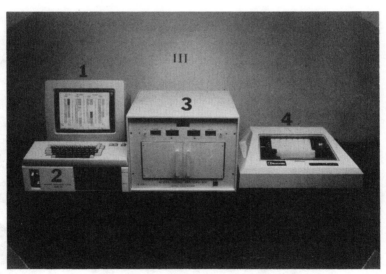

Fig. 4 — The Bactometer system. I, disposable module with its 16 separate electrode chambers: 1, electrically nonconductive base; 2, electrical contact strip; 3, plastic cover. From Cady *et al*. [68,69]. II, sections A (IIa) and B (IIb) of a single chamber, showing the pair of stainless-steel electrodes (↓) arising from the stainless-steel lead frame imbedded in the nonconductive support. From Cady *et al*. [68]. III, general view of the Bactometer 120 model: 1, video display unit; 2, microcomputer; 3, two incubators in which the modules are plugged; 4, printer.
Photograph provided by Bactomatic Ltd, Henley-on-Thames, England.

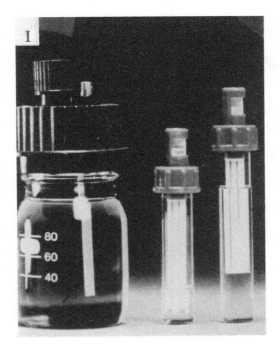

Fig. 5 — The Malthus system. I, individual reusable conductance-measuring cells with different volumes. II, incubator containing 16 racks of 8 conductance-measuring tubes. III, general view of a one-incubator system: 1, incubator; 2, computer control system and data display unit. Photographs provided by Malthus Instruments Ltd, Stoke-on-Trent, UK.

Fig. 5

groups of 8 and placed in a water bath, the temperature of which is controlled to 0.001°C. Each incubator contains 16 racks of 8 tubes, i.e. 128 tubes (Fig. 5, II), or 28 bottles. Each cell is scanned with a 10 kHz signal every 6 minutes. Different detection computer programs control the system. Conductance data are stored on disks and may be displayed as text or graphics. Malthus instruments (Fig. 5, III) have been extensively described recently [58,74].

2.3   Electrical impedance and microbial metabolism

2.3.1   Conductance

In early studies, different workers investigated the effects of microbial metabolism on the conductance of the culture medium [47–53]. They tried to relate the observed increase in conductance to changes in the chemical composition of the broth: metabolically active microorganisms are able to convert uncharged or weakly charged components of the broth (e.g. carbohydrates) to ionized end products with increased mobilities. For example, Parsons & Sturges [48,49] followed the time evolution of the culture broth conductivity during the growth of anaerobic bacteria (Clostridia); they were able to relate changes in conductivity to the bacterial production of ammonia. In the same way, Allison et al. [52], and later Kazal [53], reported that changes in the specific conductivity of Pseudomonas fluorescens cultures grown in skim milk [52] or in a synthetic glutamic acid medium [53] were directly proportional to the production of ammonia by this organism. During recent years, however, such investigations have been abandoned. The latest published literature does not report any noticeable attempt to identify metabolites involved in conductance variations or to propose a quantitative analysis of the conductance–time curves from the pathways followed by these metabolites. Priority has been given to more practical applications of conductance (or impedance) techniques to the detection and enumeration of bacteria in clinical or environmental samples, and workers have focused their interest mainly on the relations between conductance or impedance changes and microbial growth.

Thus Richards *et al.* [73], testing their conductance-measuring instruments with a culture of *Escherichia coli*, have observed that during the growth phase of bacteria the increase of the broth conductance is proportional to that of the cell population (Fig. 6). From the kinetic equations expressing exponential bacterial growth, they

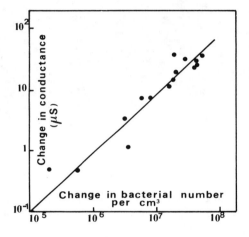

Fig. 6 — Linear relationship between change in conductance and change in bacterial number in growing cultures of *E. coli*. From Richards *et al.* [73].

have derived relations between the changes in conductance and the classical growth parameters (e.g. generation time of bacteria). In particular, they have obtained a linear relationship between the logarithm of the initial bacterial concentration of the broth and the time that elapses between the inoculation of the broth and the moment when a given change in conductance is reached:

$$\log(x_0) = \log(G_d/g) - 0.3\,(t_d - t_1)/p \qquad (1)$$

where x_0 is the initial concentration of viable bacteria in the culture broth, t_d is the time taken for the conductance to change by G_d, t_1 is the lag time preceding the exponential growth phase of bacteria, p is the generation time of bacteria, and g is the proportionality constant between the change in conductance and the change in bacterial number (Fig. 6).

In the same way, Eden & Eden [57] have proposed a simplified 'electrobacteriological' model to explain changes in the conductivity of culture broths during bacterial growth. Assuming that each bacterial cell generates the same amount K_b of conductive ions per unit time in the culture medium, these authors have expressed the concentration $C(t)$ of conductive ions in the broth at the time t:

$$C(t) = (K_b x_0 p/\ln 2) \exp [(t - t_1)\ln 2/p] \ . \tag{2}$$

Measurable increase in the broth conductivity can be detected when $C(t)$ reaches a certain level C_d: a linear relationship between $\log(x_0)$ and the corresponding detection time t_d, very similar to that reported by Richards *et al.*, can be derived from the preceding equation:

$$\log(x_0) = \log(C_d \ln 2/K_b p) - 0.3 \, (t_d - t_1)/p \ . \tag{3}$$

Equations (1) and (3) show that, for a given bacterial species cultured in given incubation conditions, it is possible to deduce x_0 from t_d, which demonstrates that conductance–time measurements may be applied to the counting of viable bacteria in an inoculum.

2.3.2   *Capacitance*
Metabolizing microorganisms are likely to modify the capacitive reactance of the electrodes by changing the composition of the double layer of charged species at the electrode/broth interface. However, relatively few studies have been devoted to the relations between microbial metabolism and the reactive (capacitive) component of impedance.

   To explain why they chose to design instruments that measure only the resistive part of impedance, Richards *et al.* [73] have reported that the capacitance changes occurring in microbial cultures are erratic, that is, they cannot be correlated with any experimental parameter; moreover, these random fluctuations are of the same order of magnitude as the changes in conductance that can be ascribed to bacterial growth. On the other hand, Tsuchiya & Terashima [66] and Cady *et al.* [69] have monitored at different frequencies the time evolutions of the impedance of culture media during the growth of *E. coli* cultures. Both teams have found that impedance curves are strongly dependent on the frequency, the slope and the amplitude of the impedance shift increasing with decreasing frequencies (Fig. 7): this is due to an increase of the frequency-dependent capacitive reactance $1/C\omega$. Tsuchiya & Terashima [66] have also shown that variations in the ionic strength of the medium are insufficient to explain the impedance changes. In agreement with Cady *et al.* [69], these authors conclude that microbial activity mainly affects the capacitance. Similar results have been obtained by Hause *et al.* [75], who have shown, moreover, that the respective parts of capacitance and conductance in impedance are strongly dependent on the electrode design.

   Faced with these contradictions, Firstenberg-Eden & Zindulis have examined in an exhaustive study [76] the relationship between microbial growth and relative changes in both the capacitance and the conductance of culture media, measuring separately these two parameters by means of a modified Bactometer system. They

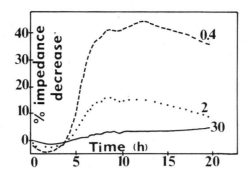

Fig. 7 — Influence of the frequency on the impedance responses of *E. coli* cultures growing at 35°C in trypticase soy broth. The numbers refer to the frequency (kHz). From Cady *et al*. [69].

have found that in low-conductivity media the conductance–time curves are clearly correlated to bacterial growth. In more conductive broths, however, the relative changes in conductance are lower than those in capacitance. Capacitance measurements are of particular interest for detecting and monitoring the growth of yeasts, which produce no noticeable increase of the broth conductance (Fig.8). As with

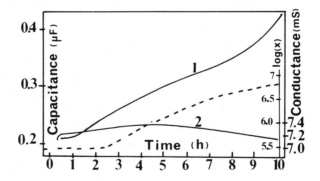

Fig. 8 — The effect of growth of *Saccharomyces cerevisiae* on the capacitive (1) and on the conductive (2) components of impedance. The yeasts were cultured in a high-conductivity medium (trypticase soy broth). Dotted line: time evolution of $\log(x)$, where x is the cell number per cm^3 in the broth. From Firstenberg-Eden & Zindulis [76].

conductance changes [57,73], it is possible to relate the detection time of the threshold capacitance change to the logarithm of the initial concentration of yeasts in the culture medium (Fig. 9).

2.3.3 Overall impedance
From the preceding works, it can be assumed that the relative importance of the resistive and the reactive part in the recorded overall impedance depends on

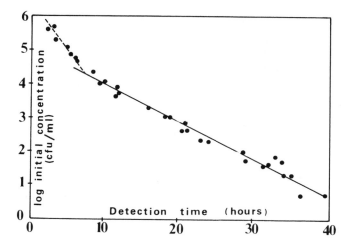

Fig. 9 — Linear relationship between the time of detectable capacitance change (detection time) and the logarithm of the initial cell concentration of *S. cerevisiae* (in yeast carbon base). From [76].

physical, chemical, and metabolic parameters such as the nature and configuration of the electrode, the frequency of the sinusoidal signal applied to the electrodes, the nature of the culture broth, and the bacterial species cultured in the broth.

It is thus not surprising that investigations of the relationships between overall impedance changes and microbial growth have led to results very similar to those obtained when only conductance or capacitance changes were monitored. In particular, these studies have highlighted noticeable similitudes between the time evolution of impedance and the time-course of the number of viable organisms in the culture broth, and also significant differences between the two curves [64,69]: the authors have concluded that impedance changes reflect bacterial metabolic activity rather than growth. Linear relationships between $\log(x_0)$ and the detection time of a significant impedance change have also been reported, for example by Cady *et al.* [69]. In agreement with equations (1) and (3), the slope of the straight line reflects the generation time of organisms (Fig. 10).

To briefly conclude this paragraph, it appears that changes in conductance or impedance of culture broths have been related both experimentally and theoretically to microbial growth parameters (inoculum size, generation time of organisms). These relationships are of obvious interest in microbiological analysis, since they may allow the enumeration of viable organisms in contaminated samples. More fundamental investigations of the mechanisms controlling the impedance changes at the cellular level must now be undertaken.

2.4   *Detection of microbial infections in clinical samples*
Impedance measurements were extensively applied to the detection of bacteria in clinical samples during the 1970s [77–82], most work during this period being performed with the Bactometer 32 device. The results obtained with the instrumen-

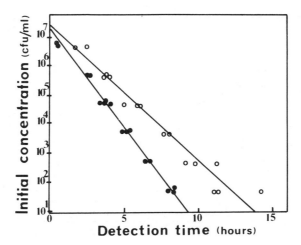

Fig. 10 — Experimental calibration curves relating the impedance detection times to the logarithms of initial concentrations of microorganisms for two bacterial species with different generation times: (●), *E. coli* (generation time of 26 min); (○), *Staphylococcus aureus* (generation time of 35 min). From Cady *et al*. [69].

tal technique were compared with those of the conventional (manual) methods routinely used in clinical microbiology laboratories, and it is perhaps necessary to give a brief account of these conventional procedures.

2.4.1 *Conventional procedures for detecting bacteremia and bacteriuria*

Blood normally contains small numbers of bacterial pathogens. The detection of organisms in blood samples thus requires an incubation step allowing bacteria to multiply: a relatively large volume of blood (10–20 cm^3) is added to a culture broth, and the mixture is periodically examined during incubation for macroscopic evidence of bacterial growth (e.g. turbidity) [83]. Most positive blood cultures are detected within 48 h of incubation [83]. Since blood contains a variety of factors likely to inhibit bacterial growth (drugs, antibodies, ...), some procedures include a preliminary step of inactivation or removal of these antimicrobial agents before blood inoculation [83].

In urine, however, the number of bacteria may be large: the commonly agreed criterion for urinary tract infection is more than 10^5 organisms per cm^3 of urine. The conventional procedure for determining bacteriuria is in fact a manual enumeration method: a low volume of urine (*c*. 0.01 cm^3) is taken from the urine sample under test with a calibrated bacteriological loop and plated onto solid culture medium (agar) by streaking the plate with the loop; colonies developing on the plate are enumerated after a proper incubation period (24 or 48 h) [4]. Since bacteria proliferate readily in urine, contamination of urine specimens during urine sample processing frequently occurs, and it is necessary to separate insignificant and clinically important bacterial growth [4].

The detection step of bacteremia and bacteriuria is followed by the isolation and identification of the organisms responsible and by susceptibility tests of isolates to chemotherapeutic agents.

2.4.2 Impedance-based procedures

The enumeration function of impedimetric (or conductimetric) techniques, arising from equations (1) or (3), applies only to pure cultures of a single organism or to mixed cultures in which a single bacterial species dominates and generates the detection signal: the number of bacteria of a given species, present in a sample submitted to the impedimetric procedure, can be deduced from the impedance detection time by referring to a calibration curve relative to this species (which is known to be detected). This enumeration function is obviously of no use in the case of blood or urine cultures where bacterial pathogens are unidentified and differ from one culture to another: impedance (conductance) procedures have been evaluated as tools for the rapid detection of the growth of a total bacterial flora.

Kagan *et al.* [78] have associated a blood lysis–filtration technique with impedance detection of bacterial activity to diagnose bacteremia. The initial lysis–filtration step in the processing of blood cultures ensures the removal from the blood samples of both antimicrobial agents and metabolically active nonbacterial cells which might affect impedance changes. The impedance-measuring circuit uses a single test cell, and impedance changes are measured at very low frequency (2 Hz). The combined lysis–filtration–impedance procedure, tested with 264 blood samples from 107 hospital patients, exhibits 36% greater detection efficiency than the conventional technique. The major part of this gain in detection efficiency, however, is attributable to the lysis–filtration step, allowing the elimination of antibiotics and larger inocula (filter concentration of blood bacteria). Using artificially seeded blood cultures, Specter *et al.* [79] have reported that the impedance technique (Bactometer 32 apparatus) detects most of the significant aerobic pathogens within 6 h (*Klebsiella pneumoniae*, *E.coli*) to 18 h (*Staphylococcus aureus*) of incubation. In these experiments, the initial bacterial concentrations were very low, about 1–5 bacteria per cm^3. Specter *et al.*, however, have found that the electrical impedance procedure is no more rapid than the radiometric technique (Bactec device; see *Introduction*).

The Bactometer impedance-monotoring system has also been applied to the screening of urine samples for detection of bacteriuria [70,80–82]. For instance, Cady *et al.* [82] have defined impedance-positive urine specimens as those whose culture generates a 0.8% change in impedance ratio (see 2.2.1) within 2.6 hours incubation. This impedimetric criterion is compared to the commonly admitted standard for bacteriuria, i.e. more than 10^5 organisms per cm^3: of 1133 urine samples, 96% are correctly classified as positive or negative by the impedance detection test (Fig. 11). The impedance test misses 13.5% of the samples found to be positive when using the conventional techniques. Most of these false-negative results are probably due to the presence of antimicrobial agents in the relevant urine specimens [82].

Despite these promising early results, no additional evaluation of the Bactometer device as a laboratory instrument for detecting bacteremia and bacteriuria has been

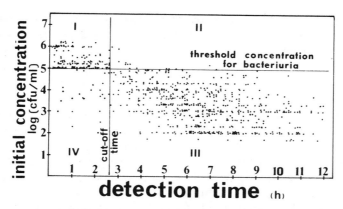

Fig. 11 — 'Scattergram' [72] relating the impedance detection times of clinical urine cultures to the initial bacterial concentration (colony-forming units per cm^3) determined by a conventional manual procedure (plate counts). A total of 1133 randomly selected clinical urine samples were screened for bacteriuria. The scattergram does not involve urine specimens with impedance detection times greater than 12 hours. I: urine cultures positive both by impedance and by the conventional procedure (162 samples). III: urine cultures negative by both techniques (923 samples). II: false negative samples (26, i.e. 2.3%). IV: false positive samples (22, i.e. 1.9%). From Cady *et al.* [82].

undertaken. The application area of the impedance technique has turned, instead, towards alimentary microbiology, namely the detection of bacterial contaminants in foods (see Part 3, Chapter 2).

2.4.3 Conductance-based Malthus system

More recent work in the field of clinical microbiology has been performed by means of the Malthus device. As specified in 2.2.2, this system selectively measures changes in the conductance of culture media during microbial growth.

Brown *et al.* have compared a Malthus prototype with conventional methods for examining blood samples from hospital patients [84]. The prototype monitored 112 (4×28) electrode cells of $100 \ cm^3$ (Fig. 5, I) incubated in four water baths. The electrical conductivity of the blood culture medium was scanned every 30 minutes for 9 days, while conventional blood cultures were examined for turbidity twice a day for six days. Organisms in positive blood cultures were isolated and identified by routine laboratory methods. Among the 651 blood cultures tested, 100 were found to be significantly positive (i.e. not due to bacterial contaminants) either by conductance or by conventional procedures, or by both techniques. There is no statistically significant difference between the number of positive cultures detected either by the Malthus device or by conventional methods. The conductance-based system detects 83.6% of positive cultures earlier and 7.3% later than the conventional procedures. Brown *et al.*, however, have highlighted some difficulties inherent to the Malthus prototype under test, in particular a number of false-positive changes in conductivity (175, i.e. 26.9%) induced by the instability of some electrodes. In addition, they have emphasized that the improvement of detection delays gained by automated systems such as the Malthus instrument is dependent on working practice in laboratories, since positive cultures may be displayed at any time during the day or night.

The Malthus system has also been applied to the detection of bacteriuria [85,86]. An evaluation relating to 500 clinical specimens of urine [86] has given less promising results than those previously reported by Cady *et al.* using a Bactometer 32 device [82]. According to standard enumeration and identification procedures, 40 of the 500 urine specimens under test contained more than 10^5 organisms of a single species per cm^3, and 44 contained a mixed bacterial flora whose total cell number was higher than 10^5 cells/cm^3. The Malthus instrument detected within 2.5 hours 32 (80%) of the pure cultures and 8 (18%) of the mixed cultures, i.e. 47.6% of the samples containing initially more than 10^5 organisms/cm^3: the percentage of positive cultures missed by the conductance procedure is much higher than that reported by Cady *et al.* using the Bactometer apparatus with an identical cut-off time (13.8%) [82]. Smith *et al.* have attributed most of their false-negative results to the presence of boric acid in urine cultures (boric acid is frequently used in hospitals to inhibit the multiplication of contaminants in urine samples) [86]. They have concluded their study by stating that the lack of a suitable culture medium, ensuring rapid growth of all urinary pathogens, limits the usefulness of the Malthus growth analysers for the screening of significant bacteriuria. Brown & Warner had previously expressed such reserves [85]. The discrepancies between the results of the two evaluations [82,86], however, were mainly due to differences in the inoculum size which is a key factor for controlling the detection time: while Cady *et al.* have inoculated 0.5 cm^3 of urine into 0.5 cm^3 of broth [82], Smith *et al.* have chosen a much lower proportion of urine in the mixture: 0.02 cm^3 added to 2 cm^3 of culture medium [86].

2.5 *Antimicrobial susceptibility testing*

The *in vitro* susceptibility of bacteria to antibiotics is classically expressed by the Minimal Inhibitory Concentration (MIC), that is, the minimal concentration of drug inhibiting bacterial growth. Usually, MICs are obtained from visual examination of growth (turbidity or colony formation) after incubation of organisms (initial concentration: *c.* 10^6 cells per cm^3) for 18 h or 24 h in liquid or solid medium provided with serial antibiotic concentrations (two-fold dilutions). Since this procedure is time-consuming, however, the current laboratory practice is to use rapid susceptibility tests, such as the disk agar diffusion technique [87], which classify the organisms under test as susceptible, intermediate resistant or resistant to a given antibiotic.

The application of impedance techniques to the determination of bacterial susceptibility to antibiotics, already evoked by Ur & Brown [63], has been studied in particular by Colvin & Sherris [88] using a Bactometer 32 device. These authors have compared MIC values from conventional broth dilution tests with MICs determined from impedance readings. The electrical impedance MIC was defined as the lowest concentration of antibiotic required to prevent the impedance ratio increasing by 3.2% or more during a given incubation time; different cut-off times (overnight or 6 h incubation) and inoculum sizes (initial bacterial concentration: 5×10^5 or 5×10^6 organisms per cm^3) were tested. The best correlation between the two methods was observed for overnight impedance readings using the lower inoculum (93% of the results within one twofold dilution). However, the 6 h electrical impedance test, which represented a noticeable gain of time, was greatly improved by using the larger inoculum (74% correlation against 34%). Colvin & Sherris noted that most discrepancies appeared in the presence of tetracycline, polymyxin E, and ampicillin. The impedance–time curves recorded in *E. coli* cultures exposed to increasing concentrations of, for example, polymyxin E and ampicillin, were markedly different, and could explain the discrepancies between the 6 h impedance MIC values and the conventional overnight visual readings (Fig. 12). Thus, the determination of impedance MICs of beta-lactam antibiotics, in particular, should be preceded by an impedimetric study of the kinetics of the antimicrobial action of each antibiotic under test, in order to find a suitable cut-off time. To assess the MIC of a wide range of antibiotics against strains of *E. coli* and *K. pneumoniae*, Tsuchiya & Terashima [66] have followed the time courses of impedance of cultures exposed to serial two-fold increasing concentrations of beta-lactams (ampicillin, cefazolin): the impedance–time signals were strongly dependent on the inoculum size. These results, taken together with those of Colvin & Sherris [88], show that it is unrealistic to attempt the standardization of impedance MIC tests for the family of beta-lactam antibiotics, prominent in antimicrobial chemotherapy.

Porter *et al.* [89], and more recently Baynes *et al.* [90], have investigated the ability of the Malthus conductance system (128 H model) to perform rapid antimicrobial susceptibility determinations. The latter study involved 194 fast-growing coliform bacilli (mainly *E. coli* and *K. pneumoniae*) and 6 antibiotics. To differentiate susceptible from intermediate or resistant bacterial strains, the organisms were submitted to only one or two concentrations of each antibiotic. Bacterial growth in 10 cm³ Malthus electrode cells provided with antibiotics was indicated by three successive increases in the conductance value (each of 2 μs), and the corresponding detection times were compared to that of a control culture (without antibiotic) of the tested organism. Susceptibility criteria referred to delays in detection times compared to the control. Bacterial inocula were chosen to obtain results within 6 hours. Baynes *et al.* have found quite satisfactory agreement between this conductance procedure and a reference susceptibility test (break-point inhibitory concentration method) [90]. Such good correlation between conductance assays and conventional susceptibility tests had been previously established by Porter *et al.* [89] (13 discrepancies on a total of 462 tests).

These experiments were performed with a Malthus instrument monitoring conductance changes in 128 electrode cells. Automated devices ensuring numerous susceptibility tests are, in fact, needed in specialized laboratories to obtain antibiograms of clinical isolates rapidly. The relatively rough classification of organisms as resistant, intermediate resistant, or susceptible, is a necessary step preceding the determination of MICs. More precise information on bacterial susceptibility would require continuous recording of conductance of cultures submitted to various amounts of antibiotics, as illustrated by Colvin & Sherris [88] or Tsuchiya & Terashima [66] using impedance-based devices of lower capacity (Fig. 12). This more basic application of impedance (or conductance) procedures deserves interest, as shown by comparable investigations with potentiometric techniques (see 3.5.2). Nevertheless, despite the quoted examples, they have not obtained in this field the same success as with other procedures for monitoring overall microbial activity, e.g. microcalorimetry (see *Introduction*). In their book devoted to impedance microbiology [58], Firstenberg-Eden & Eden have summarized and classified the different

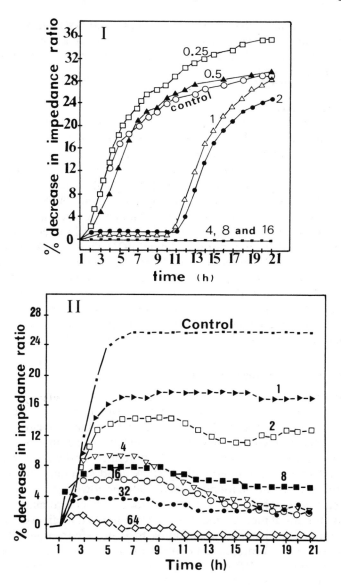

Fig. 12 — Time course of impedance in *E. coli* cultures exposed to various antibiotic concentrations (indicated in $\mu g/cm^3$ on the relevant curves). Initial cell concentration of the culture broth: 5×10^6 organisms/cm^3). Antibiotics: I, polymyxin E (visual MIC: $4 \mu g/cm^3$ with an initial cell concentration of 5×10^5 organisms/cm^3); II, ampicillin (same visual MIC as polymyxin E). From Colvin & Sherris [88].

effects exerted on impedance curves by increasing antibiotic concentrations: increase in the impedance detection time and decrease in the slope of the impedance shift (Fig. 12, I); decrease in the amplitude of the impedance shift (Fig. 12, II); a combination of these phenomena (Fig. 13). These authors have stated that such continuous impedance measurements in the presence of antibiotics do provide interesting information on bacteria–antibiotic interaction, but that much work remains to be done in this field.

2.6 Other applications

The shape of the impedance–time signal generated by a bacterial culture varies with the species cultured and with the incubation conditions: characteristic sets of impedance responses may be obtained for a given organism by varying the incubation conditions. Thus, the impedance procedures may be applied to the rapid identification of microorganisms [58,64,69,91]. This application should be of interest in clinical microbiology for identifying pathogenic organisms. Nevertheless, it has been poorly developed compared to microcalorimetry, for example.

An original application of conductance measurements in microbial cultures has recently been reported by Einarsson & Snygg [92]. These workers, using a 128-channel Malthus device, have monitored changes in conductance produced by the growth of a niacin-dependent strain of *Lactobacillus plantarum* in a broth supplemented with different amounts of niacin (nicotinic acid): conductance changes, determined after a given incubation time, were a linear function of the vitamin concentration. This conductance-based microbiological assay may be extended to the determination of antimicrobial agents in alimentary or clinical samples.

3 Redox potential (ORP) measurements

3.1 Introduction

During their incubation, metabolically active microorganisms induce changes in the oxidation–reduction potential (ORP) of the culture medium. In current laboratory practice, this ORP is most frequently estimated by using indicator dyes which undergo reversible redox reactions at specific potentials and whose colour changes can be followed spectrophotometrically [93]. A good summary of ORP determinations with redox dyes can be found in Jacob [94]. Following Potter's electrometric experiments carried out with a platinum electrode in bacterial cultures [95], however, electrode potential measurements have been applied to diverse investigations in basic and applied microbiology, including hygiene, environmental and clinical microbiology, as well as fermentation control and monitoring. Reviews on electrometric measurements of ORP in culture media and their applications have been published regularly [94,96–104].

A survey of the literature reveals various difficulties in the interpretation and consequent exploitation of such potentiometric measurements. These difficulties arise mainly from the complexity of the electrochemical phenomena that control the zero-current potential of the redox probe in fermentation broths: number and diversity of electroactive components likely to intervene in the electrode potential, improbability of equilibrium states in redox reactions and consequently of true Nernst potentials, influence of the nature, geometry, and pretreatment of the redox probes on the measured potential values. Faced with this situation, Harrison [105] concluded that 'the concept of overall redox potential seems to be of little value in studies of growing microbial cultures', while, according to Costilow [93], the

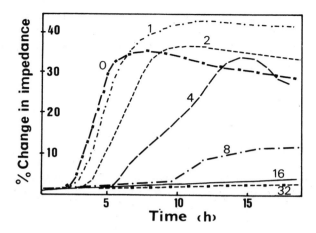

Fig. 13 — Influence of tetracycline on the impedance–time responses of *E. coli* cultures (initial cell concentration: 10^6 organisms/cm³). Antibiotic concentrations (μg/cm³) as indicated on the curves. From Firstenberg-Eden & Eden [58].

electrometric measurements of the redox potential in culture media 'are not believed to be sufficiently meaningful to the general bacteriologist' to be described in a manual devoted to laboratory bacteriological methods. Nevertheless, the importance of ORP in the life of microbial populations has been highlighted by a number of workers, including Wimpenny & Necklen [106] who assumed that ORP 'may be one of the most fundamental, though perhaps the most complex, indicators of the physiological state of microbial cultures'. The study of the time evolutions of electrode potentials in culture media has provided noticeable contributions particularly in the field of clinical microbiology, enhanced by recent developments that improve the redox measurements and their interpretation. In the next section, the difficulties inherent to ORP measurements in fermentation media, but also the achievements due to this technique, will be emphasized.

3.2   *Validity of the concept of ORP in microbial cultures*
The notion of oxidation–reduction potential has been extensively developed in many textbooks devoted to basic and applied electrochemistry. It is sufficient here to recall briefly that a reversible redox couple Ox/Red, where

$$a \ Ox + n \ e^- \rightleftharpoons b \ Red \ ,$$

sets up at equilibrium at an electrode a potential E given by the Nernst equation:

$$E \ = \ E_o + RT/nF \ \ln((Ox)^a/(Red)^b) \tag{4}$$

in which (Ox) and (Red) are the activities of the oxidized and of the reduced forms of the couple, respectively, R is the gas constant, T is the absolute temperature, F is the Faraday, and E_o is the standard potential of the couple.

Equation (4) can be rewritten in terms of concentrations (square brackets):

$$E = E_o' + RT/nF \ln([Ox]^a/[Red]^b) \tag{5}$$

where E_o' is the apparent standard potential of the system. The value of E_o' varies with the ionic strength of the medium (it equals the standard potential when the activity coefficients of each reactant are unity); for redox reactions involving protons, E_o' is also a function of the pH of the medium. In the biochemical literature, detailed tables give E_o' values at pH 7 and at 25°C or 30°C for redox couples of biological importance.

The Nernst relation is meaningful only for reversible redox systems which can reach equilibrium. This is not the case in complex media, for example natural waters [107,108], where numerous redox couples are likely to be found and where the required electrochemical reversibility is usually not found. In fermentation media more particularly, rapid time evolutions of the chemical contents of the broths occur as a consequence of microbial metabolism, and the zero-current potential indicated by an electrode is very probably not an equilibrium potential, but rather a stationary potential [109]. Jacob has stated [103] that the general term of electrode potential should be applied to such a potential instead of the term of oxidation–reduction potential.

Nevertheless, the concept of ORP in culture media has been used for a long time, in fact since the first electrometric experiments of Potter at the beginning of this century [95], and continues to be employed nowadays. In the following, the term of ORP, though often misused, will be employed with its usual meaning, i.e. a parameter that reflects the redox conditions in a culture medium.

3.3   Redox probes and reference electrodes
In an exhaustive review on the use and control of ORP in biotechnology, Kjaergaard [104], referring to Kantere [110], summarized the technical requirements that the redox electrodes must meet and selected two types of redox probes among the six main types likely to fulfil these requirements: electrodes of noble metals, either plain and massive or with large thin-layer sensing areas. In fact, commercially available electrodes of noble metals, mainly platinum, are now used for measuring ORP in fermentation media.

The shape, size, and surface state of the sensitive part of the redox electrode exert a noticeable influence on the value of the measured ORP. In a series of papers, Jacob [103,111–113] has investigated the influence of pretreatment of redox electrodes on their potentiometric responses in bacterial cultures; to gain accurate ORP values, this author recommended careful mechanical cleaning and polishing of the electrode (Fig. 14). Earlier, Zakhar'evskii [114–116] and Kovacs & Marton [117] among other workers, had highlighted the importance of the pretreatment of the electrode before ORP determination.

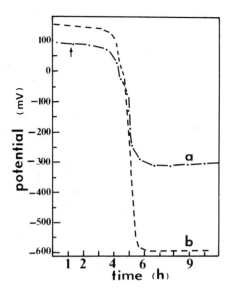

Fig. 14 — Evolution with time of ORP of a culture of *Proteus vulgaris* measured with a non-polished (a) or a polished (b) platinum electrode coupled to a saturated calomel reference. (Incubation conditions: pH 7, oxygen continuously supplied to the stirred culture broth by diffusion from the liquid/gas phase interface). (↑): inoculation. From Jacob [94].

The reference electrodes most frequently used for ORP measurements in biological media are the calomel and the silver chloride electrodes [94,104]. In commercial devices, the metal electrode may be coupled to the reference as a single compact unit (combined redox electrode). The liquid junction between the inner electrolyte of the reference electrode (saturated KCl) and the culture medium is usually ensured by a porous ceramic plug, and is a possible source of discrepancies in ORP values since it may get dirty and contaminated in the presence of proteins and microorganisms. Various attempts have been made to improve this junction or to replace the conventional liquid-junction reference electrodes by electrodes which are easier to handle and clean. In particular, different couples of metallic electrodes have been proposed to replace the conventional Pt–calomel or Pt–silver chloride associations. For instance, Wilkins [118] has tested a couple of platinum electrodes with surface area ratios of four to one, where the larger electrode served as reference, and has found that this device behaved like the platinum-standard reference electrode combination. Electrodes of the same nature but differing in size have also been employed by Holland *et al.* [119] who have monitored the potential difference between two stainless-steel electrodes immerged in blood cultures. Ackland *et al.* [120] have investigated the behaviour in various fermentation media of another bimetallic potential-sensing device, composed of a gold and an aluminium electrode. In all these metal–metal associations, however, it is obvious that the potential of both electrodes changes and that the relative potential difference recorded cannot be compared to the conventional standard potential scale.

3.4   Tentative analysis of ORP variations in microbial cultures

3.4.1   ORP and dissolved oxygen

Most workers have advocated the use of ORP measurements for monitoring and controlling the reducing activity of microbial cultures. A majority have identified this microbial reducing activity with aerobic respiration (i.e. dissolved oxygen consumption): they have considered that the redox probe reacts basically to dissolved oxygen (DO), the oxidant of an electrochemically dominant couple involved in a complex redox system, the fermentation broth. Thus, as early as 1920, Gillespie [121] showed that strongly negative ORP recorded in a culture broth rose when air was passed into the medium. Since then, the predominant role exerted by DO on ORP has been advanced by a number of authors (see e.g. [122–126]). From a quantitative viewpoint, linear relationships between ORP and the logarithm of DO concentration or partial pressure have frequently been obtained. However, the diversity of numerical values assigned to the proportionality coefficient is remarkable: from 45 mV [127] to 100 mV [128] per logarithmic unit of DO tension, with a middle value of 60 mV [129]. Ishizaki *et al.* [130] reported a coefficient of 15 mV/log. unit of DO tension for a platinum electrode dipped in a phosphate buffer solution, the coefficient value attaining 100 or 115 mV in a culture broth, depending on the composition of the broth.

Apart from microbiologists, electrochemists have also investigated the potential of metal electrodes in the presence of DO. Schuldiner *et al.* [131,132] have found a linear experimental relation between the rest potential of a platinum electrode and DO partial pressure, with a slope of 60 mV per log. unit, for DO tensions as low as 10^{-6} atm; they suggested that the potential was determined by a reversible redox reaction involving oxygen, i.e. an O_2/HO_2 exchange. Wroblowa *et al.* [133] have also shown that the rest potentials established at a Pt electrode varied linearly as a function of the logarithm of DO tension with a slope of 60 mV/log. unit; deviations from this relation occurred at DO partial pressures inferior to 10^{-2} atm. The authors stated that the potentials observed could not be attributed to a single reversible electrode reaction, and that the Pt electrode acquired instead a mixed potential due to the couple O_2/H_2O and to a couple involving impurities, most likely traces of organic substances (impurities $+ O_2 \rightleftharpoons$ oxidized impurities $+ ne^- + n\,H^+$).

The diversity in interpretations of open-circuit electrode potentials in simple electrolyte solutions highlights the inevitable difficulties in relating unambiguously these potentials to DO in fermentation media. Many results contradict the hypothesis that time variations of ORP in microbial cultures can be explained only in terms of DO consumption. Early work by Coulter & Isaacs [134] showed that ORP in a *Bacillus typhosus* culture which was given access to atmospheric oxygen continued to decrease long after the exhaustion of DO as a consequence of microbial respiration. Later, Kliger & Guggenheim [135] reported that vitamin C decreased ORP in a culture of the anaerobic bacterium *Clostridium welchii* but did not affect DO tension. More recently, Wimpenny [136] reported that, in chemostat cultures of *E. coli* at a low level of DO, an increase in the aeration rate produced a large rise in ORP, whilst DO concentration increased only slightly; conversely, only small changes in ORP occurred for large changes in DO concentration when oxygen was in excess. Shibai *et al.* [137] have investigated the effects of uncoupling agents and respiratory inhibitors (2,4-dinitrophenol, cyanide, azide) on respiration and ORP of aerobic microbial cultures: they have observed an increase in DO concentration but a decrease in ORP. A final illustrative example of the apparent lack of correlation between DO and ORP is given by Thompson & Gerson [138] who have reported that in batch *E. coli* cultures continuously aerated to keep the DO concentration at a constant level, the ORP fell by nearly 200 mV over the incubation time required to exhaust the substrate supply.

Among the various hypotheses put forward to explain the potential changes other than by DO consumption, the most frequently advocated has been to consider these potential variations as a response of the redox probe to electroactive metabolites or biosynthesized compounds excreted by the organisms during incubation [134,139,140]. The soundness of relating ORP changes to various constituents of the culture broth, namely metabolic products, has been emphasized by Wimpenny & Neklen [106]; according to these authors, the action of DO on ORP might be indirect, since DO can readily oxidize the electroactive reduced components responsible for the electrode potential. Cole [141] has also put forward the idea of an indirect influence of DO on the potential. He has advocated the role of intracellular redox systems (electron transport systems) equilibrating with the electrode through electroactive metal ions (e.g. Fe^{2+}) present in the culture broth; DO would exert a regulatory action by its ability to oxidize these metal ions.

Such relations between intracellular ORP and ORP in the culture medium had been under investigation for a long time. As early as 1929, Aurel *et al.* [142] showed that the bacterial cell was able to ajust its interior ORP to that of the environment. These first results were precursors of a number of works showing the influence of the broth ORP on the growth and metabolic activity of bacterial cultures: the ORP could be considered as a means of controlling and regulating microbial activity, and various systems have been proposed for performing this control [104,124,138,143–145], which is of obvious interest in these times of biotechnological advances.

3.4.2   ORP and specific redox couples

Electroactive substances involved in ORP variations have never been clearly identified, most authors confining themselves to hypotheses. Some workers, however, have implicated molecular hydrogen evolved by gas-producing cultures in anaerobic incubation conditions [146,147]. Wilkins *et al.* [148] have also imputed to hydrogen the large decrease in potential that they recorded with a platinum electrode during the culture of diverse *Enterobacteriaceae*; they have proposed a method for detecting bacteria based on this principle [148]. These authors do not define the oxygenation conditions of the nutrient broth, which have to be standardized to gain reproducible results: hydrogen only occurs in anaerobiosis [146,147]; this metabolic requirement is also of an electrochemical nature, since traces of oxygen strongly affect the hydrogen equilibrium (H^+/H_2 couple) on platinum electrodes [132]. Fischer *et al.* [149,150] have tried without success to relate the voltages recorded during the growth of *Thiobacillus thiooxidans* to the production of sulphate by microbial oxidation of sulphur. Brown *et al.* [151] have controlled the growth of *Desulfovibrio vulgaris*, an organism reducing sulphate into sulphide, by ORP measurements with a platinum electrode. By measuring the ORP of natural water bodies periodically over a year, Brünger [152] has identified six different redox couples, mainly nitrogen couples (NO_3^-/NO_2^-, NO_3^-/NH_4^+, NO_2^-/NH_4^+), one of which dominated and determined the ORP measured. To explain the observed time variations of ORP values, Brünger suggested that the proportions in water of the oxidized and reduced forms of the different couples changed in response to several metabolic processes involving the natural microbial flora.

These few examples are in agreement with Harrison's ideas [105]. Harrison stated that it is difficult to define the output from a redox probe immerged in a culture broth in terms of known electrochemical reactions because of both the diversity of potentially electroactive compounds and the approximate understanding of the role of DO on ORP. He suggested [105] studying specific redox couples whose interaction with the cell is known, instead of the overall ORP.

This procedure can be applied to only a few specific metabolisms (anaerobic respiration). An original technique, following Harrison's principle but of more universal nature (i.e. convenient for a number of bacteria), has been proposed and developed by Junter *et al.*: it consists in measuring the potential–time evolutions of a gold electrode in bacterial cultures which are incubated in the presence of exogenous lipoic (thioctic) acid (see [153] for a review). Lipoic acid (LA) is a coenzyme for the pyruvate dehydrogenase and α-ketoglutarate dehydrogenase cellular enzyme complexes, where it acts as a transacylating cofactor and an electron carrier; during the redox reactions of bacterial catabolism, it is reduced into dihydrolipoic acid (LAH_2) [154]. Junter *et al.* have shown [155] that the typical changes in potential recorded when LA is added to the culture broth (Fig. 15) can be assigned to the appearance of

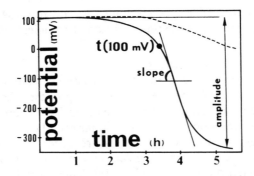

Fig. 15 — Schematic time-course of the potential of a gold electrode in bacterial cultures provided with lipoic acid (10 μM). Usually, cultures are performed in a minimal salt medium buffered at pH 7 and supplemented with glucose; a commercially available combined gold–Ag/AgCl saturated KCl electrode is used. Experimental magnitudes describing the potential–time signal are indicated on the curve. The potentiometric lag time obtained from the curve is t(100 mV), the time necessary for the potential to shift by 100 mV from its initial level (before inoculation of the broth).

LAH$_2$ in the medium. The major interest of this procedure is to confine the complex reducing activity of microbial cultures to a single electroactive phenomenon: the reduction of LA (Fig. 16). This simplification has allowed the analysis and modelling

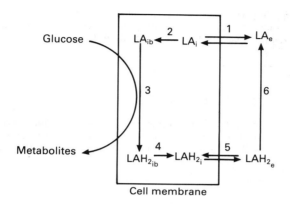

Fig. 16 — Bacterial reduction of lipoic acid: schematic pathways followed by exogenous LA at the cellular level. 1: transmembrane transport of LA. 2: covalent binding of LA to the protein fraction of enzyme complexes (pyruvate and α-ketoglutarate dehydrogenase complexes). 3: intracellular reduction of bound LA. 4: release of LAH$_2$ from its protein support. 5: excretion of LAH$_2$. 6: extracellular oxidation of LAH$_2$ by dissolved oxygen (aerobic incubation conditions). Indices: e, in the culture medium; i, free inside the cell; ib, bound to the protein fraction of enzyme complexes inside the cell. From Junter *et al*. [155].

of the time evolutions of the gold electrode potential [155,156]: the potential has been expressed as an empirical function of LAH$_2$ concentration of the broth, the time evolution of which was given by a system of kinetic equations describing LA reduction and, in aerobiosis, DO consumption (Table 1). In this procedure, DO did exert a noticeable but indirect effect on the potential by reoxidizing readily excreted LAH$_2$ into LA. Most results published by Junter's team have been obtained in *E. coli* cultures, but the technique can be applied to other bacterial species, since most organisms are able to reduce exogenous LA [157].

3.4.3   ORP and microbial growth

From the preceding examples, it appears that the electrometric measurement of ORP changes in culture media is suitable for detecting and monitoring a certain reducing activity of organisms; there is generally little or no definition of this activity with regard both to electrode reactions and to bacterial metabolism, but it deserves some interest since it may provide useful information on the physiological state of the cultures. Only a few studies (Refs. 158–160, for example) have focused on the relations between ORP changes and microbial growth. That OPR change occurs during the logarithmic phase of growth is the most frequently reported observation. In fact, ORP measurements are unsuitable for monitoring and quantitatively estimating bacterial growth. The ORP begins to fall as soon as the reducing activity

Table 1 — Modelling of the time evolutions of the gold electrode potential in bacterial cultures supplied with lipoic acid [155]

Cell concentration. In usual incubation conditions, bacterial growth follows a classical exponential law:

$$x = x_o \exp(\mu_m t) \tag{1}$$

DO concentration. Cultures are performed under controlled aerobiosis, i.e. the culture broth is continuously aerated by stirring, ensuring oxygen transfer from residual air in the flasks used for the potential–time measurements; in the culture medium, excreted LAH_2 is partly reoxidized by DO:

$$\frac{d[O_2]}{dt} = \frac{\partial[O_2]}{\partial t}\bigg|_{\text{consumption}} + \frac{\partial[O_2]}{\partial t}\bigg|_{\text{transfer}} + \frac{\partial[O_2]}{\partial t}\bigg|_{LAH_2\,\text{oxidation}} \tag{2}$$

$$= -\frac{Q_m x[O_2]}{K_o + [O_2]} + k_1 a([O_2]_0 - [O_2]) - k_{ox}[LAH_2][O_2]^{\frac{1}{4}} \tag{3}$$

LA concentration. The transmembrane transport of LA is considered as the limiting step of the reduction process schematized in Fig. 4 (binding rate ≫ transmembrane transport rate); this transport is moreover supposed to involve both a simple diffusion and an enzymatic transport due to permeases:

$$\frac{d[LA]}{dt} = \frac{\partial[LA]}{\partial t}\bigg|_{\text{diffusion}} + \frac{\partial[LA]}{\partial t}\bigg|_{\text{permeation}} + \frac{\partial[LA]}{\partial t}\bigg|_{LAH_2\,\text{oxidation}} \tag{4}$$

$$= -KA[LA]x - \frac{V_m x[LA]}{K_m + [LA]} + 2k_{ox}[LAH_2][O_2]^{\frac{1}{4}} \tag{5}$$

Mass balance. It is assumed that no noticeable intracellular accumulation or degradation of LA or LAH_2 occurs during the reduction process:

$$[LA]_o = [LA] + [LAH_2] \tag{6}$$

Expression for the gold electrode potential in the presence of electroactive species DO, LA, and LAH_2. The gold electrode potential, of the mixed type, is modelled by arbitrary functions of $\log([LAH_2])$:

$$V([O_2], [LA], [LAH_2]) = V([O_2]_o, [LA]_o, 0) + f(\log[LAH_2]) \tag{7}$$

These functions $f(\log[LAH_2]$ depend on the initial *LA* content of the broth. They have been standardized empirically in particular for $[LA]_o = 10\ \mu M$ [70].

Symbols

a	specific surface of the air–liquid interface in the flasks used for potential–time measurements under aerobic incubation conditions.
A	specific surface of effective membrane permeability to LA.
K	effective membrane permeability constant of bacteria to LA.
k_1	oxygen transfer coefficient from residual air to the culture broth (aerobic incubation conditions).
K_m	Michaelis constant relative to the enzymatic transmembrane transport of LA.
K_o	Michaelis constant relative to bacterial DO consumption.
k_{ox}	rate constant of the $LAH_2 + 1/2\ O_2 \rightarrow LA + H_2O$ oxidation reaction.
$[LA]$	concentration of LA in the culture broth at the time t.
$[LA]_o$	initial concentration of LA in the culture medium (at $t = 0$).
$[LAH_2]$	concentration of LAH_2 in the broth at t.
μ_m	maximum specific growth rate of bacteria.
$[O_2]$	concentration of DO in the medium at t.
$[O_2]_o$	initial DO concentration in the culture broth (i.e. usually, 2.1×10^{-4} M, corresponding to air saturation at 37°C).
Q_m	maximum specific rate of bacterial DO consumption.
t	time, starting from inoculation.
V	zero-current potential of the gold electrode in the presence of electroactive species LA, LAH_2 and O_2.
V_m	maximum specific rate of enzymatic transmembrane transport of LA.
x	cell concentration in the broth at the time t (cell number per unit volume).
x_o	initial concentration of organisms in the culture medium.

---

developed by the culture is sufficient to be recorded by the redox probe. During the potentiometric lag time, cultures are active and grow, and this activity and growth can indeed be followed by measuring parameters other than ORP, for example the viable cell number or the DO content of the broth. Moreover, most studies have shown that the potential shift is a relatively rapid phenomenon (see Figs 14 and 15): during the period of rapid potential change, the cellular content of the culture broth is likely to increase only slightly. Thus, the direct conversion of ORP decrease into cell population increase has not really been investigated, the workers relating their studies more readily to the notion of bacterial activity.

3.5   ORP and clinical microbiology

Historically, electrometric measurements of ORP have been applied to various investigations relevant to clinical microbiology. Several authors have reported that the time evolutions of ORP recorded in diverse bacterial cultures vary from one bacterial species to another, and have made use of these differences in ORP–time curves to characterize, identify, and differentiate microbial species or strains [161–176]. Changes in ORP recorded in cultures of pathogenic organisms have been connected with the virulence of the organisms [174,177–181]. These potentiometric measurements have also been found useful in characterizing the microbial flora of different biological samples and tissues of human or animal origin: horse muscle [182], contents of the gastrointestinal tract of normal or artificially infected animals [180,183–185], feces of normal humans and of typhoid or dysentery patients [186], saliva of healthy, coronary heart-diseased, or mentally defective males [160], dental plaque and decayed teeth [187–189].

The following paragraphs develop the application of ORP measurements to the detection of microbial contaminants in biological samples and to the study and estimation of antibacterial effects of antibiotics. In this latter field, particular reference is given to the lipoic acid technique [153], which has prompted original investigations.

3.5.1   Detection of bacteria in biological samples by using ORP measurements

The main parameters controlling ORP curves under aerobic conditions of growth have been detailed by Ward [190]: number of viable bacteria in the inoculum, DO tension, and reducing capacity of the bacterial system. Zakhar'evskii [191] suggested detecting bacterial contaminants in foods by a fall in ORP to a definite minimal level. The principle of detecting bacteria by ORP measurements was later taken up by Schmidt-Lorenz [192] and subsequently discussed by Obliger & Kraft [193] who have regretted the low sensitivity of this potential–time detection: in their experiments with pure and mixed cultures of *Salmonella* and *Pseudomonas*, bacterial concentrations in the broth approached 10^5 to 10^6 cells per cm^3 before significant changes of the redox-probe potential (platinum electrode) could be detected. The most recent detection assays using electrode potential measurements have focused on the relation between the inoculum size and the potentiometric lag time, i.e. the time that elapses between the beginning of the measurement (inoculation of the broth) and the moment when bacterial reducing activity can be detected by a significant change of the electrode potential level. By measuring specific redox phenomena, better defined than overall ORP, different teams have elicited linear relationships between the potentiometric lag time and the inoculum size, allowing the enumeration of bacteria [148,194]; these workers, however, were mainly concerned with nonclinical samples, i.e. waters (see Part 3, Chapter 2).

Thus, only very few attempts at applying ORP measurements to the detection and counting of microorganisms in clinical samples (blood or urine samples of human origin) have been reported at the present time. Referring to instrumental techniques for detecting bacteria in general, and to their application to the detection of bacteremia, Murray [83] has explained the difficulties encountered in this area partly by the delays in detection due to the incubation of samples (e.g. potentiometric lag time) and partly by the ready contamination of samples during the numerous manipulations inherent to these techniques. The latter drawback is particularly opposable to electrode potential measurements in a field where, moreover, a number of samples must be analysed in a short time to allow valuable practical developments. It seems unrealistic to imagine a device which, like those based on impedance or conductance techniques, would perform parallel ORP measurements in numerous samples, using test cells that are easily sterilizable and deliver reproducible results: the very nature of the electrochemical technique under discussion does not lend itself to that type of instrumentation.

Despite these basic limitations, however, some assays merit attention. Lamb *et al.* have applied the potential–time procedure previously described by Wilkins *et al.* [148] to the detection of bacteria in urine [195]. Urine specimens from hospital patients were tested by the potentiometric technique and by conventional bacteriological procedures: all positive potentiometric responses were caused by the presence of bacteria in the urine and were recorded within 10 hours; moreover, Lamb *et al.* did not observe false negative potentiometric responses. Another study by Holland *et al.* [119] deals with the automated detection of bacteria in blood by using a couple of stainless-steel electrodes. These authors have evaluated a prototype system in which bacterial growth was recognized by an increase in the measuring-electrode potential of at least 0.1 mV/min over three consecutive 15–min intervals. The potential–time technique was compared to the conventional visual procedure for detecting microbial growth in broth. Experiments related both to blood specimens seeded with bacterial or yeast species commonly isolated from clinical blood cultures and to patients' blood cultures: 90% of the seeded blood specimens were detected by the potentiometric technique, most of the false negatives being cultures seeded with *Cryptococcus neoformans*; the detection occurred an average of 18 hours earlier than with the standard visual method. Assays on clinical specimens evidenced a complete agreement (100% correlation) between the two procedures.

These promising investigations have not been continued, and redox (electrode) potential measurements have found a more suitable field of application in the study and evaluation of antimicrobial effects of inhibitors, namely antibiotics.

3.5.2   ORP and antimicrobial agents

3.5.2.1   Historical background. The idea of electrometrically measuring ORP variations in cultures exposed to compounds acting on bacterial metabolism, in order to obtain data on the antimicrobial effects of these compounds, was applied by Vaczi as early as 1944 [196]. Vaczi studied the action of different sulfonamides on the time evolution of ORP in *Staphylococcus albus* cultures. A little later, this author and Toth applied the potentiometric procedure to investigations on the effects exerted by vitamins on various bacterial cultures [197]. Later on, several workers studied antibiotic effects on ORP changes [198–205]. Thus, Drouhet [198] and Drouhet & Kepes [199] have shown that ORP variations in *Staphylococcus* cultures exposed to penicillin were different from those recorded in the presence of streptomycin. Marked differences in effects induced by these two antibiotics on the time-courses of ORP have been confirmed by Kramli *et al.* [200,201]: penicillin induced a sudden U-turn of the platinum electrode potential during the potential shift, the amplitude of which decreased when the antibiotic concentration added to the culture broth increased (Fig. 17, I); in cultures exposed to streptomycin, however, the slope of the potential shift decreased with increasing antibiotic concentrations, and the potential remained stable at an intermediate level with no noticeable U-turn (Fig. 17, II). Kramli *et al.* have not attempted to relate the different modifications of ORP time-courses and the antibacterial action of the tested antibiotics; they have followed bacterial growth in the presence of antibiotics and claimed that ORP measurements were more rapid than classical turbidimetric tests for detecting small amounts of antibiotics in a broth or for determining the resistance of bacteria to antibiotics.

3.5.2.2   *Application of the lipoic acid technique to the study and evaluation of antimicrobial effects of antibiotics*

Introduction

The study of the effects of inhibitors on bacterial reducing activity in the presence of exogenous lipoic acid is a privileged field of investigation developed by Junter *et al.* as an application of their potentiometric procedure [153]. They approached this application with the benefit of a precise definition of the bacterial reducing activity involved in the time variations of the gold electrode potential: the three metabolic events controlling the potential changes are LA reduction, DO consumption, and bacterial growth. Their first aim was to characterize and to compare, perhaps more precisely than in previous works which exploited overall ORP changes, the anti-microbial effects of drugs acting on bacterial metabolism, by looking at the modifica-tions they induce in the time evolutions of the electrode potential. Moreover, the theoretical modelling of the potential–time phenomena was likely both to provide information on the kinetics and mechanisms of drug action and to allow a quantita-tive estimation of these antimicrobial effects.

In preliminary assays, Junter *et al.* had shown [206] that the potentiometric signals recorded in *E. coli* cultures were modified by the presence of hormones known to regulate the penetration of LA into mammalian cells. Complementary investigations have confirmed the likelihood of hormonal effects on bacterial cells [207]. However, the potentiometric LA procedure has been found particularly suitable and promising for studying the antibacterial action of antibiotics.

Classification of antibiotics by the potentiometric LA procedure

In agreement with previously published results [198–201], Junter *et al.* have shown [208] that the potentiometric responses of *E. coli* cultures provided with increasing antibiotic concentrations varied from one antibiotic family to another. The changes noted in the presence of beta-lactam antibiotics (both penicillins and cephalosporins)

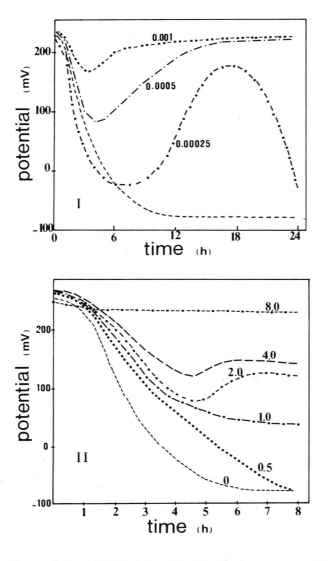

Fig. 17 — Time evolutions of ORP of a sensitive *Staphylococcus aureus* culture exposed to various antibiotic concentrations. Measurements were performed with a smooth platinum electrode coupled to a calomel reference. Antibiotic concentrations (units/cm^3) are indicated on the curves (1 unit/cm^3 corresponds to a concentration of about 1.7 μM). Initial cell concentration: 2.75×10^6 organisms/cm^3. Dotted line: control without antibiotics. Antibiotic: I, penicillin; II, streptomycin. From Kramli *et al.* [201].

were quite different from those observed with antibiotics that affect protein synthesis, e.g. aminoglycosides, tetracyclins, macrolides, or chloramphenicol.

Beta-lactam antibiotics, for example amoxicillin (penicillin : Fig. 18, I) or cefotaxime (semisynthetic cephalosporin : Fig. 18, II) induced changes quite similar to those already reported by Kramli *et al.* [201] (see Fig. 17, I). A noticeable

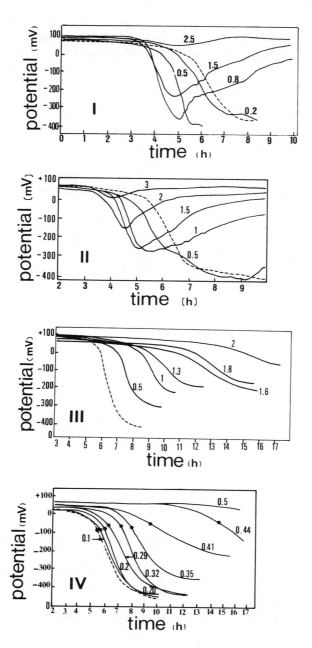

Fig. 18 — Evolution with time of the gold electrode potential in *E. coli* cultures provided with lipoic acid, dissolved oxygen, and increasing amounts of antibiotics. The numbers refer to the antibiotic concentration of the culture broth ($\mu g/cm^3$). Dotted line: control without antibiotics. (●): $t(100\ mV)$ (Fig. 18.IV). Antibiotic: I, amoxicillin [208]; II, cefotaxime [211]; III, chloramphenicol [208]; IV, kanamycin [210]. Initial cell concentration (organisms/cm^3): I, 3×10^6; II, 3.5×10^6; III, 2.5×10^6; IV, 3×10^6.

difference, however, was elicited by using the LA technique: surprisingly, the potentiometric lag times preceding the potential shift decreased with increasing antibiotic concentrations. On the other hand, all antibiotics tested other than beta-lactams showed an increase in lag times and a decrease in slopes with increasing concentrations of antibiotics (Fig. 18, III and IV).

Estimation of the bactericidal efficacity of antibiotics affecting protein synthesis
From classical turbidimetric measurements, antibiotics acting on protein synthesis were found to inhibit bacterial growth by decreasing the specific growth rate of organisms. Within the frame of the theoretical model, this decrease in the specific growth rate was sufficient to explain macroscopically the effects of antibiotics on the electrode potential variations, that is, the increase in potentiometric lag times and the decrease in slopes [209]. It was possible to assign to every tested antibiotic concentration an apparent specific growth rate of bacteria (μ_{app}), calculable from the corresponding value of the potentiometric lag time by using the system of equations modelling the potential–time signal (see Table 1); then, the relation between μ_{app} and the antibiotic concentration was used in the following way to estimate the bactericidal efficacity of the tested antibiotic [210]:

Bacterial growth in the presence of a concentration [A] of the tested antibiotic was given by equation (6) replacing equation (1) of Table 1:

$$x = x_o \exp(\mu_{app} t) \tag{6}$$

in which $$\mu_{app} = \mu_m - k \tag{7}$$

where k is the inactivation rate constant of the antibiotic.
The constant k can be connected to [A] by an empirical relation:

$$k = a\,[A]^n \tag{8}$$

or else
$$\log(k) = \log(a) + n\,\log([A]) \tag{9}$$

in which a and n are constants characterizing the bactericidal activity of the antibiotic.
Thus μ_{app} was estimated from the potentiometric lag times, k was calculated from equation (7), and a and n could be obtained by least-square analysis of $\log(k)$ as a function of $\log([A])$ (Fig. 19).
Junter *et al.* have applied these semi-empiric calculations to the family of aminoglycosides [210]. Specific values for a and n have been assigned to each antibiotic experimented. The obtained relations (8) or (9) described the bactericidal effects of aminoglycosides for any given drug concentration. Moreover, potentiometric MICs were calculated from these relations in which k was made equal to μ_m: aminoglycosides were ranged according to their potentiometric MIC in the same order of bactericidal efficacity as that given by the conventional susceptibility tests of bacteria to antibiotics.

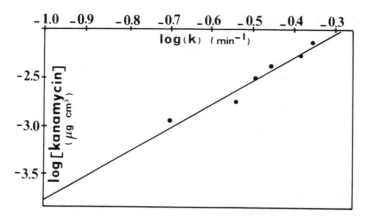

Fig. 19 — Logarithm of the inactivation rate constant k as a function of the logarithm of kanamycin concentration. ($\bullet$) experimental values of $\log(k)$ calculated from the potentiometric $t(100 \text{ mV})$ data of Fig. 18.IV; (——), correlation line [equation (9)]: $\log(k) = -1.278 + 2.5\log([A])$ (correlation coefficient: 0.992). From Bogaczyk et al. [210].

This procedure is being extended, with industrial support, to the family of macrolides.

Information provided by the potentiometric LA technique on the kinetics and mechanisms of antibacterial action of beta-lactams

The surprising decreases in the potentiometric lag times recorded in the presence of beta-lactam antibiotics may be due to the stimulation by these antibiotics either of bacterial growth or of the reducing activity of the culture. By combining the LA potentiometric technique and classical measurements of microbial growth and DO consumption, Junter & Bogaczyk-Hébert [211] have described qualitatively the effects of cefotaxime on the growth and reducing activity of *E. coli* cultures. They have shown that cefotaxime stimulates both bacterial respiration and LA reduction but inhibits the growth of organisms. In addition, theoretical calculations performed using the model have proved that the antibiotic exerts its stimulative effect more particularly on the transmembrane transport of LA (the limiting step of LA reduction in Junter's model), the rate of which was considered to be a function of the duration of cell-antibiotic contact [212]. From these works [211,212], a detailed commentary of the typical potential time-courses observed in the presence of beta-lactams can be proposed in terms of the antibacterial action of these antibiotics:

● During the first stages of incubation (period corresponding to the potentiometric lag phase), abnormal forms of bacteria appear in the broth as a consequence of the well-known fixation of the antibiotics on various membrane proteins involved in bacterial morphogenesis (the reader may refer to Refs. [213–216] for reviews on the mechanisms of antibacterial action of beta-lactams). Abnormal bacteria rapidly prevail among the cell population of the culture medium; they are nonviable, i.e. they cannot initiate colonies on nutrient agar plates, but their reducing activity is much higher than that of viable cells, and the potentiometric lag time is decreased.

- The sudden U-turn in potential that appears during or after the potentiometric shift is clearly attributable to the lysis of abnormal bacteria. This lysis produces a sudden and drastic decrease of the overall reducing activity of the culture: the broth is reoxygenated by oxygen transfer through the air/liquid interface in the flasks used for potential–time experiments, excreted LAH_2 is reoxidized into LA by DO, and the potential stops falling and reincreases.
- The later decrease in potential is due to the reducing activity developed by surviving bacteria.

These results show that the potentiometric LA technique may be a useful tool for fundamental studies on the antimicrobial action of beta-lactam antibiotics, providing original data on cell membrane phenomena. In particular, they highlight interesting questions on the mechanisms of action exerted by these antibiotics at the cellular level. Further investigations are being performed to explain the cellular origins of the drug-induced increase in the bacterial reducing activity. At the present time, it seems plausible to evoke an indirect mechanism of the antibiotics, affecting bacterial energy-transducing processes, and constraining the organisms to more intense fluxes of electron acceptors [211].

On the other hand, it is possible, as for other families of antibiotics, to make use of the potentiometric data for quantitative evaluation of these antibacterial effects of beta-lactams.

Contribution of the potentiometric LA procedure to the estimation of the in vitro *antibacterial activity of beta-lactams*
From the foregoing, two significant chronometric parameters can be obtained from the potential–time signals recorded in bacterial cultures exposed to beta-lactam antibiotics. Besides the potentiometric lag time, which reflects the stimulative effects of beta-lactams on bacterial reducing activity, the moment when the electrode potential stops falling and reincreases (see Fig. 18, I and II) represents the lysis time and can evaluate the killing effect of the antibiotics. For different beta-lactams, Junter [217] has studied quantitative relationships between the antibiotic concentration and each of these two potentiometric parameters (Fig. 20). This author was able to range the different antibiotics according to the relative intensity of both their stimulative and their killing action: he found that the two cephalosporins tested (cefotaxime and cefoperazone) exerted more stimulative effects and more intense lytic effects than the two penicillins experimented (ampicillin and amoxicillin). This original classification was consistent with that obtained from conventional susceptibility tests (MICs). Some discrepancies appeared, however: for instance, while identical MICs have been assigned to ampicillin and cefotaxime, cefotaxime was found to exert a more intense killing effect than ampicillin.

These first results are still quite fragmentary. Nevertheless, they prove that the potentiometric LA procedure applied to investigations of antibacterial effects of beta-lactams has a practical interest since it can complement the conventional *in vitro* bacterial susceptibility tests.

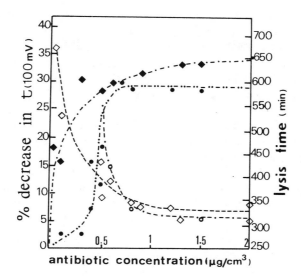

Fig. 20 — Examples of variations in potentiometric lag times t(100 mV) and lysis times as a function of the concentration of beta-lactam antibiotics. (●,♦,O,◊) experimental data recorded in *E. coli* cultures. ‒‥‒ (●) t(100 mV) in the presence of amoxicillin. ‥‒‥ (♦), t(100 mV) in the presence of cefoperazone. ‥‒‥ (O), lysis time in the presence of amoxicillin.——— (◊), lysis time in the presence of cefoperazone. From Junter [217].

4   Dissolved oxygen and pH measurements

4.1   Dissolved oxygen measurements

Although the prominent influence exerted by oxygen on microbial growth and metabolism has been recognized since the beginning of modern microbiology, quantitative investigations of this influence have for a long time been severely impeded by the lack of handy and reliable tools for measuring the concentration of dissolved oxygen (DO). It was only during the nineteen-fifties that such studies developed with the appearance of oxygen electrodes [218]. Since then, DO measurements have been applied both to fundamental investigations of cell or organelle respiration and for more practical purposes such as the control and monitoring of aerobic fermentation processes. The reader may refer to different sections of this book describing the principles of oxygen probes and their application to fermentation process control and to *in vitro* analyses in clinical biology. Additional information on DO probes and their utilization in the preceding fields may also be found in diverse reviews [219–222].

Despite the wide utilization of oxygen sensors in basic microbiology and in biotechnology, electrochemical measurements of aerobic microbial respiration have found relatively few practical applications in the field of clinical microbiology.

Jacob & Horn have investigated the kinetics of bacterial consumption of DO in the presence of antimicrobial agents of therapeutic interest [202,203,205,223]. In identical incubation conditions, they followed the time courses of redox potential and of DO tension with platinum vs reference electrodes and polarographic oxygen electrodes, respectively. Moreover, they simultaneously monitored bacterial growth by turbidimetric measurements. According to these authors (see the preceding section), polarographic oxygen electrodes may be advantageously replaced by redox probes for determining low levels of DO in culture media, that is, when the DO content of the broth falls below the detection threshold of oxygen electrodes: they state that the association of the two devices is advisable for studying antibiotic effects on aerobic bacterial cultures, and that more complete information on the kinetics and mechanisms of antibacterial action of antibiotics may be provided by this association than by classical optical monitoring of growth [205]. In connection with these statements, Junter *et al.* have reported [211] that the consumption of DO by a culture of *E. coli*, followed with a Clark-type electrode, was enhanced by cefotaxime, whereas bacterial growth, reflected by photometric measurements, was inhibited (Fig. 21). Nevertheless, Jacob & Horn have focused their interest more

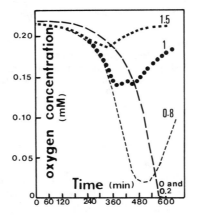

Fig. 21 — DO consumption by stirred cultures of *E. coli* exposed to cefotaxime. A Clark-type DO electrode was used. The U-turns of oxygen concentration correspond to the lysis of bacteria and to the consequent reoxygenation of the broth by air transfer through the air/liquid interface in the culture flasks. The numbers refer to the concentration of cefotaxime (μg/cm^3) in the culture medium. Initial cell concentration: 7×10^5 organisms/cm^3. From Junter & Bogaczyk-Hébert [211].

particularly on the electrometric measurement of redox phenomena, and Junter *et al.* were mainly concerned with the application of their potentiometric technique [153] to the study of antimicrobial effects of antibiotics.

Electrochemical DO measurements have found more specific applications both in environmental and clinical microbiology as rapid oxygen consumption tests for determining the toxicity of pollutants on bacteria [224–227] or the susceptibility of organisms to antibiotics [228–230], and for detecting bacteriuria [228,230,231]. The principle consists in comparing the kinetics of DO consumption by an aerobic bacterial culture exposed to an inhibitor with that of a control culture incubated without inhibitor. Different ratios, referring for example to the amount of DO consumed by bacteria or to the respiration rate of organisms during the test, express in percentages the inhibiting effect of the antimicrobial agent.

Lindemann *et al.* [228] were the first to propose applying this DO-uptake-based electrochemical procedure to the detection of bacteria (aerobes or facultative anaerobes) in urine samples and to antimicrobial susceptibility tests. The technique was developed later by Schmidt *et al.* [229] and Strassburger & Tiller [230]. Both teams followed the kinetics of DO consumption by bacterial cultures exposed to various concentrations of antibiotics. After aerobic incubation for 3 h in the presence of antibiotic, the cultures were saturated with oxygen and the time evolution of the DO content of the culture broth was followed for 1 h with an oxygen probe (Fig. 22).

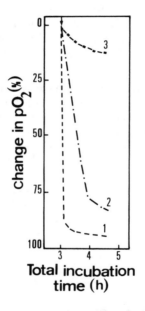

Fig. 22 — Decrease in pO_2 (in percentage) in *E. coli* cultures exposed to various concentrations of oxytetracycline. (↓) Oxygen saturation of the broth. Antibiotic concentration ($\mu g/cm^3$): 1, 0 (control culture) and 0.06; 2, 0.25; 3, 1.0. From Schmidt *et al.* [229].

The inhibiting effect of the antibiotics on DO consumption was expressed quantitatively by an 'oxygen index' OI:

$$OI = [\log(O_s)\text{-}\log(O_r)]/[\log(O_o)\text{-}\log(O_r)] \qquad (10)$$

where O_o is the partial pressure of oxygen in the culture broth at the beginning of the measurements, O_r and O_s are, respectively, the partial pressure of oxygen at the end of the test in the control culture and in the culture exposed to the antibiotic. Values of

OI vary from 0 ($O_s = O_r$, i.e. no inhibition of DO consumption) to 1 ($O_s = O_o$, i.e. total inhibition of DO consumption). Bacteria were considered to be susceptible to antibiotic concentrations yielding *OI* values superior to 0.7 and resistant to antibiotic concentrations giving *OI* values lower than 0.4. An electrochemical MIC could be defined as the lower drug concentration leading to an *OI* value higher than 0.7 [229]. The results of these electrochemical susceptibility tests satisfactorily correlated with those obtained by conventional procedures for most antibiotics tested [230].

Strassburger & Tiller have extended the procedure to the rapid detection of bacteriuria [230]. They have related the amount of DO consumed by urine cultures to the initial concentration of organisms in urine specimens as determined by standard plate count methods. In the same way as for susceptibility tests, urine cultures were incubated aerobically for 3 h and then saturated with oxygen. The pO_2 decrease was evaluated 30 min after oxygen saturation. A total of 577 urine specimens were screened by this procedure: it was found in particular that all cultures of urine specimens exhibiting significant bacteriuria according to the conventional technique (i.e. more than 10^5 organisms per cm^3 of urine) had consumed at the end of the test 60% or more of the initial DO content of the broth (Table 2). Comparable

Table 2 — Distribution of 577 urine specimens as a function of pO_2 decrease (%) recorded 1 h after broth saturation in oxygen and of the concentration of organisms in the specimens (colony-forming units per cm^3) as determined by conventional plate counts. From Strassburger & Tiller [230]

Organisms per cm^3	No. detected by conventional plate culture	No. detected electrochemically[†]				
		pO_2 decrease (%)				
		0–20	20–40	40–60	60–80	80–100
$\geq 10^5$	21	0	0	0	1	20
10^4–10^5	190	21	22	13	26	108
$\leq 10^4$	203	144	33	15	6	5
sterile	114	108	4	1	1	0
contaminated[‡]	49	38	6	1	3	1
	577	311	65	30	37	134

†Incubation time 3.5 h.
‡Samples yielding coagulase-negative staphylococci and/or Gram-negative bacteria less than 10^3 organisms per cm^3.

changes in pO_2, however, were also recorded in many cultures of urine samples containing between 10^4 and 10^5 organisms per cm^3 (Table 2): this probably induced false-positive results when assigning, for example, a 60% change in pO_2 to the detection threshold of bacteriuria.

4.2 pH measurements

Like the DO tension and the ORP level, the concentration of hydrogen ions (pH) of a culture medium exerts a strong influence on bacterial metabolic activity which, conversely, is able to generate changes in the pH of the broth.

The appearance of hydrogen ion-selective glass electrodes and their use for pH determination in fermentation media date from the beginning of this century. Since then, glass electrodes have been continuously improved (e.g. ease of handling and of sterilization, miniaturization), more particularly for pH measurements in biological samples or tissues and for pH control and regulation in the fermentative industry. Plentiful data on pH and other ion-selective electrodes can be found in the medical and biological literature (see for example [232] and references therein). The reader may refer, as for DO probes, to different chapters of the present book for information *in vitro* and *in vivo* pH determinations in clinical samples (Part 1) or the application of pH glass electrodes in the fermentation process control (Part 2).

Changes in the pH of fermentation broths mainly arise from the degradation of nitrogenous compounds leading to the release of alkaline products in the culture broth and from the catabolism of carbohydrates yielding organic acids [233]. Since the nature and the relative proportions of acid or alkaline metabolites excreted by active organisms depend on a number of factors, including the bacterial species cultured in the broth, the incubation conditions (aerobiosis/anaerobiosis), and the substrate(s), the measurement of pH changes in unbuffered culture media allows only indirect monitoring of the overall activity of bacterial cultures. On the other hand, only large populations of metabolizing organisms can produce noticeable pH variations, the amplitude of which is also controlled by the buffering capacity of the incubation broth. This relatively low sensitivity explains the lack of applications of pH measurements in analytical microbiology, more particularly with regard to the detection of bacterial contaminants. Mitz [234], however, has described an auto-mated system monitoring changes in acidity and turbidity for microbial detection on Mars (Wolf Trap, developed in the Nasa Space Program). Since then, Harrison *et al.* [235] have studied the time evolution of pH of a culture broth inoculated with low amounts of microorganisms; they have suggested that such pH–time measurements might be applied to the rapid determination of the bacterial inoculum size (Fig.23). Nevertheless, these works have revealed the limited sensitivity of pH-based methods for detecting microorganisms as compared to other techniques emerging at that time (e.g. radiometric or impedimetric procedures).

An original application of pH measurements was illustrated twenty years ago by Faine & Knight [236] who proposed a rapid pH-based microbiological assay of blood levels of aminoglycoside antibiotics. Using a sensitive pH meter, these workers measured small changes in the pH of a culture medium containing a heavy inoculum of selected test bacteria and different antibiotic concentrations. The fermentative activity of organisms (overall acidity production) was inhibited by the antibiotics, and the degree of inhibition, reflected by the amplitude of the pH shift, was related to the antibiotic concentration by referring to dose-response standard curves.

To conclude, the scope of pH measurements in the field of quantitative medical microbiology has been mainly restricted to applications where both detection times and thresholds are not determining parameters. For instance, for a long time these measurements have found an elective application field in dental research: to follow bacterial colonization of teeth (formation of bacterial plaque and caries), changes in pH of tooth surfaces are periodically or continuously monitored by means of microglass pH electrodes [189,237–244].

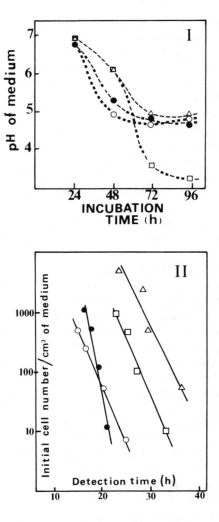

Fig. 23 — Detection and enumeration of microorganisms by monitoring the pH of the culture medium. I, changes in pH of a low-buffering-power culture medium (glucose/yeast extract broth) during incubation of various microbial cultures (initial cell concentration: 2 organisms/cm^3): (○), *Saccharomyces cerevisiae*; (●), *Lactobacillus fermenti*; (△), *Zymomonas anaerobia*; (□), *Acetobacter rancens*. II, corresponding detection times for the tested microorganisms. A change of 0.5 pH unit was chosen as the instrumental detection threshold. From Harrison *et al.* [235].

5   Conclusion

This section has presented the main physical or chemical parameters of overall activity of microbial cultures that can be detected and monitored by electrochemical techniques and that have found current applications in clinical microbiology.

Commercially available systems based on impedance or conductance measurements are suitable for the screening of large numbers of samples, to detect and enumerate pathogenic organisms or to test the susceptibility of bacteria to antibiotics. Nevertheless, these instruments are at present mainly applied to the control of the microbiological quality of foods (rapid estimation of microbial levels), and this may be partly explained by the following remarks. First, the area of susceptibility tests and rapid diagnosis of urinary tract infections is well covered by automated systems with spectrophotometric detection [4]. In addition, the recent attempt to apply the Malthus conductance system to the screening of bacteremia has highlighted some difficulties inherent in the technique (instability of electrodes in large-volume measuring cells) [84]. Other difficulties arise from the nature of the samples under test: as shown by Kagan *et al.* [78], automated devices for the rapid detection of bacteremia should include preliminary steps (centrifugation, filtration, ...) to concentrate bacteria and to remove the inhibitory agents present in blood samples [83].

Because of the nature of redox and reference electrodes, it is difficult to build at a relatively low cost a number of disposable or sterilizable measuring cells as is the case in impedance or conductance devices. While recently-developed redox procedures have been applied to the bacteriological analysis of water (i.e. to the detection and counting of fecal coliforms in water samples), an existing automated detector prototype has only two channels and is intended for determining *in situ* the fecal-coliform content of water, not for ensuring multiple analyses in the laboratory (see Part 3, Chapter 2). The major strength of redox-based procedures, namely the potentiometric measurement of lipoic acid reduction [153], lies certainly both in more basic studies of antimicrobial effects of drugs and in the practical evaluation of *in vitro* drug efficacity by referring to original criteria.

So far, susceptibility tests based on the electrochemical measurement of bacterial consumption of dissolved oxygen have mainly been applied to environmental microbiology as microbial toxicity assays for the rapid detection and quantification of pollutants. The rapidity of the test, which is achieved within 4 h [230], while impedance or conductance procedures, for example, require 6 h incubation times [88,90], seems a definite advantage of this technique for the determination of antibiograms. As with redox measurements, however, further practical developments of the procedure imply the construction of systems able to perform a number of simultaneous analyses.

Finally, the detection of bacterial activity from variations in pH of the culture broth suffers from its lack of sensitivity, since low bacterial populations induce only minor pH changes.

Other electrochemical procedures have been proposed for monitoring microbial activity. For instance, Allen has described a 'coulokinetic' technique which measures the electrical current developed by metabolizing microbial systems [245]. He has stated that his technique is more suitable than ORP measurements for investigating metabolic pathways of microorganisms. The coulokinetic procedure was found to be particularly useful for elucidating the effects of antimicrobial agents. Thus, coulombic output data obtained in the metabolism of glucose by *E. coli* cultures submitted to chlorpromazine revealed either an inhibitory or a potentiating pharmacological effect of the drug, according to its concentration range [245]: these results can be compared to those obtained by studying the antibacterial action of beta-lactam antibiotics with the potentiometric LA procedure [211].

Furthermore, besides oxygen and pH electrodes, other ion- and gas-selective electrodes or enzyme sensors, which have been increasingly developed during the last twenty years, may be used for measuring the concentration of specific ions, gaseous metabolites (ammonia or carbon dioxide), or substrates in fermentation broths: these probes are likely to be used to monitor microbial activity, reflected by precise metabolic pathways followed by the organisms. Thus, an ammonia electrode method has been used to separate *Campylobacter* strains on the basis of D-asparaginase activity: the ability of the different strains to break down D-asparagine has been tested by measuring bacterial production of ammonia with a commercial electrode [246,247].

All these promising electrochemical sensors and biosensors, however, are only at the beginning of their practical utilization in microbial culture media, the sterilization of the probes remaining one of the major difficulties to be overcome. Most work performed up to the present time relates to the control and regulation of culture broth contents during fermentation processes (Part 2). Nevertheless, these sensors and biosensors, whose applications to analyses in nonfermentation media are widely presented throughout the book, will certainly in the near future become potent investigation tools both in basic and applied microbiology, including clinical microbiology.

References
[1] Leclerc, H., Izard, D., Husson, M. O., Wattre, P. & Jacubczak, E. (1983) *Microbiologie générale*. Paris, Doin Editeurs p. 177.

[2] *Coulter Counter: medical and biological bibliography* (1984) Luton, Coulter Electronics Ltd.

[3] Koch, A. L. (1981) In: Gerhardt, P., Murray, R. G. E., Costilow, R. N., Nester, E. W., Wood, W. A., Krieg, N. R. & Phillips, G. B. (eds) *Manual of methods for general bacteriology*. Washington, American society for microbiology p. 179–207.

[4] Martin, W. J. (1983) In: *The direct detection of microorganisms in clinical samples*. New York, Academic Press p. 353–361.

[5] Mackett, D., Kessock-Philip, S., Bascomb, S. & Eastmon, C. S. F. (1982) *J. Clin. Pathol.* **35**, 107–110.

[6] Drow, D. L., Baum, C. H. & Hirschfield, G. (1984) *J. Clin. Microbiol.* **20**, 797–801.

[7] Schifman, R. B., Wieden, M., Brooker, J., Chery, M., Delduca, M., Norgard, K., Palen, C., Reis, N., Swanson, J. & White, J. (1984) *J. Clin. Microbiol.* **20**, 644–648.

[8] Welch, W. D., Thompson, L., Layman, M. & Southern, P. M., Jr (1984) *J. Clin. Microbiol.* **20**, 1165–1170.

[9] Martin, E. T., Cote, J. A., Perry, L. K. & Martin, W. J. (1985) *J. Clin. Microbiol.* **22**, 19–22.

[10] Mitruka, B. M. (1975) *Gas chromatographic applications in microbiology and medicine*. New York, John Wiley & Sons.

[11] Colve, P. J. (1978) *J. Clin. Pathol.* **31**, 365–369.

[12] Borgan, T. & Sorheim, K. (1984) *Methods Microbiol.* **15**, 345–362.

[13] Carlier, J. P. (1985) *Bull. Inst. Pasteur* **83**, 57–69.

[14] Deland, F. H. & Wagner, H. N., Jr (1970) *J. Lab. Clin. Med.* **75**, 529–534.

[15] Deblanc, H. J., Jr, Deland, F. & Wagner, H. N., Jr (1971) *Appl. Microbiol.* **22**, 846–849.

[16] Renner, E. D., Gatheridge, L. A. & Washington, J. A., II (1973) *Appl. Microbiol.*, **26**, 368–372.

[17] Brooks, K. & Sodeman, T. (1974) *Am. J. Clin. Pathol.* **61**, 859–866.

[18] Smith, A. G. & Little, R. R. (1974) *Ann. Clin. Lab. Sci.*, **4**, 448–455.

[19] Randall, E. L. (1975) In: Schlessinger, D. (ed.) *Microbiology–1975*. Washington, American Society for Microbiology p. 39–44.

[20] Thiemke, W. A. & Wicher, K. (1975) *J. Clin. Microbiol.*, **1**, 302–308.

[21] LaScolea, L. J., Jr, Dryja, D., Sullivan, T. D., Mosovich, L., Ellerstein, N. & Neter, E. (1981) *J. Clin. Microbiol.* **13**, 478–482.

[22] Carey, R. B. (1984) *J. Clin. Microbiol.* **19**, 599–605.

[23] Arpi, M., Lester, A. & Frederiksen, W. (1985) *Acta Path. Microbiol. Immunol. Scand. Sect. B* **93**, 263–271.

[24] Mitchell, M. J., Zwahlen, A., Elliott, H. L., Ford, N. K., Charache, P. & Moxon, E.R. (1985) *J. Clin. Microbiol.* **22**, 225–228.

[25] Smaron, M. F., Boonlayangoor, S. & Zierdt, C. H. (1985) *J. Clin.Microbiol.*, **21**, 298–301.

[26] Forrest, W. W. (1972) In: Norris, J.R. & Ribbons, D.W. (eds) *Methods in microbiology*. Vol. 6B. New York, Academic Press p. 285–318.

[27] Spink, E. & Wadsö, I. (1976) In: Glick, D. (ed.) *Methods in biochemical analysis*. Vol. 23. New York, Wiley–Interscience p. 1–159.

[28] Lamprecht, I. & Schaarschmidt, B. (eds.) (1977) *Application of calorimetry in life science*. Berlin, Walter de Gruyter.

[29] Beezer, A. E. (ed.) (1980) *Biological microcalorimetry*. London, Academic Press.

[30] Luong, J. H. T. & Volesky, B. (1983) *Adv. Biochem. Eng. Biotechnol.* **28**, 1–40.

[31] Binford, J. S., Binford, L. F. & Adler, P. (1973) *Am. J. Clin. Pathol.* **59**, 86–94.

[32] Beezer, A. E., Miles, R. J., Shaw, E. J. & Willis, P. (1979) *Experientia* **35**, 795–796.

[33] Beezer, A. E., Miles, R. J., Shaw, E. J. & Vickerstaff, L. (1980) *Experientia* **36**, 1051–1052.

[34] Beezer, A. E. & Chowdhry, B. Z. (1981) *Experientia* **37**, 828–831.

[35] Perry, B. F., Beezer, A. E., Miles, R. J., Smith, B. V., Miller, J. & De Graza Nascimento, E. (1986) *Microbios* **45**, 181–192.

[36] Mardh, P. A., Andersson, K. E., Ripa, T. & Wadsö, I. (1976) *Scand. J. Infect. Dis.* **9** (Suppl.), 12–16.

[37] Mardh, P. A., Ripa, T., Andersson, K.E. & Wadsö, I. (1976) *Antimicrob. Agents Chemother.* **10**, 604–609.

[38] Arhammer, M., Mardh, P.A., Ripa, T. & Andersson, K.E. (1978) *Acta Path. Microbiol. Scand.* **86**, 59–65.

[39] Mardh, P. A. & Arhammer, M. (1978) *J. Antimicrob. Chemother.* **4**, 73–78.

[40] Semenitz, E. (1978) *Immun. Infekt.* **6** 260–266.

[41] Mardh, P. A., Arhammer, M. & Andersson, K. E. (1979) *Chemotherapy* **25**,

106–116.

[42] Semenitz, E. (1979) *Zbl. Bakt. Hyg.* **243**, 372–380.

[43] Beezer, A. E. & Chowdhry, B. Z. (1980) In: Beezer, A.E. (ed.) *Biological microcalorimetry*. New York, Academic Press p. 195–246.

[44] Semenitz, E. (1980) *Thermochim. Acta* **40**,99–107.

[45] Semenitz, E., Casey, P. A., Pfaller, W. & Gstraunthaler, G. (1983) *Chemotherapy* **29**, 192–207.

[46] Steward, G. N. (1899) *J. Exp. Med.* **4**, 235–243.

[47] Oker–Blom, M. (1912) *Zbl. Bakteriol. Abs.* **65**, 382–389.

[48] Parsons, L. B. & Sturges, W. S. (1926) *J. Bacteriol.* **11**, 177–188.

[49] Parsons, L. B. & Sturges, W. S. (1926) *J. Bacteriol.* **12**, 267–272.

[50] Parsons, L. B., Drake, E. T. & Sturges, W. S. (1929) *J. Am. Chem. Soc.* **51**, 166–171.

[51] Sierakowski, S. & Leczycka, E. (1933) *Zbl. Bakt. Parasitenk., Abt.* 1, *Reihe B* **127**, 486–492.

[52] Allison, J. B., Anderson, J. A. & Cole, W. H. (1938) *J. Bacteriol.* **36**, 571–586.

[53] Kazal, L. A. (1943) *J. Bacteriol.* **45**, 277–292.

[54] Schwan, H. P. (1963) In: Nastuk, W. L. (ed.) *Physical techniques in biological research*. Vol. 6. New York, Academic Press p. 323–407.

[55] Schwan, H. P. (1968) *Ann. N. Y. Acad. Sci.* **148**, 191–209.

[56] Pollak, V. (1974) *Med. Biol. Eng.* **12**, 460–464.

[57] Eden, G. & Eden, R. (1984) *IEEE Trans. Biomed. Eng.* **31**, 193–198.

[58] Firstenberg–Eden, R. & Eden, G. (1984) *Impedance microbiology*. Letchworth, Research Studies Press.

[59] Ur, A. (1970) *Nature (Lond.)* **226**, 269–270.

[60] Ur, A. (1970) *Biomed. Eng.* **5**, 342–345.

[61] Ur, A. (1971) *Am. J. Clin. Pathol.* **56**, 713–718.

[62] Ur, A. & Brown, D. F. J. (1973) *J. Int. Res. Commun.* **1**, 37.

[63] Ur, A. & Brown, D. F. J. (1974) *Biomed. Eng.* **9**, 18–25.

[64] Ur, A. & Brown, D. F. J. (1975) *J. Med. Microbiol.* **8**, 19–28.

[65] Ur, A. & Brown, D. F. J. (1975) In: Héden, C. G. & Illéni, T. (eds) *New approaches to the identification of microorganisms*. New York, John Wiley & Sons p. 61–71.

[66] Tsuchiya, T. & Terashima, E. (1978) *Nihon Univ. J. Med.* **20**, 319–330.

[67] Cady, P. & Welch, W. J. (1973) *U.S. Patent* No. 3,743,581.

[68] Cady, P., Omura, T., Rudin, M. & Fleming, J. H. (1978) *U.S. Patent* No. 4,072,578.

[69] Cady, P., Dufour, S. W., Shaw, J. & Kraeger, S. J. (1978) *J. Clin. Microbiol.* **7**, 265–272.

[70] Hadley, W. K. & Senyk, G. (1975) In: Schlessinger, D. (ed.) *Microbiology–1975*. Washington, American Society for Microbiology p. 12–21.

[71] Cady, P. (1977) *Food Prod. Dev.* **11**, 80–85.

[72] Eden, R. (1983) *Food Technol.* **37**, 64–70.

[73] Richards, J. C. S., Jason, A. C., Hobbs, G., Gibson, D. M. & Christie, R. H. (1978) *J.Phys. E: Sci. Instrum.*, **11**, 560–568.

[74] Baynes, N. C., Comrie, J. & Prain, J. H. (1983) *Med. Lab. Sci.* **40**, 149–158.

[75] Hause, L. L., Komorowski, R. A. & Gayon, F. (1981) *IEEE Trans. Bio-*

med.Eng. **28**, 403–410.

[76] Firstenberg–Eden, R. & Zindulis, J. (1984) *J. Microbiol. Meth*. **2**, 103–115.

[77] Kahn, W., Friedman, G., Rodriguez, W., Controni, G. & Ross, S.(1976) In: *Proceedings of the Second International Symposium on Rapid Methods and Automation in Microbiology*, *Cambridge (England)*, 19–25 September 1976. Oxford, Learned Information (Europe) p. 14–15.

[78] Kagan, R. L., Schuette, W. H., Zierdt, C. H. & MacLowry, J. D. (1977) *J.Clin. Microbiol*. **5**, 51–57.

[79] Specter, S., Throm, R., Strauss, R. & Friedman H. (1977) *J. Clin. Microbiol*. **6**, 489–493.

[80] Throm, R., Specter, S., Strauss, R. & Friedman, H. (1977) *J. Clin. Microbiol*. **6**, 271–273.

[81] Zafari, Y. & Martin, W. J. (1977) *J. Clin. Microbiol*. **5**, 545–547.

[82] Cady, P., Dufour, S. W., Lawless, P., Nunke, B. & Kraeger, S. J. (1978) *J. Clin. Microbiol*. **7**, 273–278.

[83] Murray, P. R. (1983) In: *The direct detection of microorganisms in clinical samples*. New York, Academic Press p. 343–351.

[84] Brown, D. F. J., Warner, M., Taylor, C. E. D. & Warren, R. E. (1984) *J. Clin. Pathol*. **37**, 65–69.

[85] Brown, D. F. J. & Warner, M. (1982) In: Tilton, R. C. (ed.), *Rapid methods and automation in microbiology*. Washington, American Society for Microbiology p. 171–175.

[86] Smith, T. K., Eggington, R., Pease, A. A., Harris, D. M. & Spencer, R. C. (1985) *J. Clin. Pathol*. **38**, 926–928.

[87] Bauer, A. N., Kirby, W. M. M., Sherris, J. C. & Turk, M. (1966) *Am. J. Clin. Pathol*. **45**, 493–496.

[88] Colvin, H. J. & Sherris, J. C. (1977) *Antimicrob. Agents Chemother*. **12**, 61–66.

[89] Porter, I. A., Reid, T. M. S., Wood, W. J., Gibson, D. M. & Hobbs, G. (1983) In: Russell, A.D. & Quesnel, L.B. (eds) *Antibiotics: assessment of antimicrobial activity and resistance*. London, Academic Press p. 49–60.

[90] Baynes, N. C., Comrie, J. & Harper, I. A. (1986) *Med. Lab. Sci*. **43**, 232–240.

[91] Cady, P. (1975) In: Hedén, C.G. & Illéni, T. (eds) *New approaches to the identification of microorganisms*. New York, John Wiley & Sons p. 73–99.

[92] Einarsson, H. & Snygg, B. G. (1986) *J. Appl. Bact*. **60**, 15–19.

[93] Costilow, R. N. (1981) In: Gerhardt, P., Murray, R. G. E., Costilow, R. N., Nester, E. W., Wood, W. A., Krieg, N. R. & Phillips, G. B. (eds) *Manual of methods for general bacteriology*. Washington, American Society for Microbiology p. 66–78.

[94] Jacob, H. E. (1970) In: Norris, J.R. & Ribbons, D. W. (eds) *Methods in microbiology*. Vol.2. London and New York, Academic Press p. 91–123.

[95] Potter, M. C. (1911) *Proc. R. Soc. London Ser. B* **84**, 260–276.

[96] Hewitt, L. F. (1931) *Oxidation-reduction potentials in bacteriology and biochemistry*. London County Council.

[97] Michaelis, M. (1933) *Oxydations-Reduktion-Potentiale mit besonderer Berücksichtigung ihrer physiologischen Bedeutung*. Berlin, Springer-Verlag.

[98] Kanel, E. S. (1941) *Microbiology (USSR)* **10**, 595–620.

[99] Tizzano, A. (1946) *Experientia* **2**, 86–99.

[100] Hewitt, L. F. (1950) *Oxidation-reduction potentials in bacteriology and bio-chemistry*. 6th ed. Edinburgh, Livingstone .

[101] Clark, W. M. (1960) *Oxidation-reduction potentials of organic systems*. Baltimore, Williams and Wilkins.

[102] Rabotnowa, I. L. (1963) *Die Bedeutung physikalisch-chemischer Faktoren (pH und rH₂) für die Lebenstätigkeit der Mikroorganismen*. Jena, Fischer-Verlag.

[103] Jacob, H. E. (1971) *Z. Allg. Mikrobiol.* **11**, 691–734.

[104] Kjaergaard, L. (1977) In: Ghose, T. K., Fiechter, A. & Blakebrough, N. (eds) *Advances in biochemical engineering*. Vol. 7. Berlin, Springer-Verlag p. 131–150.

[105] Harrison, D. E. F. (1972) *J. Appl. Chem. Biotechnol.* **22**, 417–440.

[106] Wimpenny, J. W. T. & Necklen, D. K. (1971) *Biochim. Biophys. Acta* **253**, 352–359.

[107] Morris, J. C. & Stumm, W. (1967) *Adv. Chem. Ser.*, No. 67, 27–85.

[108] Stumm, W. (1967) In: Jaag, O. (ed.) *Advances in water pollution research. Proceedings of an International Conference*, 1966, vol. 1. Washington, Water Pollut. Contr. Fed. p. 283–308.

[109] Balakireva, L. M., Kantere, V. M., & Rabotnova, I. L. (1974) *Biotechnol. Bioeng. Symp.*, No. 4, 769–780.

[110] Kantere, V. M. (1971) In: Milazzo, G., Jones, P. E. & Rampazzo, L. (eds) *Biological aspects of electrochemistry*, Basel, Birkhäuser Verlag p.355–366.

[111] Jacob, H. E. (1967) *Z. Allg. Mikrobiol.* **7**, 407–408.

[112] Jacob, H. E. (1970) *Path. Microbiol.* **36**, 57–62.

[113] Jacob, H. E. (1974) *Biotechnol. Bioeng. Symp.*, No. 4, 781–788.

[114] Zakhar'evskii, M. S. (1940) *Microbiology (USSR)* **9**, 872–878.

[115] Zakhar'evskii, M. S. (1941) *Z. Mikrobiol. Epidemiol. Immunitätsforsch.*, 111–114.

[116] Zakhar'evskii, M. S. (1944) *Zh. Mikrobiol. Epidemiol. Immunobiol.* **4–5**, 78–82.

[117] Kovacs, E. & Marton, K.P. (1958) *Naturwissenschaften* **45**, 285–286.

[118] Wilkins, J. R. (1978) *Appl. Environ. Microbiol.* **36**, 683–687.

[119] Holland, R. L., Cooper, B. H., Helgeson, N. G. P. & McCracken, A. W. (1980) *J. Clin. Microbiol.* **12**, 180–184.

[120] Ackland, M. R., Manvell, P. M. & Bean, P. G. (1984) *Biotechnol. Lett.* **6**, 137–142.

[121] Gillespie, L. J. (1920) *Soil Sci.* **9**, 199–216.

[122] Okey, R. W., Cohen, R. L., Monteith, L. E., Chapman, D. D., Proctor, C. M. & Wedemeyer, C. A. (1963) *Develop. Ind. Microbiol.* **4**, 223–234.

[123] Jacob, H. E. & Horn, G. (1965) *Z. Allg. Mikrobiol.* **5**, 33–41.

[124] Lengyel, Z. L. & Nyiri, L. (1965) *Biotechnol. Bioeng.* **7**, 91–100.

[125] Dahod, S. K. (1982) *Biotechnol. Bioeng.* **24**, 2123–2125.

[126] Chung, I. S. & Lee, Y. Y. (1986) *Enzyme Microb. Technol.* **8**, 503–507.

[127] Tengerdy, R. P. (1961) *Biochem. Microbiol. Technol. Eng.* **3**, 241–253.

[128] Matkovics, B., Kovacs, E. & Sipos, G. (1959) *Z. Path. Bakt.* **22**, 214–224.

[129] Squires, R. W. & Hosler, P. (1958) *Ind. Eng. Chem.* **50**, 1263–1266.

[130] Ishizaki, A., Shibai, H. & Hirose, Y. (1974) *Agric. Biol. Chem.* **38**, 2399–2406.
[131] Warner, T. B. & Schuldiner, S. (1965) *J. Electrochem. Soc.* **112,** 853–856.
[132] Schuldiner, S., Piersma, B. J. & Warner, T. B. (1966) *J. Electrochem. Soc.* **113**, 573–577.
[133] Wroblowa, H., Rao, M. L. B., Damjanovic, A. & Bockris, J.O'M. (1967) *J. Electroanal. Chem.* **15**, 139–150.
[134] Coulter, C. B. & Isaacs, M. L. (1929) *J. Exp. Med.* **49**, 711–725.
[135] Kliger, J. & Guggenheim, K. (1938) *J. Bacteriol.* **35**, 141–156.
[136] Wimpenny, J. W. T. (1969) *Biotechnol. Bioeng.* **11**, 623–629.
[137] Shibai, H., Ishizaki, A., Kobayashi, K. & Hirose, H. (1974) *Agr. Biol. Chem.* **38**, 2407–2411.
[138] Thompson, B. G. & Gerson, D. F. (1985) *Biotechnol. Bioeng.* **27**, 1512–1515.
[139] Kluyver, A. J. & Hoogerheide, J. C. (1936) *Enzymologia* 1–21.
[140] Vennesland, B. & Hanke, M. E. (1940) *J. Bacteriol.* **39**, 139–169.
[141] Cole, J. A. (1976) In: Rose, A. H. & Tempest, D. W. (eds) *Advances in microbial physiology*. Vol. 14. London, Academic Press p. 1–92.
[142] Aurel, E., Aubertin, E. & Genevois, L. (1929) *Ann. Physiol. Physicochim. Biol.* **5**, 1–11.
[143] Hanke, H. E. & Katz, Y. J. (1943) *Arch. Biochem.* **2**, 183–200.
[144] Dobson, A. J. (1964) *J. Gen. Microbiol.* **35**, 169–174.
[145] Rippon, J. W. (1968) *Appl. Microbiol.* **16**, 114–121.
[146] Johnstone, K. I. (1942) *J. Path. Bact.* **54**, 25–38.
[147] Douglas, F. & Rigby, G. J. (1974) *J. Appl. Bact.* **37**, 251–259.
[148] Wilkins, J. R., Stoner, G. E. & Boykin, E. H. (1974) *Appl. Microbiol.* **27**, 949–952.
[149] Fischer, D. J., Herner, A. E., Landes, A., Batlin, A. & Barger, J. W. (1965) *Biotechnol. Bioeng.* **7**, 471–490.
[150] Fischer, D. J., Landes, A., Sanford II, M. T., Herner, A. E. & Wiegand, C. J. W. (1965) *Biotechnol. Bioeng.* **7**, 491–506.
[151] Brown, D. E., Groves, G. R. & Miller, J. D. A. (1973) *J. Appl. Chem. Biotechnol.* **23**, 141–149.
[152] Brünger, R. (1982) *Radiat. Environ. Biophys.* **21**, 141–154.
[153] Junter, G. A. (1984) *Trends Anal. Chem.* **3**, 253–259.
[154] Reed, L. J. (1966) In: Florkin, M. & Stotz, E. H. (eds) *Comprehensive biochemistry*. Vol.14. Amsterdam, Elsevier p. 99–126.
[155] Junter, G. A., Sélégny, E. & Lemeland, J. F. (1982) *Bioelectrochem. Bioenerg.* **9**, 679–697.
[156] Junter, G. A., Lemeland, J. F. & Sélégny, E. (1982) *Bioelectrochem. Bioenerg.* **9**, 699–709.
[157] Charrière, G., Jouenne, T., Lemeland, J. F., Sélégny, E. & Junter, G. A. (1984) *Appl. Environ. Microbiol.* **47**, 160–166.
[158] Jacobs, L. (1950) *Am. J. Trop. Med.* **30**, 803–815.
[159] Chodat, F. & Naves, R. (1954) *Schweiz. Z. Allg. Pathol. Bakteriol.* **17**, 492–497.
[160] Turner, N. C., Gertler, M. M. & Garn, S. M. (1954) *Science* **119**, 847–849.
[161] Hewitt, L. F. (1930) *Biochem. J.* **24**, 669–675.

[162] Hewitt, L. F. (1930) *Biochem. J.* **24**, 676–681.
[163] Tuttle, C. D. & Huddleson, I. (1934) *J. Infect. Dis.* **54**, 259–272.
[164] Kholchev, N. (1935) *J. Microbiol. Epidemiol. Immunobiol.*, **15**, 52–55.
[165] Burrows, W. & Jordan, E. O. (1935) *J. Infect. Dis.* **56**, 255–263.
[166] Burrows, W. & Jordan, E. O. (1936) *J. Infect. Dis.* **58**, 259–262.
[167] Gillespie, R. W. H. & Rettger, L. F. (1938) *J. Bacteriol.* **36**, 605–620.
[168] Gillespie, R. W. H. & Rettger, L. F. (1938) *J. Bacteriol.* **36**, 621–631.
[169] Gillespie, R. W. H. & Porter, J. R. (1938) *J. Bacteriol.* **36**, 633–637.
[170] Burrows, W. (1941) *J. Infect. Dis.* **69**, 141–147.
[171] Zador, S. (1961) *Acta Biol. Acad. Sci. Hungaricae* **11**, 387–392.
[172] Zador, S. (1961) *Acta Biol. Acad. Sci. Hungaricae* **12**, 35–46.
[173] Gimranov, M. G. (1965) *Zh. Mikrobiol. Epidemiol. Immunobiol.* **42**, 58–62.
[174] Blagova, N. V. & Belozerova, A. V. (1970) *Zh. Mikrobiol. Epidemiol. Immunobiol.* **47**, 59–62.
[175] Shikova, L., Payanotova, K. & Deskova, G. (1972) *Ser. Sci. Med. Annu. Sci. Pap.* **10**, 123–135.
[176] Walden, W. C. & Hentges, D. J. (1975) *Appl. Microbiol.* **30**, 781–785.
[177] Aksianzew, M. I. (1933) *Z. Tuberk.* **68**, 249–253.
[178] Messing, V. & Semich, A. (1935) *J. Microbiol. Epidemiol. Immunobiol.* **15**, 665–667.
[179] Moretti, G. F. & Catros, P. (1950) *Compt. Rend. Soc.Biol.* **144**, 1500–1503.
[180] Sawada, T., Oshima, T. & Suzuki, I. (1953) *Gunma J. Med. Sci.* **2**,127–135.
[181] Jayawardene, A. & Goldner, M. (1975) *Antonie van Leeuwenhoek* **41**, 553–568.
[182] Barnes, E. M. & Ingram, M. (1955) *J. Sci. Food Agr.* **6**, 448–455.
[183] Bergeim, O., Kleinberg., J. & Kirsch, E. R. (1945) . *Bacteriol.* **49**, 453–458.
[184] Moretti, G. F. (1950) *C. R. Soc. Biol.* **144**, 1499–1500.
[185] Celesk, R. A., Asano, T. & Wagner, M. (1976) *Proc. Soc. Exp. Biol. Med.* **151**, 260–263.
[186] Lipsits, D. V. (1953) *Voprosy Pitaniya* **12**, 43–50.
[187] Kurokochi, T. (1950) *J. Japan. Biochem.* **22**, 1–6.
[188] Kenney, E. B. & Ash., M. (1969) *J. Peridont.* **40**, 630–633.
[189] Russell, C. & Coulter, W. A. (1975) *Appl. Microbiol.* **29**, 141–144.
[190] Ward, W. E. (1938) *J. Bacteriol.* **36**, 337–355.
[191] Zakhar'evskii, M.S. (1939) *Z. Mikrobiol. Epidemiol. Immunitätsforsch.*, No. 2–3, 87–97.
[192] Schmidt–Lorenz, W. (1960) *Naturwissenschaften* **47**, 208–209.
[193] Oblinger, J. L. & Kraft, A. A. (1973) . *Food Sci.* **38**, 1108–1112.
[194] Junter, G. A., Lemeland, J. F. & Sélégny, E. (1980) *Appl. Environ. Microbiol.* **39**, 307–316.
[195] Lamb, V. A., Dalton, H. P. & Wilkins, J. R. (1976) *Am. J. Clin. Path.* **66**, 91–95.
[196] Vaczi, L. (1944) *Z. Immunitäts.* **105**, 178–188.
[197] Vaczi, L. & Toth, I. (1946) *Z. Immunitäts.* **105**, 285–296.
[198] Drouhet, E. (1949) *Ann. Inst. Pasteur*, **76**, 168–173.
[199] Drouhet, E. & Kepes, A. (1949) *Ann. Inst. Pasteur* **76**, 174–177.
[200] Kramli, A., Stur, J. K. & Turay, P. (1954) *Acta Physiol. Acad. Sci. Hung.* **5**,

549–551.

[201] Kramli, A., Stur, J. K. & Turay, P. (1955) *Acta Physiol. Acad. Sci. Hung.* **8**, 15–24.

[202] Jacob, H. E. & Horn, G. (1964) *Zbl. Bakt. Parasitenk. Infekt.– Krankh. Hyg.*, I, **192**, 315–319.

[203] Jacob, H. E. & Horn, G. (1964) *Abh. Dtsch. Akad. Wissensch. Berlin (Kl. Chem. Geol. Biol.)* **1**, 320–326.

[204] Kovacs, E. & Kokai, K. (1965) *Path. Microbiol.* **28**, 454–459.

[205] Jacob, H. E. & Horn, G. (1966) *Abh. Dtsch. Akad. Wissensch. Berlin (Kl. Med.)* **4**, 93–101.

[206] Junter, G. A., Lemeland, J. F. & Sélégny, E. (1980) *Path. Biol.* **28**, 435–439.

[207] Sélégny, A., Junter, G. A., Sélégny, E. & Lemeland, J. F. (1984) *C. R. Soc. Biol.* **178**, 722–729.

[208] Junter, G. A., Bogaczyk, M. C., Lemeland, J. F. & Sélégny, E. (1984) In: Allen, M. J. & Usherwood, P. N. R. (eds), *Charge and field effects in biosystems*, Tunbridge Wells, Abacus Press p. 195–200.

[209] Junter, G. A. (1984) *Bioelectrochem. Bioenerg.* **12**, 81–92.

[210] Bogaczyk, M. C., Sélégny, E. & Junter, G. A. (1984) *Bioelectrochem. Bioenerg.* **12**, 93–103.

[211] Junter, G. A. & Bogaczyk–Hébert, M.C. (1985) *J. Antimicrob. Chemother.* **15**, 685–693.

[212] Junter, G. A. (1987) *Bioelectrochem. Bioenerg.* **17**, 43–57.

[213] Tomasz, A. (1979) *Rev. Infect. Dis.* **1**, 434–467.

[214] Spratt, B. G. (1983) *J. Gen. Microbiol.* **129**, 1247–1260.

[215] Waxman, D. J. & Strominger, J. L. (1983) *Ann. Rev. Biochem.* **52**, 825–869.

[216] Hoover, J. R. E. (1983) *Handbook Exper. Pharmacol.* **67**, 119–247.

[217] Junter, G. A. (1986) *Path. Biol.* **34**, 549–553.

[218] Clark, L. C., Jr, Wolf, R., Granger, D. & Taylor, Z. (1953) *J. Appl.Physiol.* **6** 189–193.

[219] Hitchman, M. L. (1978) *Chem. Anal. Ser. Monogr. Anal. Chem.* **49**, 1–255.

[220] Lee, H. Y. & Tsao, G. T. (1979) *Adv. Biochem. Eng.* **13** 35–86.

[221] Pungor, E., Feher, Z. & Varadi, M. (1980) *CRC Crit. Rev. Anal. Chem.* **9** 97–165.

[222] Havas, J. (1985) *Ion– and molecule–selective electrodes in biological systems* Berlin-Heidelberg-New York-Tokyo, Springer-Verlag p. 168–187.

[223] Jacob, H. E. & Horn, G. (1964) *Abh. Dtsch. Akad. Wissensch. Berlin (Kl. Med.)* **4**, 87–92.

[224] Robra, K. H. (1976) *Gas-Wasserfach, Wasser-Abwasser* **117**, 80–86.

[225] Boyles, D. T. & Rose, H. J. (1977) *French patent* No. 77 31941.

[226] Bauer, N. J., Seidler, R. J. & Knittel, M. D. (1981) *Bull. Environ. Contam. Toxicol.* **27**, 577–582.

[227] Paga, U. (1981) *Vom Wasser* **57**, 263–275.

[228] Lindemann, J., Gernand, K. & Blume, H. (1977) *Dtsch. Gesundh.-Wes.* **32**, 1476–1478.

[229] Schmidt, E., Strassburger, J., Tiller, F. W. & Thilo, W. (1982) *Dtsch. Gesundh.-Wes.* **37**, 977–981.

[230] Strassburger, J. & Tiller, F. W. (1984) *Zbl. Bakt. Hyg.* **A 256**, 466–474.

[231] Stradtmann, H. & Kleint, V. (1978) *Z. Ges. Inn. Med.* **33**, 407–409.
[232] Havas, J. (1985) *Ion– and molecule–selective electrodes in biological systems*. Berlin–Heidelberg–New York–Tokyo, Springer–Verlag p. 91–116.
[233] Munro, A. L. S. (1970) In: Norris, J. R. & Ribbons, D. W. (eds) *Methods in microbiology*. Vol 2. New York, Academic Press p. 39–89.
[234] Mitz, M. A. (1969) *Ann. N. Y. Acad. Sci.* **158**, 651–664.
[235] Harrison, J., Webb, T. J. B. & Martin, P. A. (1974) *J. Inst. Brew.* **80**, 390–398.
[236] Faine, S. & Knight, D.C. (1968) *Lancet* **7564**, 375–378.
[237] Stephan, R. M. (1940) *J. Am. Dent. Assoc.* **27**, 718–723.
[238] Stephan, R. M. (1944) *J. Dent. Res.* **23**, 257–266.
[239] Eastoe, J. E. & Bowen, W. H. (1967) *Caries Res.* **1**, 59–68.
[240] Graf, H. & Mühlemann, H. R. (1966) *Helv. Odontol. Acta* **10**, 94–101.
[241] Newman, P., Macfadyen, E. E., Gillespie, F. C. & Stephan, K. W. (1979) *Arch. Oral Biol.* **24**, 501–507.
[242] Brown, J. P., Huang, C. T., Oldershaw, M. D. & Bibby, B. G. (1981) *J. Dent. Res.* **60**, 724.
[243] Schactele, C. F. & Jensen, M. E. (1982) *J. Dent. Res.* **61**, 1117–1125.
[244] Hudson, D. E., Donoghue, H. D. & Perrons, C. J. (1986) *J. Appl. Bacteriol.* **60**, 301–310.
[245] Allen, M. J. (1966) *Bacteriol. Rev.* **30**, 80–93.
[246] Karmali, M. A., Williams, A., Fleming, P. C., Krishnan, C. & Wood, M. M. (1984) *J. Hyg.* **93**, 189–196.
[247] Karmali, M. A., Roscoe, M. & Fleming, P. C. (1986) *J. Clin. Microbiol.* **23**, 743–747.

2

In vivo determinations

2.1   CARBON FIBRE ELECTRODES

J. L. Taboureau and P. Blond

It was during the last decade that Adams [1–3] showed the importance for neurochemistry of the electrochemical study of aromatic compounds. He especially drew attention to the group of catechols which is easy to oxidize electrochemically (Fig. 1).

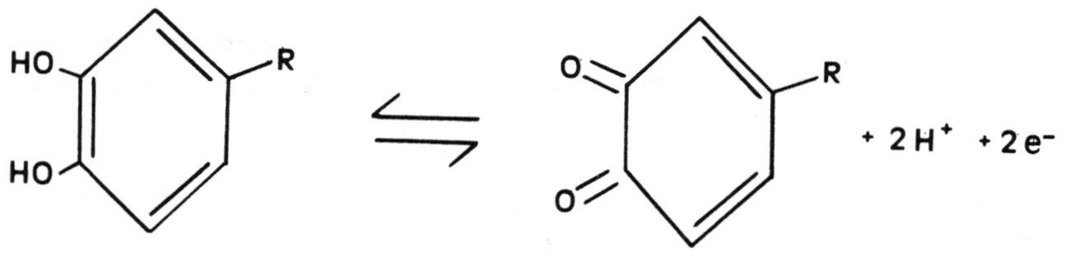

Fig. 1 — Oxidation of catechols.

At that time, Adams hoped that these compounds were neurotransmitters and that it would be possible to study them *in vivo* in nerve tissue. In fact, the extracellular fluid is, for numerous reasons, a good electrolyte support (weak ohmic resistance, sufficient electrolyte concentrations to be conductive, a strong buffer). One of the most difficult problems has been to obtain electrodes which can be implanted *in vivo*. This is the neurochemical context in which the technology of carbon fibre electrodes has been developed. The properties of mechanical resistance, the properties of electrical conductivity, and finally, the weak construction factor (diameter = 8 μm) make the carbon fibres the chosen material for obtaining microelectrodes which can be implanted *in vivo* and used in neurochemistry.

1   Nerve tissue and neurotransmitters

Nerve tissue is different from other living tissues, as the nerve cells (neurones) serve to communicate with other cells. The neurone is composed of a cellular body (pericaryon) and of cytoplasmic extensions which favour information exchanges. There are two types of extensions: axon and dendrite; the dendrite types are short extensions which branch out in tree-like formations; the axons branch out a little and are longer (Fig. 2).

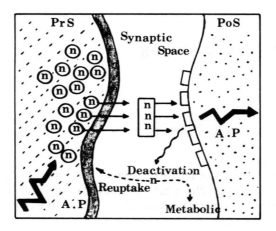

Fig. 2 — 'Typical' neurone in a mammalian brain.

Communication between neurones takes place by means of exact structures: synapses. In classical neurophysiology, the dendrites receive information, and axons transmit it. Usually, the synapses are represented by the axon of a primary neurone and the dendrites of a secondary neurone.

Also, in some neurones where synaptic transmission takes place directly by an electrical signal going through the synapse, e.g. in mammalian brains, the synaptic transmission is of a chemical nature and unidirectional (Fig. 3). On the structural level, the synapse possesses a synaptic cleft between 200 and 800 Å bordered by a presynaptic membrane belonging to the first neurone and a postsynaptic membrane belonging to the second neurone (Fig. 3). The electron microscope shows the accumulation of vesicles at the presynaptic structure. These vesicles contain molecules of weak molecular mass: neurotransmitters or neuromediators. In addition, the presynaptic structure possesses a range of cellular organites which allow the synthesis of the vesicles and of the neurotransmitters. The neurotransmission is initiated by an electrical signal: the action potential. Action potentials have a value of 100 mV and last 1 to 2 ms.

When an action potential arrives at the presynaptic end of the first neurone, the contents of some synaptic vesicles in the synaptic cleft are released; thus, the neurotransmitters move by convection or diffusion towards the membraneous receptors situated on the postsynaptic membrane. The formation of a neurotransmitter–receptor complex causes a modification in the membraneous permeability of

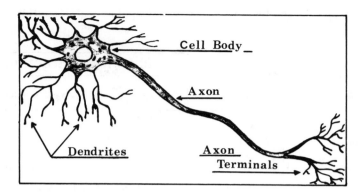

Fig. 3 — Outline representation of a chemical synapse (Prs: presynaptic membrane; Pos: postsynaptic membrane; n: neurotransmitter).

Na^+ ions and other ionic species. The entry of Na^+ ions in the postsynaptic structure induces the creation of a postsynaptic action potential. This new potential can be inhibiting or activating; its amplitude registers several millivolts and lasts from several milliseconds to several seconds. The amplitude of the postsynaptic action potential depends on the quantity of neurotransmitters released in the synaptic cleft.

After the formation of the neurotransmitter–receptor complex, the neurotransmitter is inactivated so that the stimulation or inhibition of the postsynaptic structure stops. Inactivation can occur through the enzymatic deterioration of the molecule or through the recapture of the neurotransmitter by means of the presynaptic structure. Recapture is the most frequent type because in this way the neurotransmitter can be re-used. For a long time, the generally accepted idea was that one type of neurone uses one single mediator; it is now known, however, that certain neurones use two or more neurotransmitters. The chemical structures of the most common neurotransmitters are summarized in Table 1.

It is easy to understand the importance of measuring these neurotransmitters *in vivo*, both for our knowledge of the physiology of the nervous system and for the study of certain drugs and their effects on the nervous system. Since the catechols are easily oxidizable, their electrochemical detection *in vivo* has been the subject of a number of studies. This type of detection encounters three types of major difficulty: firstly, because the signal sought must not be masked by other oxidizable molecules present in heavy concentrations (notably, ascorbic acid), the electrodes implanted in the brain of an animal must exhibit an excellent selectivity and preserve this property long enough in a complex and poorly controlled milieu; secondly, the detection limits must be very low (5 nM for dopamine); finally, the working electrodes, and therefore the reference and auxiliary electrodes, must all be miniaturized. We will briefly look at the different voltammetric techniques most frequently used in neurochemistry, as well as the different types of working electrodes implanted *in vivo*.

2   Voltammetric techniques and electrodes
2.1   *Voltammetry*
Polarography was discovered in 1924 by Heyrovsky who showed [4] that the current obtained at a mercury electrode was not directly proportional to the applied tension,

Table 1 — Chemical structure of principal neurotransmitters

name	abbreviation	chemical structure
Acetylcholine	Ach	$(CH_3)_3\text{-}N^+\text{-}CH_2\text{-}CH_2\text{-}O\text{-}\overset{\overset{\displaystyle O}{\|}}{C}\text{-}CH_3$
Dopamine	DA	HO–(benzene ring)–$CH_2\text{-}CH_2\text{-}NH_2$, with HO substituents
Noradrenaline	NE	HO–(benzene ring)–$CHOH\text{-}CH_2\text{-}NH_2$, with HO substituents
Serotonin	5_HT	HO–(indole ring)–$CH_2\text{-}CH_2\text{-}NH_2$, with HO substituents
ɣ-Aminobutyric acid	GABA	$HOOC\text{-}CH_2\text{-}CH_2\text{-}CH_2\text{-}NH_2$

but that it was the function of different oxidizable chemicals in the solution. This technique measures oxidizable or reducible compounds at the electrode surface. We will speak of polarography when a mercury electrode is used, and of voltammetry for other types of electrodes.

On the practical side, a montage of three electrodes is used: the working electrode, the reference electrode, and the auxiliary electrode. The voltage from a generator is applied to the working electrode by the intermediary of an auxiliary electrode (generally of a metallic nature, e.g. platinum wire). The reference electrode is stabilized; the electrochemical theory shows that the oxidation current (i) decreases with time (t) as stated in Cottrell's equation:

$$i = (n\,F\,A\,D^{\frac{1}{2}}C)/(t^{\frac{1}{2}}\,\pi^{\frac{1}{2}}).$$

The same number of electrons is kept during oxidation ($n=2$ for catechols). F is the Faraday, A is the electrode surface, D is the diffusion coefficient of the oxidized substance, and C is its concentration in the solution.

The current flowing through the working electrode is composed of a charging current and a faradic current. The charging current is a result of the physicochemical nature of the electrode and does not appear in the oxido-reduction phenomena, while the faradic current is due exclusively to the processes of oxidation and reduction. The majority of voltammetric techniques are designed in such a manner that they favour the faradic current and minimize the charging current.

2.2 *Different voltammetric techniques used* in vivo
Chronoamperometry (CH)
This is the simplest and most frequently used method for carbon paste electrodes. The method gives higher faradic to charging current ratios. It possesses a weak aptitude to measure more than one electrode species. Measurements can be performed every five seconds, and they show good sensitivity [5].

Linear Sweep Voltammetry (LSV)
This technique examines the potential range and gives weak peaks for each compound encountered. It exhibits a solid resolving power for the different compounds. The measurement interval is 10 min. The technique has a moderate sensitivity [6].

Cyclic Voltammetry
This technique is similar to LSV but with a reduced time for response after oxidation. It is not currently used *in vivo* [7].

Differential Pulse Voltammetry (DPV)
This technique is very widespread, and is primarily used with electrodes of carbon fibres which have been electrochemically treated. The interval between two measurements is 2 min; the method exhibits a great sensivity [8–11].

Normal Pulse Voltammetry (NPV)
This technique is derived from CH; it supplies a series of pulsations of increasing size to the working electrode. It is primarily used with non-treated carbon fibre electrodes. It exhibits moderate resolving power and sensitivity; the interval between 2 measurements is 2 min [12,13].

Differential Normal Pulse Voltammetry (DNPV)
This technique, derived from NPV, uses carbon fibre electrodes and exhibits good resolving power. Also, there is no progressive poisoning of the electrode. The measurement interval is 60 s, and this method is very sensitive [14].

Differential Pulse Amperometry (DPA)
This technique is derived from CH and also uses carbon fibre electrodes. Better selectivity is noted than in CH, and a very high sensitivity is also exhibited. The interval between two measurements is 400 ms [15].

High Speed Cyclic Voltammetry (HSCV)
This technique is used only with carbon fibre microelectrodes of very small size. Resolving power is limited and sensitivity is moderate. Measurements are possible at intervals of 25 ms [16].

2.3 *Electrodes used* in vivo

To determine dosages *in vivo*, notably in the brain, working electrodes and also the reference and auxiliary electrodes must be miniaturized. The electrodes available in the 1970s in electrochemistry laboratories could not be used in studies *in vivo* (the smallest diameters were several millimeters). At present, most electrodes used *in vivo* are made of carbon compounds: carbon paste, graphite–resin, glassy–carbon, and of course, carbon fibre. It can be noted that certain researchers have used iodinated platinum electrodes in some experiments. A brief review of the different types of electrodes follows, before an in-depth study of carbon fibre electrodes:

Carbon paste electrodes

These electrodes are very popular owing to the simplicity of their use [17–19]. They consist of a Teflon-sheathed silver wire. The Teflon sheath extends over the end of the silver wire by several millimetres and forms a well that is filled with a paste made of graphite powder and either silicone oil or Nujol. The resulting electrode has a diameter of about 200 μm and exhibits a weak background current and good flexibility which makes it more useful than rigid electrodes, especially in long-term implantations.

Carbon epoxy electrodes

These are a variation of the carbon paste electrode in which the silicone oil or Nujol is replaced by an epoxy resin [20]. At the end of the Teflon-sheathed silver wire, a pulled glass micropipette is packed with a carbon-epoxy mixture and then heat-cured. The resulting electrode has a diameter of 50 to 100 μm. This type of electrode is used for chronic implantation, and it withstands dissolution in nerve tissue, but being rigid, it is more fragile than carbon paste electrodes.

Glassy carbon electrodes and platinum electrodes

Glassy carbon electrodes are rarely used *in vivo* because of manufacturing difficulties and their very limited resolving powers *in vivo*. Their principal disadvantage is the loss of electron transfer rates between ascorbic acid and catechols on the implantation of these electrodes in nerve tissue. Their use is very limited.

2.4 *Modification of the electrode surface*

Modification of the state of the electrode surface is an area of great interest for electrochemistry *in vivo*. The different processes used have been studied by Snell & Keenan [21]. For electrochemical detection *in vivo*, a very high selectivity of working electrodes is needed. The extracellular cerebral milieu is complex and poorly regulated. The detection of monoamines is complicated by the presence of numerous metabolites, notably high concentrations of ascorbic acid. All the electrodes previously cited, in their application *in vivo*, are incapable of separating directly the catechols from ascorbic acid: the solution to this problem has been attempted by various treatments of electrode surfaces in order to obtain this separation.

The first treatments were developed by Lane *et al.* [8,9] in 1976 on iodinated platinum electrodes. Yamamoto *et al.* [22] modified carbon paste electrodes. They added stearic acid to the carbon paste of a conventional carbon paste electrode to inhibit electrostatically the access of anions (notably of ascorbic acid). Gonon and coworkers have tried different electrochemical treatments on carbon fibre electrodes; their results will be discussed later in section 3.3.2.

Many criteria have been established for the evaluation of the quality of surface treatment of electrodes and its efficiency. Firstly it is necessary for the electrode to be able to separate catechols from ascorbic acid *in vitro* before being used *in vivo*. Secondly, tests *in vitro* must be carried out in a solution containing both ascorbic acid and catechols. The voltammograms obtained from a solution of ascorbic acid and a solution of catechols show obvious differences in comparison with voltammograms obtained from a mixture; this can bring about a nonreproducible detection of catechols *in vivo*.

3   Carbon fibre electrodes

Carbon fibre electrodes [23–26] are formed by one or more carbon fibres of 8 μm diameter (mono or multifibre electrode). The great advantage of this type of electrode is its small diameter; it works with elements which have the same size as the neuroneic structure, and implantation causes only minimum damage to nerve tissue. The primary disadvantage is fragility and probable deterioration over a long period *in vivo* which limits chronic implantations. Dayton *et al.* [27,28] have designed a carbon fibre microelectrode whose active surface is about 100 times smaller than that of carbon fibre electrodes. The difficulty with this type of electrode is the measurement of the oxidation currents, which are very weak (around pA). Forni [29] has made a multifibre electrode whose active surface corresponds to the section of a bundle of carbon fibres. Some studies *in vivo* have been carried out with this type of electrode.

3.1   Carbon fibres
3.1.1   Origin and use of carbon fibres
Carbon fibres have been known for about 100 years. They were first used by Edison, Swan and others in incandescent electric lamps. At that time these carbon fibres were prepared by means of carbonized bamboo or cellulose (rayon). This use was replaced by tungsten filaments. Later, carbon fibres were used as thermic insulation, but they have been rapidly replaced by glass fibres.

Since 1960, considerable technical advances have been made with carbon fibres. There has been a diversification of raw materials and of manufacturing processes. Among the raw materials most often used at present are polyacrylonitriles, the forerunners of cellulose, the derivatives of asphalt, and aromatic heterocyclic polymers. Manufacturing processes today use high temperatures ranging from 1000°C to 2800°C. By means of these different techniques, carbon fibres can be obtained which exhibit great solidity, low mass, and strong electric conductivity.

The uses and present applications of carbon fibres are very diverse, notably on the level of heterogeneous materials. Table 2 summarizes briefly the diverse applications of carbon fibres.

For the specific use of carbon fibres in electrochemistry, notably for the construction of electrodes implanted *in vivo*, it is necessary to know the structure and the state of the surface of this material.

Table 2 — Applications of carbon fibres. From Donnet [31]

Physical characteristics	Applications
Specific strength, specific toughness, light weight	Aerospace, road and marine transport, sports goods
High dimensional stability, low coefficient of expansion, low abrasion	Missiles, aircraft brakes, aerospace antenna
Good vibration damping, strength, toughness	Audio equipment
Electrical conductivity	Automobile hoods
Biological inertness, X-ray permeability	Medical application in prosthesis, surgery, X-ray equipment
Fatigue resistance, self-lubrication, high damping	Textile machinery, general engineering
Chemical inertness, high corrosion resistance	Chemical industry, nuclear field

3.1.2   Structure and state of the surface of carbon fibres

The exact structure of carbon fibres has not yet been entirely clarified, owing to the diversity of the raw materials and of the techniques used to obtain them. Carbon fibres can be classified in categories according to the degree of crystalline orientation in the macrostructure of the fibre. This orientation is nonetheless dependent on the processing temperature of the fibre (Table 3).

Table 3 — Different types of carbon fibres. From [31]

Carbon fibre type	Heat treatment temperature (°C)	Crystallite orientation	Long distance order
Type I High modulus	>2000	Mainly parallel to the fibre axis	High
Type II High strength	≈1500	Mainly parallel to the fibre axis	Low
Type III Isotropic	<1000	Random	Very low

The macroscopic structure proposed by Johnson & Bennet [30] is generally accepted today. This structure shows a difference in the arrangement and the density of the 'shreds' of carbon and of carbon microcrystals between the peripheral zone close to the fibre surface layer and its centre (Fig. 4). This model illustrates the make-

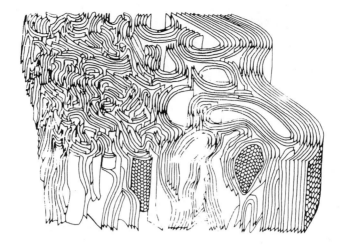

Fig. 4 — Three dimensional model of a carbon fibre, according to Johnson & Bennet [30].

up of a 'skin' of the fibre, which is in agreement with the behaviour of carbon fibres in many processes, e.g. electrochemical oxidation, as pointed out by Donnet [31].

So far as the microstructure is concerned, great heterogeneity can be observed among the models proposed. In general, the investigators agree in giving a high ratio of graphite carbon [32–34] in fibres treated at high temperatures (>2000°C).

Graphite carbon consists of layers in which each carbon atom is surrounded by three other carbon atoms (the structure is comparable to a giant hydrocarbon). This surface has only weak chemical and electrochemical reactivity (it is non-ionic, weakly polar, and hydrophobic). Diverse oxygenated functions are found; notably, carboxylic acid, hydroxide, aldehyde, etc., are found at the fibre periphery. These are the various functions responsible for the chemical and electrochemical reactions of carbon fibres (Fig. 5).

The concentration of different chemical species at the carbon fibre surface can be evaluated by chemical, electrochemical, or spectroscopic methods. In spite of all the studies carried out, results remain approximate; the density of different functional groups at the surface of carbon fibres is unknown. The state of the surface can be modified by different chemical or electrochemical oxidation treatments. These different treatments are carried out in order to increase the fibre reactivity.

3.2 Carbon fibre electrodes
3.2.1 Preparation of the electrode

This type of electrode is composed of a carbon fibre with an 8 μm diameter extending 0.5 mm over a glass micropipette (Fig. 6) [23,24]. The manufacturing process is as follows: a tube of borosilicated glass (external diameter: 1.5 mm, internal diameter: 1.05 mm) is stretched with the help of a pipette drawer in order to obtain a diameter of several micrometres (Fig. 7A). A carbon fibre (e.g. Carbon–Lorraine AGT 10 000) 20 mm in length is inserted in the micropipette until it seals the end (Fig. 7B), which is then cut back at this point, allowing the fibre to push through and extend

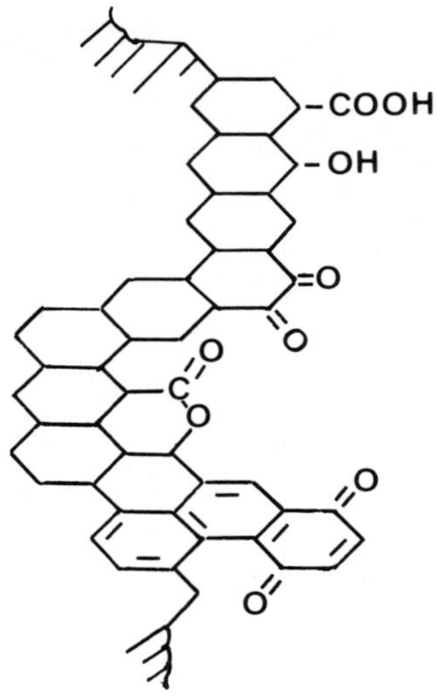

-COOH
-OH
=O
O
C
O
O
O
O
O

Fig. 5 — Various functions attributed to carbon fibre surface.

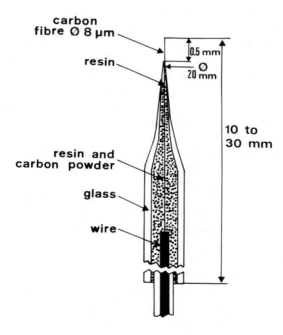

carbon
fibre Ø 8 µm
resin
0.5 mm
Ø
20 mm
10 to
30 mm
resin and
carbon powder
glass
wire

Fig. 6 — Sketch of a standard carbon fibre electrode. From Ponchon *et al.* [24].

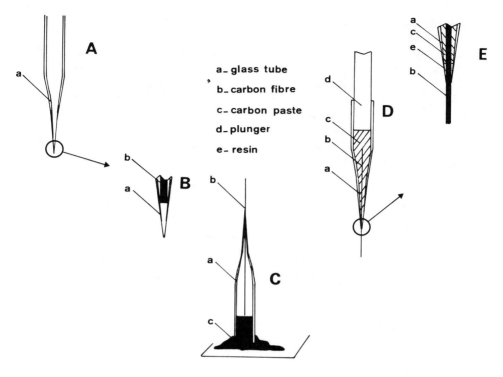

a – glass tube
b – carbon fibre
c – carbon paste
d – plunger
e – resin

Fig. 7 — A,B,C,D,E: Different stages in the manufacturing of a carbon fibre electrode. From [24].

beyond the end by several millimetres. This method minimizes intersticial space between the pipette and the fibre.

A paste is prepared from a carbon and polyester resin powder. The powder is mixed thoroughly for 5 min to obtain a highly conductive homogeneous paste, which remains sufficiently malleable for 30 min to allow electrode preparation. Several millimetres of carbon paste are introduced into the open end of the micropipette by pushing it into the preparation (Fig. 7C). A special plunger 'forces' the carbon paste into the interior of the micropipette (Fig. 7D), at the extremity of which, if the intersticial space is minimum, the resin separates from the carbon paste; this separation ensures perfect insulation in the non-active part of the electrode (Fig. 7E). A copper wire is inserted into the carbon paste to connect the measuring instrument. To ensure that the resin dries out, the electrodes are left for 24 h at ambient temperature. With this method, six to ten electrodes are produced in one hour with 90% success. Before use *in vitro* or *in vivo*, the electrodes are cut back to the desired dimensions (generally 0.5 mm long per fibre); they have a resistance of around 1 kΩ.

3.2.2 *Characteristics of the carbon fibre electrode*
This type of electrode was used by Gonon *et al.* in NPV [23,24]. The electrochemical set-up is one of 3 electrodes: the working electrode is a carbon fibre electrode; for use

in vivo, the reference electrode is a microelectrode Ag/AgCl, i.e. a silver wire of 0.2 mm in diameter covered by silver chloride and inserted into a glass micropipette filled with a gelatine salt solution (1 g of gelatine in 9 ml of NaCl 3M); the auxiliary electrode, used as successfully *in vivo* as *in vitro*, is a 0.5 mm diameter platinum wire.

Electrochemical measurements are carried out with a commercial polarograph (e.g. Model PRG 5 from Tacussel, Lyon, France). NPV is more successful than direct voltammetry; in particular, the charging current is minimized and the reactions to the electrode are also simpler because the potential returns to its initial value between each impulse. The analytical use of the impulse with solid electrodes has been greatly developed by Söderhjelm [35]. Fig. 8 shows voltammetric curves

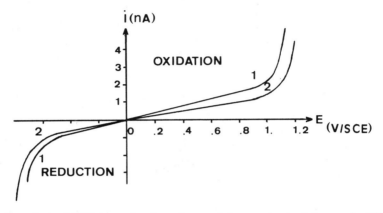

Fig. 8 — Experimental NPP (normal pulse polarography) current-potential curve for PBS solution. NPP parameters: $T=1$ s, $t_1=88$ ms, $V=10$ mV s^{-1}. Curve 1: first sweep; Curve 2: second sweep (and following ones). From [24].

obtained by a carbon fibre electrode in a phosphate buffered saline (PBS) solution [24].

The potential domain of a carbon fibre electrode (-0.8 V to +1.2 V vs SCE) is comparable to that of a carbon paste electrode, but it is definitely superior to that obtained with a chemically treated platinum electrode [1,9]. The residual current is weaker for the second sweep and for the following sweeps; this phenomenon is apparently due to an electrochemical clearing of the fibre or to conditioning of the electrode surface. The residual current is between 0.5 and 1 nA (0.6 V vs SCE) for a standard carbon fibre electrode.

3.3   Chemical and electrochemical treatments of carbon fibre electrodes
The major problem for the dosage of catechols in cerebral tissue is the presence of a high quantity of ascorbic acid which oxidizes at nearby potentials. Different treatments have been elaborated by workers to increase the sensitivity of detection methods.

3.3.1   Chemical treatments
Kuwana and coworkers [36–38] have treated pyrolitic graphite carbon electrodes by oxygen plasma and have grafted electroactive molecules, such as benzidine or other

quinonic molecules. The improvement of the oxidation of ascorbic acid and other compounds has been shown by the same workers [39]. The grafting of identical molecules, in particular dopamine and benzidine, has been accomplished on carbon fibres by Blond & Allemand [40].

3.3.2 Electrochemical treatments

Gonon *et al.* [41–43] have developed an electrochemical treatment of carbon fibres which produces successful separation of catechols from ascorbic acid. The treatment applied is as follows: the carbon fibre electrodes are submerged in a PBS solution and a triangular-shaped alternating potential is introduced. Under normal conditions, the parameters of the treatment are the following: frequency of the applied potential, 70 Hz; potential minimum limit, 0 V; potential maximum limit, +3.0 V; application timespan, 20 s. Many of these parameters have been studied separately, more particularly the potential maximum limit (Fig. 9) and the timespan (Fig. 10) on

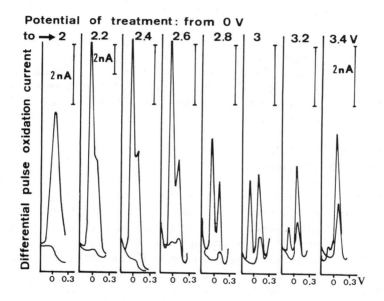

Fig. 9 — Effect of various upper potential limits on the AA + DOPAC signal. For each treated electrode, voltammograms are recorded after equilibration first in AA + DOPAC (higher traces) and secondly in PBS solution (lower traces). From Gonon *et al.* [41].

the dopamine–ascorbic acid mixture. Since the mechanism by which this type of treatment reacts on carbon fibres is known, it is possible to obtain the same type of response for the separation of catechols from ascorbic acid (AA). The separation is carried out by obtaining a reading for AA from a peak of -80 mV and for catechols from peaks of +70 mV. These measurements have been carried out by DPV with an Ag/AgCl electrode as reference. In addition, this treatment increases the sensitivity of the electrode to dopamine (DA) (20- to 400-fold) while that of AA is not modified.

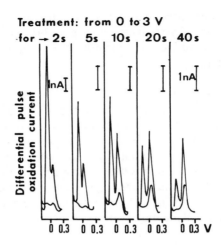

Fig. 10 — Effect of various treatment durations on the AA + DOPAC signal. From [41].

This property allows the detection of very low concentrations of DA in the presence of high concentrations of AA.

Along the same lines, Plotsky *et al.* [44,45] submerged carbon fibre microelectrodes in concentrated dichromic acid and also carried out an electrochemical treatment. The results obtained on the separation of DA and AA are similar to those obtained by Gonon *et al.* [42].

4   Results obtained *in vivo*†
4.1   *Implantation techniques*
The techniques developed for electrode implantation (Fig. 11) are quite numerous. Although this aspect is less critical than others, one point must be stressed. To properly identify and measure a monoamine derivative in nerve tissue, it is necessary to verify *in vitro* that the material and electrochemical techniques to be used *in vivo* are capable of separating this compound from AA. These tests *in vitro* must be carried out before and after *in vivo* experiments. Gonon's team observed that a non-treated carbon fibre exhibited a positive response to DA *in vitro* before implantation. After use of the electrode *in vivo*, *in vitro* tests showed a 20% reduction of DA [43]. Similar observations have been reported by Ewing *et al.* [46]. Likewise, O'Neill *et al.* [47], using a carbon paste electrode in semi-derived voltammetry, observed that the oxido-reduction potential of some compounds was greatly modified by the implantation of electrodes *in vivo*.
There are two types of preparation for animals:

Acute animal preparation
Animals are commonly anaesthetized and fixed on a stereotaxic frame. Commercially available reference electrodes and auxiliary electrodes (a platinum or stainless

† To facilitate the reader's comprehension of the experimental results reported in this section, the metabolism of catecholamines (dopamine and noradrenaline) and serotonin is summarized in Appendix A1.

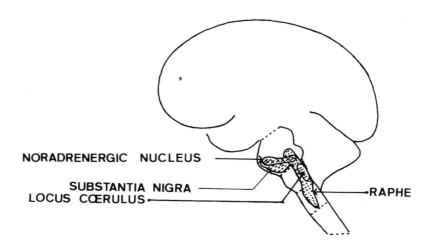

Fig. 11 — Main areas for implantation of carbon fibre electrodes in brain.

steel wire) are easily placed in contact with the skull. Electrochemical contact is accomplished by means of a semiliquid junction. The placing of the working electrode in the brain is done stereotaxically. The main advantage of this preparation is that the working electrode can be easily removed after *in vivo* experimentation to test it *in vitro*.

Unanaesthetized freely moving animals
Pharmacological studies are greatly facilitated by this type of preparation. The most widely used technique was first reported by Adams [2] and described in detail by Conti *et al.* [48]. Auxiliary and reference electrodes, as well as the working electrode (in this case, a graphite–resin electrode) are cemented on the skull. While implantation is easy to perform, the major disadvantage of such a preparation is that the working electrode cannot be tested *in vitro* after the *in vivo* experiment. To solve this problem, Gonon *et al.* [49,50] developed a double-implantation technique which allowed the working electrode to be changed. The first improvement of this technique offers the possibility of changing the working electrode without the need for anaesthesia (3 min for a carbon fibre electrode); also, it is possible to adjust the vertical localization of the electrode by means of an implanted micrometric screw.

4.2   *Compounds synthesized by dopaminergic neurones*
Considering the major interest of catecholaminergic neurones for neurobiologists, numerous electrochemical studies have been carried out to detect and measure catechols. In this respect the nigro-striatal system represents a good model because striatal DA levels are very high and stereotaxic implantation is easy to perform. Moreover, unilateral dopaminergic denervation is made possible routinely by means of 6-hydroxydopamine (6-OHDA) injection in the substantia nigra. Difficulties in detection arise from the fact that dopaminergic neurones synthesize not only DA but also several catechol and O-methylated catechols.

4.2.1 Detection of dopamine

Since the first tests were performed for detecting and measuring DA in nerve tissue, high AA tissue levels (0.2 to 0.4 ng per g of wet tissue) have complicated these studies [51,52]. Lane & Hubbard [8,9] approached the problem by means of an iodine surface-modified platinum electrode used in combination with DPV. Voltammograms recorded *in vivo* from the striatum exhibited two distinct peaks which were attributed to AA and the other to DA (+0.2 V). These conclusions were supported by the following arguments: firstly, separate injections of AA or DA brought about increases in the corresponding peaks; secondly, the injection of powerful DA-releasing agent (amphetamine) slightly elevated the presumed DA peak but did not change that of AA. A criticism of this test is the absence of *in vitro* experiments and of anatomical study. Moreover, the workers themselves observed a rapid loss in electrode stability.

On the other hand, referring to Lane's experimental data [9], numerous workers have considered that an amphetamine injection favours an increase in DA levels in dopaminergic neurones [53–58]. Many other arguments, not quoted here, support this view: the amphetamine effect was removed if the electrode was implanted in a striatum previously denervated of its dopaminergic terminals; the inhibition of catechol synthesis by α-methyl-*p*-tyrosine resulted in a decrease in the presumed DA peak and removed the effect of amphetamines; finally, the lack of AA response to amphetamine was cited by Conti *et al.* [20] from striatal brain slices.

The observations of Gonon *et al.* [10,42] using electrochemically pretreated carbon fibre electrodes showed two quite distinct peaks: one at −50 mV, the other at +100 mV. The latter was attributed to catecholamines because this same peak was obtained *in vitro*. Pharmacological studies using amphetamines conclude that the main constituent of this peak is 3,4-dihydroxyphenylacetic acid (DOPAC). A minimal quantity of DA is always detectable, in any case, because its release is stimulated by the amphetamines. The peak of −50 mV is attributed to AA (tests *in vitro*); it is not modified by striatal dopaminergic denervation, and it disappears in scorbutic guinea pigs.

Another interesting observation concerning DA is the anterior pituitary stem where DA is a prolactin inhibitor [44,45].

4.2.2 DOPAC and homovanillic acid

As stated earlier, electrochemically treated carbon fibres enabled workers to record a catechol peak related to dopaminergic neurones. Inhibition by pargyline of monoamine oxidase (MAO), which catalyses the transformation of DA into DOPAC by oxidative removal of the DA amino group, caused a total suppression of this peak in all structures studied (nucleus accumbens, striatum, substantia nigra, ventral tegmental area...) [10,59]. *In vitro* tests performed with electrodes which were used previously *in vivo* show that carbon fibres are about 100 times less sensitive to DA than to DOPAC; and there is no doubt that catechol peaks obtained in regions already mentioned are due to DOPAC. Various tests using numerous drugs show that haloperidol, a dopaminergic antagonist (DA receptor blocker), increased the DOPAC peak in the striatum and the nucleus accumbens [59,60]. On the other hand, haloperidol does not noticeably increase the DOPAC peaks in the substantia nigra and the ventral tegmental area [59]. Maindment & Marsden show that an injection of haloperidol increases the DOPAC signal by 48% in the nucleus accumbens ipsolateral, while an injection of DA reduces it by 36% [61].

Adams and coworkers [3,18,62] were the first to report a peak at +0.8 V attributed to homovanillic acid (HVA: 4-hydroxy-3-methoxyphenylacetic acid) in cerebrospinal fluid; in addition, the small surrounding peaks of +0.1 V would correspond to DOPAC formed by HVA electrolysis [63]. Lane *et al.* [7,54] recorded a peak at +0.5 V at the neostriatum and also attributed it to HVA.

4.3 *Compounds synthesized by noradrenergic neurones*

In contrast to dopaminergic neurones, very few studies have been carried out on noradrenergic neurones. Experimental study has been complicated by the fact that noradrenaline (NA) oxidizes at the same potential as DA [50,63]. The 3,4-dihydroxyphenylglycol (DOPEG), however, which is the direct NA metabolite, cannot be resolved with other catechols [50]. Regions chosen for these studies are rich in noradrenergic terminals and not in dopaminergic terminals (anteroventral nucleus of the thalamus, lateral geniculate nucleus). Unfortunately, no peak has been observed in these regions. These negative results can be interpreted in the following manner: as with DA, the concentrations of noradrenaline in the extracellular space are below the detection limit; Nielsen & Braestrup believe that in the rat brain, DOPEG is entirely in sulfate form and that oxidation peaks are at a high potential in an unexplored domain [64]. Plotsky *et al.* [44] using a microelectrode of carbon fibre in combination with DPV, observed a peak at +130 mV from the locus coeruleus (in rats) that workers attribute to catecholamines other than DOPAC (+0·4 mV) and AA (−30 mV). Buda *et al.* [50,65,66], by means of different pharmacodynamic studies, showed that noradrenaline is below the detection limit (<20 nM) in the extracellular space, and that it can be detected only through a powerful pharmacological treatment. Other studies by the same group of workers showed that noradrenergic neurones synthesize DOPAC [50,65,66].

4.4 *Compounds synthesized by serotoninergic neurones*

Electrochemical monitoring of serotonin (5-hydroxytryptamine: 5-HT) has been facilitated by the fact that 5-HT is easily separated from AA and other catecholamines. These compounds oxidize at about +300 mV vs Ag/AgCl. As with DA, numerous experiments focus on the stimulation of serotoninergic neurones. Wightman *et al.* [18] report that the stimulation of the median raphe nucleus released 5-HT in the lateral ventricles.

Brazell & Marsden [67,68] obtained an oxidation peak at the 5-hydroxyindole potential in striatum and frontal cortex. This peak was abolished by *p*-chlorophenylalanine (PCPA, inhibitor of 5-HT biosynthesis), indicating that it is due to 5-hydroxyindoles. Linear sweep voltammetry at carbon paste electrodes in the hippocampus has shown a peak at the 5-hydroxyindole potential. The peak is increased by *p*-chloro-amphetamine and by tryptophan (5-HT precursor) and reduced by PCPA [69–71]. However, recent studies have shown that with carbon paste electrodes this peak is due to uric acid in the striatum [72]. Carbon fibre electrodes are less sensitive to uric acid. Most (60–70%) of the striatal 5-hydroxyindole peak is due to 5-hydroxyindole-3-acetic acid (5-HIAA, metabolite of 5-HT) and only 30% to uric acid [73,74].

The technique of Cespuglio *et al.* (DPV at electrochemically treated carbon fibre electrodes) has been successfully employed for measuring 5-HIAA in the striatum [75], the cortex [76], spinal cord [77], cerebrospinal fluid [78], suprachiasmatic nuclei [79], hypothalamus [80],and in raphe nuclei [81].

Cespuglio *et al.* registered a peak at +0·3 V which they attribute to 5-hydroxyindoles synthesized by the serotoninergic terminals of the striatum [11]. Their view is based on the following arguments: firstly, the peak appeared at the same potential as 5-hydroxyindole *in vitro*; secondly, catechols and AA did not contribute to the peak since they oxidized at a lower potential; thirdly, *in vivo* inhibition of 5-hydroxyindole synthesis by *p*-chlorophenylalanine decreased the peak by about 72%; and lastly, selective denervation of the 5-HT striatal terminals was followed by a total suppression of the signal. Moreover, workers observed that detection of neuronal 5-hydroxyindole could be contaminated by blood 5-HT. To avoid this, they developed an appropriate implantation technique which prevents haemorrhage. Use of DPV has also enabled anatomic investigation and determination of innervation of serotoninergic neurones in mammalian brains. Lamour *et al.* [82] showed a laminar distribution of the 5-HIAA peak in the rat somatosensory cortex. Crespi *et al.* [83] have mapped the raphe nuclei by the same method. Still concerning the raphe nuclei, Echizen & Freed [84,85] have shown two oxidation peaks (using LSV, carbon paste electrodes), one of the two peaks being almost certainly due to 5-HIAA. These peaks responded to drug-induced alterations in blood pressure. Hypotension, induced by nitroprusside, caused no change. In contrast, phenylephrine-induced hypertension doubled the 5-HIAA peak, which continued to increase even after cessation of the phenylephrine infusion. In the striatum, the oxidation peak for 5-HIAA can be eliminated by an electrical lesion of dorsal raphe nuclei [86]. However, transplantation of neonatal mesencephalic raphe nuclei into the lateral ventricles adjacent to the striatum restores the striatal 5-HIAA peak, indicating that metabolism of 5-HT occurred in the transplants [87].

5   Conclusions concerning the detection of monoamine derivatives

Objectively, the easiest compound to detect *in vivo* in brain tissue is AA. It is of interest in dopaminergic functions in the striatum [10,27,42]; however, the detection of monoamine derivatives is complicated by the presence of AA. The identification and measurement of this type of compound must comply with the strict criteria listed by Gonon and his team [43]:

(i) the potential of the *in vivo* peak must be identical to that of the X peak *in vitro*;
(ii) the peak should be increased by local *in vivo* injections of X in the vicinity of the electrode;
(iii) the separation of X from ascorbic acid is absolutely necessary:
• the validity of this separation must be ascertained *in vitro* in solutions containing both AA and X:
• these *in vitro* tests must be performed before and after *in vivo* implantation with the same electrode;
• *in vitro* calibration in terms of X concentration must not be influenced by variations in AA levels.

(iv) anatomical data (including specific lesioning) must demonstrate that the peak is related to the presence of neuronal components responsible for the synthesis of X;

(v) pharmacological inhibitions of X synthesis at different steps should induce peak height decreases which should be consistent with those of X levels as measured post mortem;

(vi) drugs which are known to act on X levels should induce corresponding variations on the peak height.

In a general review, Stamford [88] shows in conclusion that neither ideal voltammetric methods nor perfect electrodes exist in reality. In most instances, experimentation must choose the voltammetric technique and working electrode which are the most adequate with regard to experimentation needs; it seems at the moment that electrochemically treated carbon fibres used in DPV or DNPV and carbon paste electrodes used in LSV are the most frequently used techniques in neurochemistry.

6   Other applications of carbon fibres

It has already been seen that the major use of carbon fibre electrodes has been the detection and measuring of catechols in neurochemistry. Owing to the mechanical, chemical, and electrochemical properties of carbon fibres, other areas of research are developing, especially with chemically modified electrodes [89] or enzymatic electrodes. The chemical modification of the state of the surface of carbon fibres (oxidation) makes it possible to contemplate grafting electroactive molecules [90–92], and the grafting of spacers [93,94] or of enzymes, in particular dehydroge-nases using NADH/NAD$^+$ as the enzymatic cofactor. Some studies on the electro-chemical regeneration of NAD$^+$ have been performed with carbon fibre electrodes; but few *in vivo* studies have used an enzymatic carbon fibre electrode. It is nevertheless necessary to mention the work of Suaud-Chagny & Gonon [95] who have immobilized lactate dehydrogenase on a microelectrode of carbon fibre. The electrode test was performed to measure pyruvate in rat cerebrospinal fluid. In fact, this electrode is the juxtaposition of two systems.

The first, a biochemical system, in which lactate dehydrogenase is included in a coating of bovine serum albumin to form a sheath around the carbon fibre, brought about the following reaction:

$$\text{lactate} + \text{NAD}^+ \underset{}{\overset{\text{lactate dehydrogenase}}{\rightleftharpoons}} \text{Pyruvate} + (\text{NADH} + \text{H}^+).$$

The second is an electrochemical system in which the carbon fibre oxidizes the NADH during the reaction:

$$\text{NADH} \rightarrow \text{NAD}^+ + 2\,\text{e}^- + \text{H}^+.$$

The oxidation current obtained is proportional to the concentration of NADH between 10^{-5} and 10^{-3} M. The detection limit is 10 μM. This type of electrode is highly specific and perhaps gives an encouraging response in favour of improvement of carbon fibre electrodes. It should be remembered that the design of this type of electrode is delicate, especially when the immobilization of an enzyme on a carbon fibre microelectrode is concerned. At present, each enzyme requires a specific and unique immobilization procedure and exhibits a low yield, which brings about difficulties in electrochemical detection.

A1   Metabolism of catecholamines and serotonin
Biological amines responsible for electrochemical responses recorded *in vivo* in the central nervous system can be determined with precision by biological and neuropharmacological investigations which require preliminary knowledge of the biosynthesis and catabolism of catecholamines and serotonin.

Catecholamine biosynthesis
The term 'catecholamine' generally designates dopamine, noradrenaline, and adrenaline. The distribution of these amines in the central nervous system is quite heterogeneous and depends on the neuronal distribution of the enzymes participating in their biosynthesis. This biosynthesis (Fig. 12) involves an initial step where the amino acid precursor is caught by catecholaminergic terminals, and then several enzyme reactions successively catalysed by (i) tyrosine hydroxylase, specifically located in catecholaminergic neurones, (ii) 3,4-dihydroxyphenylalanine (DOPA) decarboxylase, located in diverse neurones, in particular catecholaminergic neurones, (iii) dopamine β-hydroxylase, specific of noradrenergic and adrenergic neurones, and (iv) phenylethanolamine N-methyltransferase, located exclusively in adrenergic neurones.

Since tyrosine hydroxylase is inhibited by both catecholamines and DOPA, this enzyme acts as a regulator for catecholamine production.

Catabolism of catecholamines
A certain amount of catecholamine is degraded in neuronal cytoplasm by oxidative deamination implying mitochondrial monoamine oxidase (MAO) (Figs 13 and 14). This degradation is achieved by aldehyde oxidases or reductases. Thus, DA and NA are mainly deaminated in DOPAC (3,4-dihydroxyphenylacetic acid) and in DOPEG (3,4-dihydroxyphenyl glycol), respectively.

Deaminated derivatives can be O-methylated in the extraneuronal space by catechol-O-methyltransferase (COMT).

The successive action of MAO, aldehyde oxidases or reductases and COMT mainly leads to an acid derivative, HVA (homovanillic acid), in the case of DA, and to an alcohol derivative, MOPEG (3,4-dihydroxyphenyl glycol), for NA.

Biosynthesis and catabolism of serotonin
Serotonin (5-hydroxytryptamine) is synthesized from tryptophan (Fig. 15), an essential amino acid bound at 90% to serum albumin, by two enzyme steps successively catalysed by tryptophan hydroxylase, a specific enzyme of serotoniner-

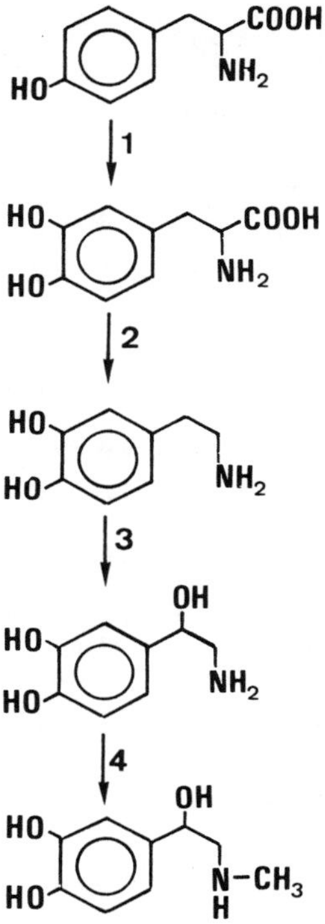

Fig. 12 — Pathway for the biosynthesis of catecholamine neurotransmitters. Enzymes: 1, tyrosine hydroxylase; 2, DOPA decarboxylase; 3, DA β-hydroxylase; 4, phenylethanolamine N-methyltransferase.

gic neurones, and by 5-hydroxytryptophan decarboxylase which is not specific since it can be involved in the decarboxylation of DOPA into DA.

Serotonin is metabolized into 5-hydroxyindolacetic acid by both MAO and aldehyde oxidase at the presynaptic and glial level (Fig. 15).

References
[1] Adams, R. N. (1969) *Electrochemistry at solid electrodes*. New York, Marcel Dekker.
[2] Adams, R. N. (1976) *Anal. Chem.* **48**, 1126A–1137A.
[3] Adams, R. N. (1978) *Trends Neurosci.* **1**, 160–164.
[4] Heyrovsky, J. (1924) *Trans. Faraday Soc.* **19**, 785–788.
[5] Schenk, J. O., & Adams, R. N. (1984) In: Marsden, C. A. (ed.) *Measurement of neurotransmitter release* in vivo. New York, Wiley p. 193–208.
[6] O'Neill, R. D., Grunewald, R. A., Fillenz, M., & Albery, W. J. (1982) *Neuroscience* **7**, 1945–1954.
[7] Lane, R. F., Hubbard, A. T., & Blaha, C. D. (1978) *Bioelectrochem. Bioenerg.* **5**, 504–525.

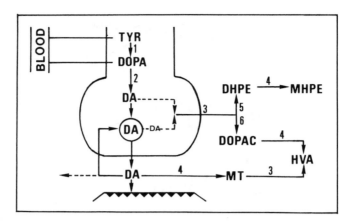

Fig. 13 — Biosynthesis and catabolism of dopamine at the level of a central dopaminergic nerve terminal. Enzymes: 1, tyrosine hydroxylase; 2, DOPA decarboxylase; 3, monoamine oxidase; 4, catechol-O-methyltransferase; 5, aldehyde reductase; 6, aldehyde oxidase. Abbreviations: TYR, tyrosine; DOPA, 3,4-dihydroxyphenylalanine; DA, dopamine; DHPE, 3,4-dihydroxy-phenylethanol; DOPAC, 3,4-dihydroxyphenylacetic acid; HVA, homovanillic acid; MT, 3-methoxytyramine; MHPE, 3-methoxy-4-hydroxyphenylethanol. A, tyrosine; B, 3,4-dihyd-roxyphenylalanine (DOPA); C, dopamine; D, noradrenaline; E, adrenaline.

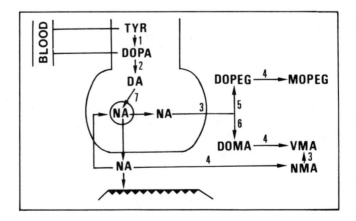

Fig. 14 — Biosynthesis and catabolism of noradrenaline at the level of a central noradrenergic nerve terminal. Enzymes: 1, tyrosine hydroxylase; 2, DOPA decarboxylase; 3, monoamine oxidase; 4, catechol-O-methyltransferase; 5, aldehyde reductase; 6, aldehyde oxidase; 7, dopamine β-hydroxylase. Abbreviations: TYR, tyrosine; DOPA, 3,4-dihydroxyphenylala-nine; DA, dopamine; NA, noradrenaline; DOPEG, 3,4-dihydroxyphenyl glycol; MOPEG, 3-methoxy-4-hydroxyphenylethyleneglycol; DOMA, 3,4-dihydroxymandelic acid; VMA, vanillylmandelic acid (4-hydroxy-3-methoxymandelic acid); NMA, normethoxyadrenaline.

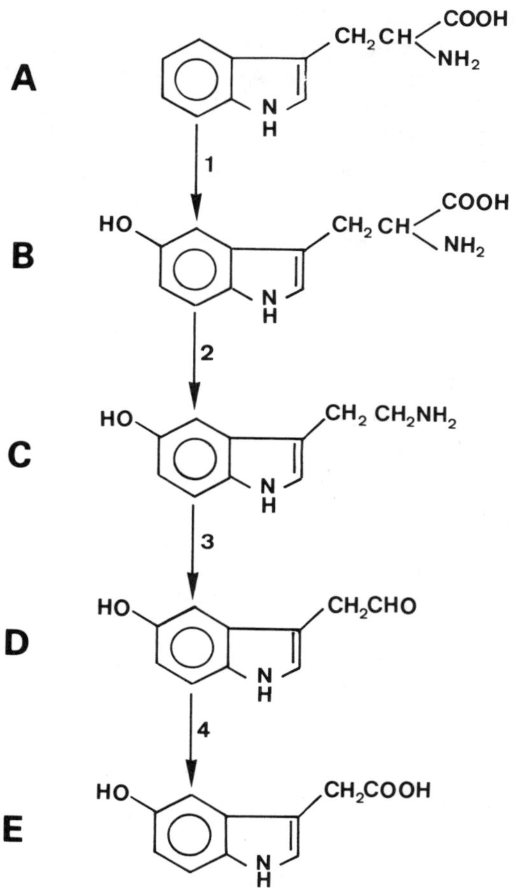

Fig. 15 — Pathway for the biosynthesis and catabolism of serotonin. A, tryptophan; B, 5-hydroxytryptophan; C, serotonin; D, 5-hydroxyindole acetaldehyde; E, 5-hydroxyindole acetic acid. Enzymes: 1, tryptophan hydroxylase; 2, 5-hydroxytryptophan decarboxylase; 3, monoamine oxidase; 4, aldehyde dehydrogenase.

[8] Lane, R. F., & Hubbard, A. T. (1976) *Brain Res.* **114**, 346–352.

[9] Lane, R. F., & Hubbard, A. T. (1976) *Anal. Chem.* **48**, 1287–1292.

[10] Gonon, F., Buda, M., Cespuglio, R., Jouvet, M., & Pujol, F. (1980) *Nature (London)* **286**, 902–904.

[11] Cespuglio, R., Faradji, H., Riou, F., Buda, M., Gonon, F., Pujol, F., & Jouvet, M. (1981) *Brain Res.* **223**, 287–311.

[12] Ewing, A. G., Dayton, M. A., & Wightman, R. M. (1981) *Anal. Chem.* **53**, 1842–1847.

[13] Ewing, A. G., Wightman, R. M., & Dayton, M. A. (1982) *Brain Res.* **249**, 361–370.

[14] Gonon, F. G., & Buda, M. (1985) *Neuroscience* **14**, 765–774.

[15] Marcenac, F. M., & Gonon, F. G. (1985) *Anal. Chem.* **57**, 1778–1779.

[16] Armstrong–James, M., & Millar, J. (1984) In: Marsden, C. A. (ed.) *Measurement of neurotransmitter release* in vivo. New York, Wiley p. 209– 224.

[17] Kissinger, P. T., Hart, J. B., & Adams, R. N. (1973) *Brain Res.* **55**, 209–213.

[18] Wightman, R. M., Strope, E., Plotsky, P. M., & Adams, R. N. (1976) *Nature (London)* **262**, 145–146.

[19] Wightman, R. M., Strope, E., Plotsky, P. M., & Adams, R. N. (1976) *Brain Res.* **159**, 55–58.

[20] Conti, J. C., Strope, E., Adams, R. M., & Marsden, C. A. (1978) *Life Sci.* **23**, 2705–2713.

[21] Snell, K. D., & Keenan, A. G. (1979) *Chem. Soc. Rev.* **8**, 259–282.

[22] Yamamoto, B. K., Lane, R. F., & Freed, C. R. (1982) *Life Sci.* **30**, 2155–2162.

[23] Gonon, F., Cespuglio, R., Buda, M., Ponchon, J. L., Jouvet, M., Adams, R. N., & Pujol, J. F. (1978) *C.R. Acad. Sci. (Paris) Sér. D* **286**, 1203–1206.

[24] Ponchon, J. L., Cespuglio, R., Gonon, F., Jouvet, M., & Pujol, J. F. (1979) *Anal. Chem.* **51**, 1483–1486.

[25] Armstrong-James, M., & Millar, J. (1979) *J. Neurosci. Methods* **1**, 279–287.

[26] Wightman, R. M. (1981) *Anal. Chem.* **59**, 1125A–1134A.

[27] Dayton, M. A., Brown, J. C., Sutts, K. H., & Wightman, R. M. (1980) *Anal. Chem.* **52**, 946–950.

[28] Dayton, M. A., Ewing, A. G., & Wightman, R. M. (1980) *Anal. Chem.* **52**, 2392–2396.

[29] Forni, C. (1982) *J. Neurosci. Methods* **5**, 167–171.

[30] Johnson, D. J., & Bennet, S. C. (1978) *Soc. Chem. Ind.* **1**, 377–381.

[31] Donnet, J. B. (1982) *Carbon* **20**, 267–282.

[32] Johnson, D. J. (1980) *Phil. Trans. Soc. (London)* **A294**, 443–449.

[33] Johnson, D. J. (1982) *J. Chem. Ind.* **6**, 692–698.

[34] Donnet, J. B., Voet, A., Marsh, P. A., Erhburger, P., & Danksch, H. (1973) *Carbon* **11**, 430–431.

[35] Söderhjelm, P. (1976) *J. Electroanal. Chem.* **71**, 109–115.

[36] Evans, J. F., & Kuwana, T. (1977) *Anal. Chem.* **49**, 1632–1635.

[37] Evans, J. F., & Kuwana, T. (1978) *J. Electroanal. Chem.* **80**, 409–416.

[38] Chi-Sing, D., & Kuwana, T. (1978) *Anal. Chem.* **50**, 933–936.

[39] Ueda, C., Tse, D. C., & Kuwana, T. (1982) **54**, 850–856.

[40] Blond, P., & Allemand, P. H. (1986) In: *Deuxième Rencontre de Biotechnologies de la Région Rhône-Alpes, Grenoble, April 1986.*

[41] Gonon, F. G., Fombarlet, C. M., Buda, M., & Pujol, J. F. (1981) *Anal. Chem.* **53**, 1386–1389.

[42] Gonon, F. G., Buda, M., Cespuglio, R., Jouvet, M., & Pujol, J. F. (1981) *Brain Res.* **223**, 69–80.

[43] Gonon, F. G., Buda, M., & Pujol, J. F. (1984) In: Marsden, C. A. (ed.) *Measurement of neurotransmitter release* in vivo. New York, Wiley p. 153–171.

[44] Plotsky, P. M., de Greef, W. J., & Neill, J. D. (1982) *Brain Res.* **235**, 179–184.

[45] Plotsky, P. M., & Neill, J. D. (1982) *Endocrinology* **110**, 691–696.

[46] Ewing, A. G., Dayton, M. A., & Wightman, R. M. (1981) *Anal. Chem.* **53**, 1842–1847.

[47] O'Neill, R. D., Gründewal, R. A., Fillenz, M., & Albery, W. J. (1982) *Neuroscience* **7**, 1945–1954.

[48] Conti, J. C., Strope, E., Adams, R. N., & Marsden, C. A. (1978) *Life Sci.* **23**, 2705–2716.

[49] Gonon, F. G., Buda, M., Cespuglio, R., Jouvet, M., & Pujol, J. F. (1981) *Brain Res.* **223**, 69–80.

[50] Gonon, F. G., Buda, M., de Simoni, M. G., & Pujol, J. F. (1983) *Brain Res.* **273**, 207–216.

[51] Kissinger, P. T., Hart, J. B., & Adams, R. N. (1973) *Brain Res.* **55**, 209–213.

[52] Subramanian, N. (1977) *Life Sci.* **20**, 1479–1484.

[53] Lane, R. F., Hubbard, A. T., & Blaha, C. D. (1978) *Bioelectrochem. Bioenerg.* **5**, 506–507.

[54] Lane, R. F., Hubbard, A.T., & Blaha, C. D. (1979) *J. Electroanal. Chem.* **95**, 117–122.

[55] Lane, R. F., Hubbard, A. T., & Blaha, C. D. (1979) In: Usdin, E. (ed.) *Catecholamines – Basic and clinical frontiers.* Vol. 1. New York, Pergamon Press p.883–885.

[56] Huff, R., Adams, R. N., & Rutledge, C. (1979) *Brain Res.* **173**, 369–372.

[57] Lindsay, W. S., Kizzort, B. L., Justice, J. B., Salamone, J. D., & Neill, D. B. (1979) *J. Electroanal. Chem.* **95**, 117–122.

[58] Huff, R., & Adams, R. N. (1980) *Neuropharmacol.* **19**, 587–590.

[59] Buda, M., Gonon, F., Cespuglio, R., Jouvet, M., & Pujol, J. F. (1981) *Eur. J. Pharmacol.* **73**, 61–68.

[60] Gonon, F., Buda, M., Cespuglio, R., Jouvet, M., & Pujol, J. F. (1981) *Brain Res.* **223**, 69–80.

[61] Maindment, M. T., & Marsden, C. A. (1983) *Brit. J. Pharmacol.* **80**, 647.

[62] Wightman, R. M., Strope, E., Plotsky, P. M., & Adams, R. N. (1978) *Brain Res.* **159**, 55–68.

[63] Sternson, A. W., McCreery, R., Feinberg, B., & Adams, R. N. (1973) *J. Electroanal. Chem.* **46**, 313–330.

[64] Nielsen, M., & Braestrup, C. (1976) *J. Neurochem.* **27**, 1211–1217.

[65] Buda, M., de Simoni, G., Gonon, F., & Pujol, J. F. (1983) *Brain Res.* **273**, 197–206.

[66] Quintin, L., Buda, M., Hilaire, G., Bardelay, C., Ghignone, M., & Pujol, J.F. (1986) *Brain Res.* **375**, 235–245.

[67] Brazell, M. P., & Marsden, C. A. (1981) *Brit. J. Pharmacol.* **74**, 219

[68] Brazell, M. P., & Marsden, C. A. (1981) *Brit. J. Pharmacol.* **75**, 539–547.

[69] Joseph, M. H., & Kennet, G. A. (1980) *Neuropharmacol.* **20**, 1361–1364.

[70] Joseph, M. H., & Kennet, G. A. (1984) In: Schlossberger, H. G., Kochen, W., Linzen, B., and Steinhart, H. (eds.) *Progress in tryptophan and serotonin research.* New York, Walter de Gruyter p.124–153.

[71] Keller, R. W., Zigmond, M. J., & Stricker, E. M. (1981) *Soc. Neurosci. Astr.* **7**, 53–54.

[72] O'Neill, R. D., Fillenz, H., Grünewald, R. A., Blomfield, M. R., Albery, W. J., Janieson, C. M., Williams, J. H., & Gray, J. A. (1984) *Neurosci. Lett.* **45**, 39–46.

[73] Cespuglio, R., Faradji, H., Ponchon, J. L., & Buda, M. (1981) *J. Physiol. (Paris)* **77**, 327–332.

[74] Cespuglio, R., Faradji, H., Riou, F., & Gonon, F. (1981) *Brain Res.* **223**,

299–311.

[75] Cespuglio, R. (1982) *J. Histochem. Cytochem.* **30**, 821–823.

[76] Cespuglio, R., Faradji, H., Crespi, F., & Jouvet, M. (1983) In: *Proceedings of the 6th European Congress on Sleep Research, Zurich*. Basel Karger p.282–284.

[77] Riley, J. P., & Gu, H. (1981) *Anal. Chim. Acta* **130**, 199–201.

[78] Hahn, Z., Cespuglio, R., Faradji, H., & Jouvet, M. (1983) *Brain Res.* **289**, 151–156.

[79] Faradji, H., Cespuglio, R., & Jouvet, M. (1983) *Brain Res.* **279**, 111–119.

[80] Baumann, P. A., & Waldmeier, P. C. (1984) *Neuroscience* **11**, 195–204.

[81] Cespuglio, R., Faradji, H., & Jouvet, M. (1983) *C. R. Acad. Sci. (Paris) Sér. D* **29(**, 611–616.

[82] Lamour, Y., Rivot, P. J., Pontis, D., & Ory-Lavollée, L. (1983) *Brain Res.* **259**, 163–166.

[83] Crespi, F., Cespuglio, R., & Jouvet, M. (1983) *Brain Res.* **270**, 45–54.

[84] Echizen, H., & Freed, C. R. (1983) *Brain Res.* **277**, 55–62.

[85] Echizen, H., & Freed, C. R. (1984) *J. Neurochem.* **42**, 1483–1486.

[86] Scatton, B., Serrano, A., Rivot, J. P., & Nishikawa, T. (1984) *Brain Res.* **305**, 353–356.

[87] McRae-Degueurce, A., Serrano, A., Sandillon, F., Privat, A., & Scatton, B. (1984) *Neurosci. Lett.* **48**, 97–102.

[88] Stamford, J. A. (1986) *J. Neurosci. Methods* **17**, 1–29.

[89] Theodoridou, E., Besenhard, J. O., & Fritz, H. P. (1981) *J. Electroanal. Chem.* **124**, 87–94.

[90] Yacynych, A. L., & Kuwana, T. (1978) *Anal. Chem.* **50**, 640–645.

[91] Edmonds, T. E. (1985) *Anal. Chim. Acta* **175**, 1–22.

[92] Ueda, C., Tse, D. C., & Kuwada, T. (1982) *Anal. Chem.* **54**, 850–856.

[93] Fujisawa, T., & Yosomya, R. (1985) *Polym. Bull.* **13**, 7–14.

[94] Fujisawa, T., & Yosomya, R. (1985) *Polym. Bull.* **12**, 523–530.

[95] Suaud-Chagny, M. F., & Gonon, F. (1986) *Anal. Chem.* **58**, 412–415.

2.2   ION-SELECTIVE MICROELECTRODES
D. Michel and P. Blond

1   Introduction

Differences in concentration or activity of small inorganic ions are responsible for the production of biopotentials between the inside and the outside of the cell, between individual cell components, or between whole cell systems [1]. To understand the electrophysiological properties of excitable cells, it is highly desirable to know the intracellular ion activities rather than their bulk concentrations [2]. This is a difficult problem for two reasons [3]. Firstly, the cells, and therefore, the intracellular volumes, are very small (cell volume can be as small as about $10^{-5}\,\mu l$). Secondly, the measurements must be made without appreciable damage to cell membrane. If the membrane is damaged to the extent that the membrane potential is not normal, the intracellular ionic activities are certain to be different from their normal values [3]. So experimental life sciences are now one of the most exciting fields of application of ion-selective electrodes [4]. Many workers have contributed to the development of special types of widely applicable microelectrodes, measuring methods, and apparatus. Electrodes with measuring tips having submicrometre diameters are now frequently prepared to monitor *in vivo* intracellular ion activities. Easily made electrodes with measuring tips of a few micrometres are also regularly used for *in vivo* extracellular measurements [4].

Three kinds of microelectrodes are so far used in biological studies: solid-state, glass membrane, and liquid ion-exchanger microelectrodes [5]. The latter two are the most widely used at present, and this chapter will be essentially devoted to them.

The first microelectrodes developed consisted in ion-selective glass microelectrodes for hydrogen and alkali cations. These electrodes all have the drawback of having a large ion-sensitive surface which must be entirely within the cell, and their use is, therefore, limited to larger cells [3]. In addition to the problem of size, they are difficult to fabricate.

In the last decade, liquid ion-exchanger microelectrodes have been largely developed, because of their relatively ease of construction (see e.g. Refs. [16–10]), the commercial availability of a number of satisfactory liquid ion-exchanger solutions, and their suitability for incorporation in double-barrelled or multi-barrelled microelectrodes [5].

Walker, in 1971, first dealt with the problem of miniaturization of specific liquid ion-exchanger electrodes, for the express purpose of measuring intracellular ion activities in living cells [3]. The liquid ion-exchanger microelectrode consists, in principle, of a micropipette in which the internal wall of the tip region is made organophilic and hydrophobic so that a column of the organic ion exchanger remains held there, when the rest of the inside of the electrode is filled with a conducting aqueous solution and the tip of the electrode is in a tissue or a solution [11]. So an electrical potential difference is generated, which is a function of the ionic activities on both sides of the membrane [1].

Liquid ion-exchangers are solutions in an organic solvent of charged or neutral organic compound, which is often highly branched to prevent micelle formation. Such ion-exchanger solutions were originally developed for industrial purposes. A physical property required for both the solvent and the ion-exchanger solution is very low water solubility so that the exchanger solution can form a membrane-like phase separating two aqueous solutions. Solvents that have been used have low dielectric constants. Wright *et al.* [12] described the equilibria that occur when the organic solution contacts an aqueous electrolyte solution.

A quantitative theory has been developed which describes the electrode potential, but it is cumbersome and contains parameters which are difficult to measure [3]. It is, therefore, more convenient to use the following empirical equation (extended Nicolsky equation) [13]:

$$E = E° + (nRT/z_iF) \ln\left(a_i + \sum_{j \neq i} k_{i,j} \, a_j^{z_i/z_j}\right)$$

where E (volts) is the potential, $E°$ (volts) is a constant for a given electrode, n (dimensionless) is an empirical constant, R is the gas constant (8.3 joules/deg.mol), T is the temperature (°K), F is the Faraday number (96 500 coulomb/equivalent), a_i (respectively a_j) (mol/l) is the activity of the principal ion i (respectively, of the jth interfering ion), z_i and z_j (dimensionless) are the corresponding valences, and $k_{i,j}$ is the selectivity coefficient for the ion i over the ion j. The smaller the value of $k_{i,j}$ the greater the selectivity of the electrode for i over j. This selectivity is often expressed as $1/k_{i,j}$. If the sum is taken as zero (the perfectly selective electrode), the Nicolsky and the Nernst equations are formally identical. At 25°C, the slope S ($S = nRT/z_iF$) comes to 59.16 mV/z_i if the common logarithm is used [14].

Firstly, the electrode must be calibrated to determine the slope of the line relating activity and potential which is usually slightly less than the theoretical expected value. So the dimensionless fractional term, n, is determined [12]. The evaluation of ion-selective electrodes focuses on their selectivity coefficient, $k_{i,j}$, that is, the mathematical expression of their ability to distinguish between the principal ion i and the interfering ion j. In determination of the selectivity coefficients, two major problems have emerged [15]. Firstly, the selectivity coefficient of an ion-selective microelectrode varies, depending on the methods used (fixed interference method or separate solution method [16]). Secondly, the selectivity coefficient varies with the levels of principal ion i activity and the interfering ion j activity. It is apparent that the reported $k_{i,j}$ values vary from one investigator to another and that the $k_{i,j}$ values differ somewhat between individual microelectrodes. Although the reason for such variations is not completely clear, there are a few possible explanations [15]. The discussion about selectivity coefficients for divalent ions, however, is more complicated [17,18].

Immediately before making intracellular measurements, the electrode is calibrated in a series of solutions of known concentration which should encompass the expected concentration of the unknown solution [3]. All the solutions used should have the same ionic strength.

The electrode is then moved to the solution bathing the cell in which the measurement is to be made [3], which normally contains some of the ion to be measured. The potential of the electrode in that solution should agree with the potential predicted from the calibration curve, taking into account any interfering ions which may be present in the solution.

The tip of the electrode is then inserted into the cell. Since it is not possible to see the tip of the electrode enter the cell, the criterion used to determine the cell entry is an abrupt shift in the potential of the electrode as it is slowly advanced toward the cell. This shift in potential is due to two factors: (i) the difference in activity of the ion being measured between the extracellular and intracellular fluids; (ii) the cell membrane potential. This shift can be written as [12]:

$$\Delta E = E_m + (nRT/z_i F) \ln \left[\frac{a_i^o + k_{i,j} a_j^o}{a_i^i + k_{i,j} a_j^i} \right]$$

where ΔE (volts) is the difference in electric potential between the inside and the outside of the cell, E_m (volts) is the cell membrane potential, and the superscripts on the activities, 'o' and 'i', refer to outside and inside of the cell.

The membrane potential is determined with the reference electrode. In a first approximation, it is assumed that the potential of the reference electrode, including the liquid junction potential, is constant in respect to the sample solution. In fact, the liquid junction potential variation can be taken into account with the Henderson equation [14,19]. The membrane potential can be determined by impaling the reference micropipette into another cell if a single-barrelled ion-selective microelectrode is used. The membrane potential is more precisely determined when the reference micropipette is impaled in the same cell as the ion-selective microelectrode. So the construction of double-barrelled microelectrodes, where one barrel is the ion-selective electrode and the other barrel the reference electrode, is a better solution to this problem.

In some papers, ionic levels are expressed as concentration despite the fact that the electromotive force seen by the ion-selective microelectrode varies with the ionic activity rather than with the concentration. This is because in calibrating solutions, ionic concentrations are known, and activities would have to be calculated by using an activity coefficient [14,18,19]. Furthermore, expressing the results as ionic concentrations facilitates a comparison with data obtained by more indirect methods. In some papers, these two terms are used ambiguously, or the activity coefficient used is not satisfactorily indicated.

At the outset, it might be well to define what we meant by the term 'microelectrode'. This term, previously used by some authors for electrodes with tip diameters ranging from 50 to 100 μm, is usually taken to imply electrodes with tip diameters that do not exceed 1–2 μm [5]. As will be seen, this is not a purely pedantic distinction. Decreasing tip size can modify, in several important aspects, the operational characteristics of ion-selective microelectrodes.

Very few publications have been found on gas measurement with microelectrodes, as minielectrodes are used for *in vivo* measurements. Vaupel *et al.* [20] measured the partial pressure of oxygen (pO_2) in tumour with gold microelectrode (1–5 μm diameter). The technique used is a polarographic one. Von Ardenne [21] also used a pO_2 gold microelectrode for the determination of pO_2 in normal and in tumour tissues. Sohtell & Karlmark [22] describe a pCO_2 microelectrode, having a tip diameter of a few micrometres, to measure *in vivo* carbon dioxide pressure in a proximal tubule. The carbon dioxide outside the electrode tip penetrates a silicone membrane into an isolated bicarbonate solution which changes its pH. This pH is measured with an Sb/Sb_2O_3–$Ag/AgCl$ system.

All workers should question the reliability of the data obtained with microelectrodes. So, the proper functioning of the electrode, the stability of the system, its selectivity, limits of sensitivity, reproducibility, accuracy, response time, the influence of the temperature, of pH and total ionic strength, should be studied in detail [23]. Moreover, the physiologist is likely to ask: whether or not the system being used can give him the information being sought; how normal the cell or tissue preparation is; what reliability and confidence can he place on the data being gathered?

The purposes of this chapter are to acquaint the reader with the kinds of ion-selective microelectrodes that are currently available for making *in vivo* extracellular or intracellular measurements, with emphasis on the use of liquid ion-exchanger microelectrodes, to provide references in which details of their construction, calibration, and characteristics are presented, and to present a general outline for the experimental use of such electrodes [13].

2 Li$^+$-selective microelectrodes
Several attempts have been made to develop ion-selective electrodes for measuring lithium ion activities in biological systems, especially because of the importance of lithium salts in manic depressive psychosis. But little is known about Li$^+$ accumulation or transport by nerve cells, owing to the analytical difficulties encountered. Here, the development of several neutral carriers for lithium ions are reported.

The first synthetic Li$^+$ carrier was described by Guggi *et al.* in 1975 [24]: the ligand N,N'-diheptyl-N,N',5,5-tetramethyl-3,7 dioxanone diamide (ligand 5) (5.8%)† in tris-(2-ethylhexyl)-phosphate (62.8%) was used as a membrane component in a PVC matrix (31.4%). Fig. 1 shows the structure of the neutral carrier. The

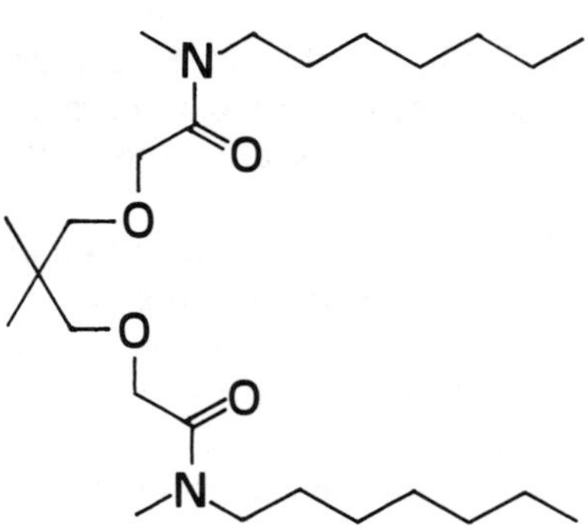

Fig. 1 — Structure of the neutral carrier ligand 5 for Li$^+$. From [24].

† Here and in the following, proportions are indicated in weight/weight.

response was linear in the 1–0.0001 M lithium concentration range with a linear regression slope of 57.5 mV (theoretical slope: 59.2 mV), the e.m.f. reading being corrected for changes in the liquid junction potential. The authors showed that this ligand 5-based Li^+ electrode had superior selectivity over Na^+ compared to glass electrodes and that its e.m.f. remained constant when pH changed in the range of physiological values (pH > 4.5).

More recently, Zhukov et al. [25] have presented several ionophores as Li^+ neutral carriers. The solvent polymeric membranes were prepared from 1–2% ligand, 33% PVC, and 65–66% o-nitrophenyloctylether. The bulky ligand 9 (Fig. 2)

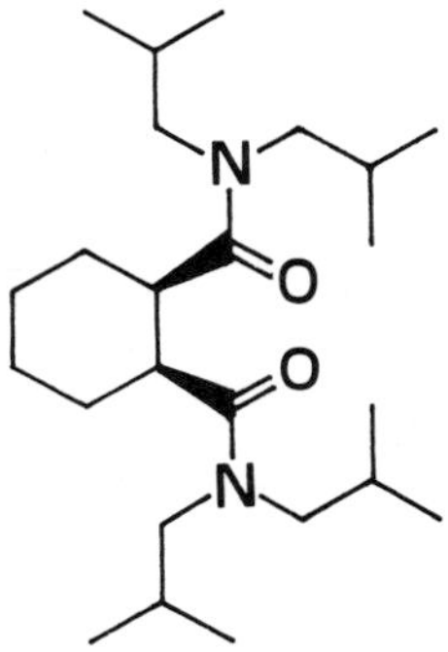

Fig. 2 — Structure of the neutral carrier ligand 9 for Li^+. From [25].

yields membranes with greatly improved selectivity compared to other membranes, e.g. for lithium ions ($k_{Li^+,Na^+} = 6 \times 10^{-3}$) and protons ($\log[k_{Li^+,H^+}] = 0.8$), which is attractive for clinical applications. Because of the fair selectivity over hydrogen ions, no H^+ interference occurs over the physiological pH range. Compared with the membrane based on ligand 5, there is a slight increase in the detection limit, but this will not create problems in monitoring lithium ion activities around 10^{-3} M. The regression line for the data obtained for pure lithium chloride solutions ($5 \times 10^{-4} - 10^{-1}$ M) has a slope of 57.7–58.1 mV per decade change in Li^+ activity, the theoretical slope being 58.2 mV at 20°C (Fig. 3). The 95% response time† of the cell assembly is less than 2 s if 1 ml of a 10^{-1} M LiCl solution is injected into 50 ml of a rapidly stirred 10^{-3} M LiCl solution. The e.m.f. stability of the electrode system in a 10^{-2} M LiCl solution over a period of 200 min is 90 μV.

Thomas et al. [26] used the neutral ligand 5 in Li^+-selective microelectrodes. These were made from borosilicate glass micropipettes and were siliconized by dipping in a solution of tributylchlorosilane and by heating for several minutes at 100°C. The last few millimeters of the micropipette were filled with the Li^+ ligand

---

† The electrode response time is theoretically defined as the period beginning with the dipping of the tip into a new solution up to the time when the e.m.f. comes to more than 95% of complete equilibrium. However, in practice, one can only estimate the rise time, i.e. the time required for the output response to rise from 10 to 95% of net change towards the final value for a given applied step input.

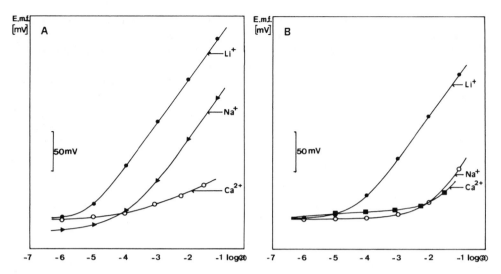

Fig. 3 — E.m.f. response of two solvent polymeric membrane electrode cell assemblies to different activities of Li^+, Na^+ and Ca^{2+} in the sample solution. A, electrode based on ligand 5; B, electrode based on ligand 9. From [25].

solution (i.e. 9.7% Li^+ carrier, 85.5% tris-(2-ethylhexyl)phosphate and 4.8% sodium tetraphenylborate). The shank of the electrode was filled with 2M LiCl. To complete the microelectrode, chloride-coated silver wire was dipped into the LiCl solution. A reasonably low resistance of about 10^{10} Ω was obtained. The selectivity coefficient of Li^+ over Na^+ was approximately 0.02 as determined by a pure solution method. The Li^+-selective microelectrodes were usable for at least 2 days after filling, and were very stable in the calibration solutions or intracellularly.

Thomas *et al.* [26] described physiological experiments done on the largest neurone at the rear of the right pallial ganglion of the snail *Helix aspersa*. The exposed suboesophageal ganglia were superfused with continuously flowing snail Ringer. According to the authors, the Li^+ -selective microelectrode described worked satisfactorily both extracellularly and intracellularly.

Li^+-selective microelectrodes based on ligand 9 have not yet been applied to *in vivo* determinations.

3 Mg^{2+}-selective microelectrodes

Magnesium ions are an important component of the cytoplasm of all living cells. These metal ions participate in many biochemical reactions where they regulate the activity of enzyme systems, in particular those involving the transfer of phosphate groups. The total magnesium content of muscular tissue is determined quite easily, but, because of compartmentalization in subcellular organelles and Mg^{2+} binding in the cytoplasm, the free cytoplasmic Mg^{2+} concentration is likely to be lower than the averaged cellular concentration.

Until 1980, only spectrophotometric determination of Mg^{2+} was a reliable technique. Direct determination of the intracellular Mg^{2+} activity had not been possible until then. The availability of neutral ligand resins, with acceptable sensitivity and selectivity properties toward Mg^{2+}, and the possibility of constructing microelectrodes with such resins, permit the measurement of free myoplasmic Mg^{2+} concentration in muscle fibres, at rest and submitted to different experimental conditions [27], and in cardiac and skeletal muscle preparations [28]. All these experiments have been performed in superfused experimental chambers.

3.1 Fabrication

Pulled glass micropipettes with tip diameters of 1–2.5 μm [29] or less (e.g. 1 μm [28] or 0.4 μm [27]) are dipped into pure hexamethyl disilazane and baked for several minutes at 200°C [29]. Another method consists in pre-heating to 200–250°C and then exposing to N-trimethylsilyldimethylamine (TMSDMA) [28] or tri-n-butyl-chlorosilane vapor [27] to make the tip hydrophobic by silanization. According to Hess *et al.* [28], the TMSDMA has the advantage over chloride-substituted silicon-izing agents that it does not produce an etching-byproduct when reacting with the glass.

The sensor can be introduced in different ways. Lanter *et al.* [29] injected a small volume of internal filling solution into the top of the micropipette shank to a height of about 5 mm; then the sensor was sucked into the tip to a height of about 700 μm. Hess *et al.* [28] indicated only that the pipette was back-filled into the tip with the sensor, and Lopez *et al.* [27] filled the microelectrode with the neutral ligand, to a length of about 6 mm, by capillarity (by application of a small drop of the resin at the back of the electrode).

The micropipettes are then back-filled with internal filling solution (10 mM $MgCl_2$ in water).

The sensor used is a solution of the Mg^{2+}-sensitive neutral carrier ETH 1117 (N,N'-diheptyl-N,N'-dimethylsuccinic acide diamide) and of sodium tetraphenyl-borate in propylene carbonate.

3.2 Characteristics

The e.m.f. experimental values, corrected with the liquid junction potential changes, were plotted against the logarithm of Mg^{2+} activity ($a_{mg^{2+}}$). In the absence of interfering ions, the slope of the response was in the range of 27.3–28.7 mV (theoretical value: 29.3 mV at 22°C) [28,29].

With an intracellular ion background (10 mM Na^+, 100 mM K^+ and 0.001 mM Ca^{2+}) (Fig. 4), the lower detection limit was for $\log[a_{Mg^{2+}}] = -3.5$. A useful e.m.f. response was obtained for the physiologically estimated free intracellular Mg^{2+} concentration of about 1 to 6 mM ($\log[aMg^{2+}] = -3.4$ to -2.7) [29]. To be valid, this response must be obtained with calibration solutions whose composition is as close as possible to the physiological concentrations (especially concentrations of K^+ and Na^+, whose selectivity coefficients are $\log[k_{Mg^{2+},K^+}] = -1.4$ and $\log[k_{Mg^{2+},Na^+}] = -1.1$, respectively): at high K^+ and Na^+ concentration, the response of the Mg^+-selective microelectrode was highly modified. Interference from Ca^{2+} and H^+ was found to be negligible [27,28]. These findings are consistent with the selectivity factors reported by Lanter *et al.* [29]. Caffeine (3 mM), oligomy-cin (0.05 mM) and ouabain (1 mM) did not induce any interference with the microelectrode function [27].

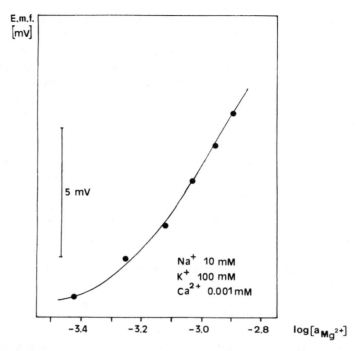

Fig. 4 — E.m.f. response of the microelectrode cell assembly to different Mg^{2+} activities in the physiological free range at constant ion background. From [29].

Owing to the relatively low resistance of this neutral carrier-based microelectrode (2 to 4×10^{10} Ω for tip diameters of 1–2.5 μm), the 90% response time was inferior to 5 s [29].

The cell assembly showed an e.m.f. drift of less than 1 mV/h over at least 4 days [27,29]. For Hess *et al.* [28], optimal Mg^{2+}-selective microelectrode responses were obtained after conditioning the electrodes by immersing their tips in a 100 mM $MgCl_2$ solution overnight. The electrodes showed no significant drift over several days. Lopez *et al.* [27] observed that the best responses of the microelectrodes were obtained when the filled microelectrodes were stored for 24–36 h; they retained their responsiveness over periods of 48–72 h.

3.3 Experiments
Mg^{2+}-selective microelectrodes have been used for continuous measurements of intracellular Mg^{2+} level in sheep cardiac Purkinje fibres [28–30], in ferret papillary muscles [28], and in frog skeletal muscles [27,28,30], placed in superfused experimental chambers.

3.4 Conclusion
The Mg^{2+}-selective microelectrode described here is suitable for measuring free intracellular magnesium concentration, in the millimolar range and in the presence

of the expected intracellular levels of interfering cations. But this technique has several limitations. Firstly, under intracellular conditions, the response of the microelectrode is noticeably influenced by Na^+ and K^+ concentrations: so it is necessary to know precisely the Na^+ and K^+ intracellular concentrations, because of the low selectivity of the microelectrodes for Mg^{2+} over these two ions. Secondly, these microelectrodes can be used for the detection of intracellular resting levels of Mg^{2+} and even for the observation of possible intracellular changes occurring over tens of seconds; with double-barrelled microelectrodes, however, it would be possible to measure rapid changes in intracellular ion activity, as the ion-selective and the reference microelectrodes are impaled in the same cell. Thirdly, measurements of intracellular activity with ion-selective microelectrodes represent localized point sampling. The question arises whether such measurements adequately reflect intracellular ionic levels. According to several authors [28], free cytoplasmic Mg^{2+} is likely to be evenly distributed and thus accurately represented by the measurements described above. So it seems reasonable to assume that there are no significant gradients of intracellular Mg^{2+} concentration within the sarcoplasm.

Another potential limitation of such measurements is the fact that the estimations of the ionic levels are based on the difference between the signals of the reference and of the ion-selective microelectrodes. Each electrode is located in a different cell and might have induced local damage of small but variable degree, so the membrane potentials recorded by the two microelectrodes may differ slightly.

It would therefore be interesting to develop double-barrelled Mg^{2+}-selective microelectrodes, with diameters of 1 μm or less, and with relatively low resistances. Up to now, all the experiments using Mg^{2+}-selective microelectrodes have been performed on preparations placed in superfused experimental chambers, and not *in vivo*. The development of such double-barrelled microelectrodes would allow some *in vivo* experiments to be undertaken.

4 Ca^{2+}-selective microelectrodes

A Ca^{2+}-selective microelectrode for the assay of Ca^{2+} activity in the presence of other interfering ions, mainly K^+, Na^+, and Mg^{2+}, should have many different applications, and would clearly be of value to the cellular physiologist, since it would provide another tool to study the involvement of Ca^{2+} in several vital cellular functions where it plays an important direct role. Several techniques have been employed to determine free Ca^{2+} intracellular concentration, each having intrinsic advantages. The photoprotein aequorin and other metallochromic and fluorescent indicators have had the widest applications. These methods have had considerable success in measuring fast Ca^{2+} transients but have not been as sensitive or reliable for determining the absolute value of free intracellular Ca^{2+} concentration. Several attempts have been made to develop Ca^{2+}-selective microelectrodes suitable for the measurement of intracellular Ca^{2+} concentration.

If an electrode is to measure the Ca^{2+} activity in living cells, it must be sensitive, selective, and have a small tip.

Though the study of monovalent ions in cellular processes with ion-selective microelectrodes has succeeded, the intracellular measurement of Ca^{2+} with a microelectrode is much more difficult, primarily because the electrode must be able to detect very low levels of Ca^{2+} (less than 10^{-6} M), in the presence of relatively high amounts of K^+ (0.2 M), Na^+ (10^{-2} M) and Mg^{2+} (10^{-3} M).

It is only since about 1977 that Ca^{2+}-selective electrodes have been made with selectivities even approaching the necessary levels.

Two classes have been reported: first those based on organophosphate anions, which appear to have inadequate selectivity against Mg^{2+} for intracellular use, although their selectivity against Na^{2+} and K^+ is excellent. Then electrodes based on a neutral ligand (ETH 1001) have been studied. The initial report on such microelectrodes showed selectivities which would be insufficient for detecting intracellular Ca^{2+} levels [31]; but since then, several modifications have been reported, so that much better performances are obtained. However, several problems occurred, especially concerning the super-Nernstian slope of the calibration curves and the potential drift, with electrodes which have tip diameters less than 1 μm.

4.1 *Alkylphosphate sensor*

Ross, in 1967 [32], and then Moody *et al.* [33], in 1970, first described a Ca^{2+} liquid ion exchanger incorporated into a poly(vinylchloride) matrix, placed on a macroelectrode. The ion exchanger used was didecylphosphoric acid dissolved in dioctylphenylphosphonate. The potential response of this electrode is linear only from 10^{-1} to 5×10^{-5} M. Ruzicka *et al.* [34] developed this calcium-sensitive macroelectrode thus enabling Brown *et al.* [35] to design and present a microelectrode with a tip diameter of about 1–2.5 μm.

The sensor is a 70%–30% mixture of the di- and monoesters of octylphenyl acid phosphate. The chemical names and abbreviations for the di- and monoesters in the mixture are di-[*p*-(1,1,3,3-tetramethylbutyl)phenyl]phosphoric acid (*t*-HDOPP) and mono-[*p*-(1,1,3,3-tetramethylbutyl)phenyl]phosphoric acid (*t*-H$_2$MOPP) (Fig. 5). Some of the diesters were converted to their calcium salts. Di-(*n*-octyl) phenylphosphonate (*n*-DOPP) was used as mediator.

The sensor was prepared by combining the calcium salt, the esters, and the mediator in a convenient molar ratio. Polyvinylchloride was added in various percentages as a function of the tip size, this ratio being crucial to obtain functioning microelectrodes.

This sensor on a 100 μm length was introduced into a silanized pipette, and then back-filled with 0.1 M $CaCl_2$ (Fig. 6). The relationship between the electrode potential and the logarithm of the Ca^{2+} concentration was reasonably linear over the 10^{-1}–10^{-8} M range of Ca^{2+} concentration, with a slope of 23 mV/decade.

The interferences with other ions, such as K^+, Na^+, Mg^{2+} and H^+ were studied by the mixed-solution method (see Chapter 1 (§1.3):

- A background of 200 mM K^+ did not affect the slope down to 2×10^{-8} M Ca^{2+}, but at lower concentrations, the response deviated (Fig. 7). Nevertheless, between pCa 7 and 8 at 200 mM K^+, the electrode response remained substantial (10–11 mV). The average k_{Ca^{2+},K^+} value was 4.5×10^{-7}.
- Sodium affected the electrode in much the same way as potassium, and the selectivity coefficients of individual electrodes were in the same range. There was no evidence that the selectivities were concentration dependent.
- The electrode response remained virtually unchanged in Ca^{2+} solutions contain-

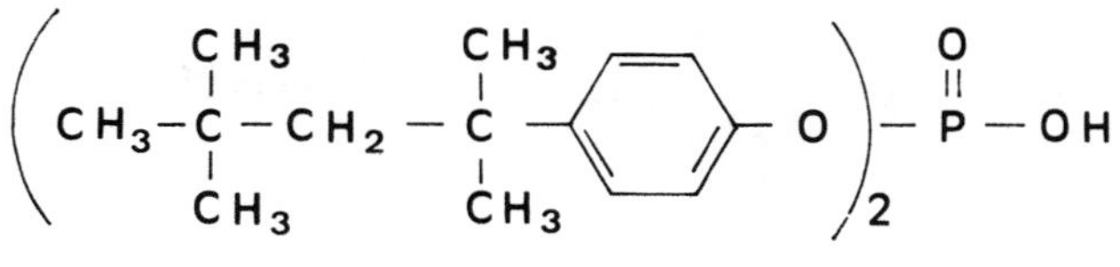

$$\left(CH_3-\underset{\underset{CH_3}{|}}{\overset{\overset{CH_3}{|}}{C}}-CH_2-\underset{\underset{CH_3}{|}}{\overset{\overset{CH_3}{|}}{C}}\left\langle\underset{}{\bigcirc}\right\rangle O \right)_2 \overset{\overset{O}{\|}}{P}-OH$$

(A)

$$CH_3-\underset{\underset{CH_3}{|}}{\overset{\overset{CH_3}{|}}{C}}-CH_2-\underset{\underset{CH_3}{|}}{\overset{\overset{CH_3}{|}}{C}}\left\langle\underset{}{\bigcirc}\right\rangle O-\overset{\overset{O}{\|}}{P}-(OH)_2$$

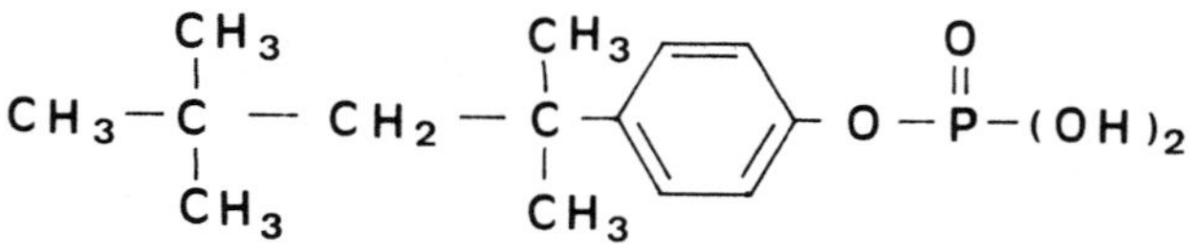

(B)

Fig. 5 — t-HDOPP (A) and t-H$_2$MOPP (B). From [35].

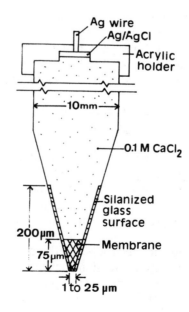

Fig. 6 — Construction features of the calcium-selective microelectrode. From [35].

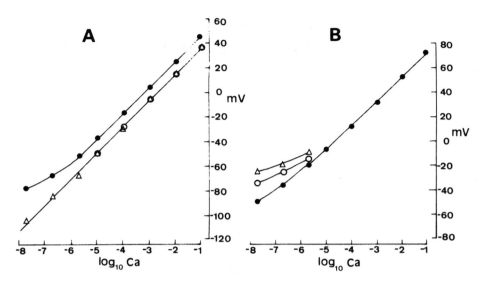

Fig. 7 — Potassium interference. A: Average e.m.f. from six electrodes calibrated in solutions containing only CaCl$_2$ (O), Ca^{2+} solutions containing 50 mM K$^+$ (△), and Ca^{2+} solutions containing 200 mM K$^+$ (●). Note that the slopes are parallel for each potassium concentration down to 2 × 10^{-6} M Ca^{2+}. B: Average e.m.f. of another group of four electrodes in Ca^{2+} solutions containing 200 mM K$^+$ (●), 500 mM K$^+$ (O), and 1000 mM K$^+$ (△). From [35].

ing a constant background of 200 mM K$^+$ and 2 mM Mg^{2+}. The $k_{Ca^{2+},Mg^{2+}}$ selectivity coefficients were about 10^{-3}.

• At all calcium concentrations, the behaviour of the electrodes indicated a preference for hydrogen ion over Ca^{2+} − . But the electrode response at constant calcium concentration of 10^{-2}–10^{-4} M remained stable within 1–2 mV from pH 9 to pH 5.

The response time in pure calcium chloride solutions was about 6 s, but in the presence of 2.5 × 10^{-2} M magnesium chloride, the response time increased to about 12 s.

Christoffersen & Johansen [36] also described, in 1976, a calcium-selective electrode based on di-(n-octylphenyl)phosphoric acid dissolved in di-octylphenyl-phosphonate and combined with polyvinyl chloride dissolved in tetrahydrofuran. The detection limit (10^{-4}–10^{-3} M), the tip diameter (10–20 μm), and the response time (from 1 to 15 min, decreasing with age) were much greater than those reported by Brown *et al.* [35].

Owen *et al.* [37] used the electrode described by Brown *et al.* [35] to measure intracellular calcium activity in the giant neuron of the marine mollusc, *Aplysia californica*, by substracting the membrane potential (measured with a reference microelectrode) from the potential recorded by the Ca^{2+}-selective microelectrode.

4.2 Neutral ligand ETH 1001

In 1975, Ammann *et al.* [18] described a liquid membrane electrode using a synthetic neutral carrier in *o*-nitrophenyloctylether as the membrane component in a PVC matrix, in the presence of sodium tetraphenylborate. This ligand is N,N'-di[(11-ethoxycarbonyl)undecyl]-N,N',4,5-tetramethyl-3,6-dioxaoctane diacid diamide (ETH 1001) (Fig. 8). The role of the sodium tetraphenylborate is to reduce to a

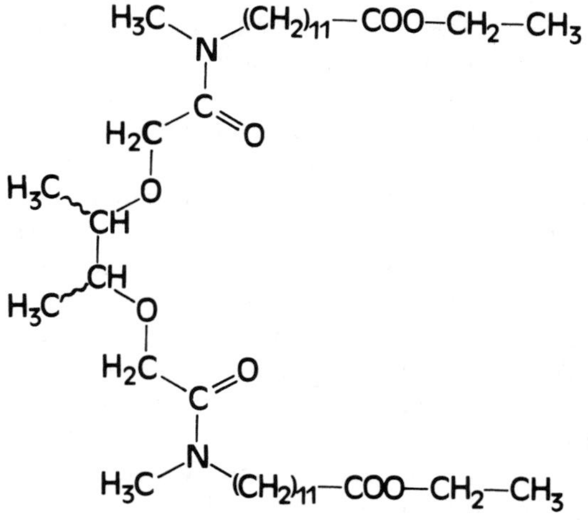

Fig. 8 — Structure of the Neutral Carrier for Ca^{2+}. The product used is a mixture of components with different configurations at the tertiary carbon atoms (70 mole-% mesoform and 30% racemate). From [18].

certain extent the interference by lipophilic anions.

A year later, Oehme *et al.* [38] used this neutral carrier in a Ca^{2+}-selective microelectrode. Double-barrelled micropipettes with a tip diameter of about 2 μm were filled with this sensor up to a height of about 200 μm. The resistance is approximatively 10^{10} Ω, a value of $2-15 \times 10^9$ Ω being reported by Heinemann *et al.* [39]. The response of the electrode assembly is linear in the range or 1×10^{-1} to at least 1×10^{-5} M $CaCl_2$, giving a slope of 31.0 mV (theoretical slope: 29.1 mV) for the linear regression of the calcium ion activity. The characteristics and performances of this type of microelectrode, in particular selectivity, have been studied in detail by several workers [15,18,39].

Later on, Tsien & Rink [31] and Weingart & Hess [40], using the same sensor, made a single-barrelled microelectrode, with a tip diameter between 0.3 and 0.8 μm [31] or slightly less than 1 μm [40]. The tip was bevelled, and the shank was painted with silver conductive paint, to reduce the capacitive artefacts produced by fluctuations in the solution level. The resistance (about $2-3 \times 10^{10}$ Ω for an electrode with 1.5 μm tip diameter in a pCa 3 solution) was not noticeably dependent on the sensor column, which was always greater than 100 μm.

These authors report that the performances of the Ca^{2+}-selective microelectrodes were quite variable, even between electrodes made in the same batch and so far as possible in the same way.

So each electrode must be tested individually for response to Ca^{2+} levels below

1 μM. Fig. 9 shows calibration curves for two electrodes. Fig. 9A gives the calibration plot for one of the best electrodes of this type, which gave a virtually Nernstian response down to pCa 7 and a further useful response down to pCa 8 and lower. Fig. 9B shows a microelectrode which presents an important hysteresis phenomenon [31].

The calibration curves presented by Weingart & Hess [40] (Fig. 10) show a Nernstian response between pCa 3 and 4 and a super-Nernstian response between pCa 4 and 6. This latter phenomenon was dependent upon the tip diameter of the electrode. The super-Nernstian response almost completely disappeared when the bevelled tip of approximately 1 μm diameter was reduced to roughly 10 μm. By the same method as Tsien & Rink [31], Barber [41] obtained double-barrelled electrodes (tip diameter of 0.5 μm or less) with a mean slope of 29 mV/decade between pCa 3 and 6. 15 mV/decade between pCa 6 and 7, and 6 mV/decade between pCa 7 and 8.

These papers [31,41] also dealt with interference from other ions: H^+, Mg^{2+}, Na^+, K^+, among which sodium caused the most significant interferences.

Dagostino & Lee [15] measured the potentials of the microelectrode with the single electrolyte solutions containing the interfering ions K^+, Na^+, and Mg^{2+}, and calculated the corresponding selectivity coefficients: $k_{Ca^{2+},K^+} = 5 \times 10^{-6}$ (150 mM KCl), $k_{Ca^{2+},Na^+} = 5 \times 10^{-5}$ (10 mM NaCl), $k_{Ca^{2+},Mg^{2+}} = 3 \times 10^{-7}$ (2 mM $MgCl_2$). These authors observed that the selectivity and slope of the Ca^{2+}-selective microelectrodes were considerably improved by breaking the tips to 1–2 μm [15].

Preliminary tests revealed [40] that the electrodes respond inconsistently to temperature changes. Therefore, both calibrations and intracellular measurements were performed at the same temperature (20–22°C).

Even with electrodes selected for good calibration slopes, the response time has been rather variable [31]. The response to changes in the range pCa 3–5 is almost always complete within the fraction of a second taken to flush the new solution through. Other workers [39] have also reported very fast responses at high Ca^{2+} levels. However, the time resolution becomes progressively worse at lower pCa, and typically many seconds are required for a 95% complete response going from pCa 5 to 6 or from pCa 6 to 7. The fastest responses are usually obtained with bevelled electrodes, but not all bevelled electrodes are fast.

Two main forms of instability have been observed: drift, that is, a shift of the entire calibration curve along the e.m.f. axis, and loss of slope, that is, a decrease in the slope at the lowest Ca^{2+} levels.

Tsien & Rink have detailed a number of mechanisms for loss of slope [31]. They observed that the drifts during calibration are approximatively linear in time, not stepwise; so the reference e.m.f. for intracellular experiments can be taken from the linear interpolation of the pre- and post-impalement values. Obviously, drift-free electrodes are desirable, but baseline shifts, typically of less than 0.5% of the full scale response per minute, are not necessarily unusable in practice. According to these authors, all electrodes suffer from loss of slope as they age (flattening in the calibration curve, specifically at low Ca^{2+} levels), but the rate of deterioration is highly variable and not obviously correlated with overall drift of e.m.f. Electrodes for intracellular use should be calibrated and used as soon as possible after they are made.

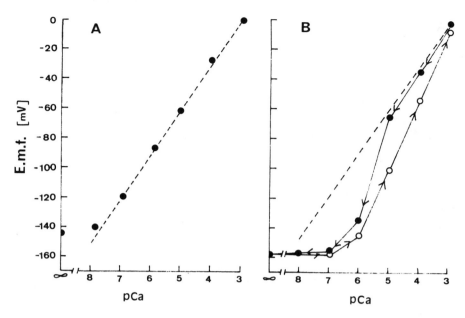

Fig. 9 — Calibration plots for two electrodes made with a sodium tetraphenylborate sensor. A: 1 μm tip electrode. The e.m.f. response is plotted against progressively lower free Ca^{2+} concentrations. This electrode showed little hysteresis. B: Another 1 μm electrode. The filled circles show the response to successively decreasing free Ca^{2+} concentrations, and the open circles to increasing concentrations. The dashed lines show the theoretical (Nernstian) response for [Ca^{2+}]. The e.m.f. has been arbitrarily offset to give O mV in pCa 3.From [31].

However, Dagostino & Lee [15] observed that the potentials of the electrodes changed considerably for about 2 h before becoming stable. They also observed that the rate of the microelectrode potential change depended on the column length of the Ca^{2+}-selective sensor: the longer the column length, the slower the rate of the microelectrode potential change.

The undesirable hysteresis (Fig. 9B) and the two-phase time course were tentatively blamed on the loading of an exchangeable cation, Na^+, into the sensor. Tetrabutylammonium tetraphenylborate has been substituted for sodium tetraphenylborate. This salt gave some electrode slopes greater than Nernstian, and a rather better response between pCa 6 and 8 was sometimes obtainable. Another substitute was tetraphenylphosphonium bis(1,3-diethyl-2-thiobarbiturate) trimethine oxonol in *o*-nitrophenyloctylether [31]. But the electrode behaved as described earlier. Tetraphenylphosphonium oxonol electrodes hardly ever show calibration slopes significantly greater than Nernstian, or hysteresis of more than 2 mV between ascending and descending calibrations in the region pCa < 7. One drawback seems to be the response reduction, particularly between pCa 7 and 8, between which no values greater than 11–15 mV can be produced (Fig. 11). Also, shown in Fig. 10, is the usual effect of tip diameter, the 0.5–μm electrode having a distinctly poorer performance than the 1.5 μm electrode. The decrease in selectivity with tip size seems to be a general phenomenon in liquid sensor microelectrodes.

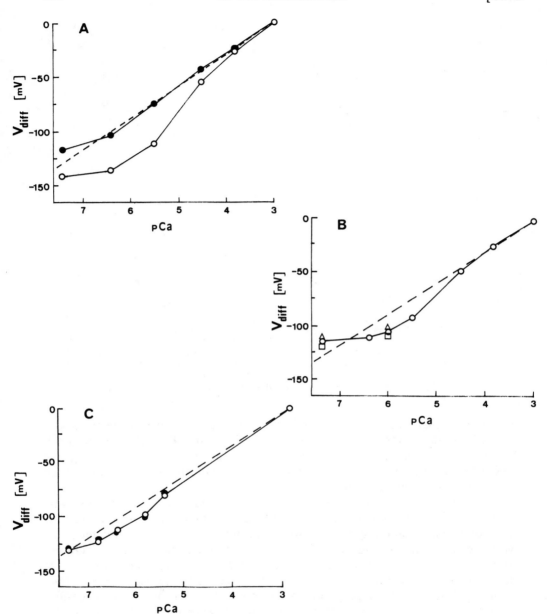

Fig. 10 — Calibration curves of different Ca^{2+} selective microelectrodes. The potential difference between a Ca^{2+} electrode and a 3 M KCl reference electrode, V_{diff}, is plotted versus pCa of the calibrating fluids. V_{diff} in the presence of 1 mM $CaCl_2$ was defined as 0 mV. The dotted line represents the Nernstian slope for an ideal Ca^{2+} response (29 mV/pCa unit). A: Influence of tip diameter. The electrode was calibrated before (○, tip diameter <1 μm) and after breaking the tip (●), tip diameter approximately 10 μm). B: Influence of Na^+. Substantial effects were observed over the physiological range of pCa when the NaCl concentration was altered from 0 mM (□) to 7 mM (○) and 20 mM (△). C: Influence of Mg^{2+}. No significant effects were seen when the $MgCl_2$ concentration was varied from 0 mM (○) to 5 mM (●). From [40].

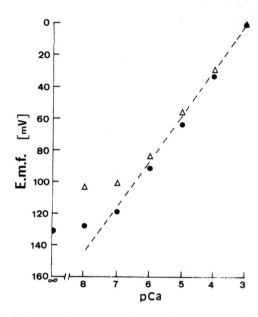

Fig. 11 — Calibration plots for two tetraphenylphosphonium oxonol microelectrodes with outer tip diameters of 1.5 ($\bullet$) and 0.5 ($\triangle$) μm. The dotted line represents the theoretical (Nernstian) response to Ca^{2+}. From [31].

Rink *et al.* [42] also replaced sodium tetraphenylborate in the sensor by tetraphenylphosphonium bis(1,3-diethyl-2-thiobarbiturate) trimethine oxonol. In the presence of 0.1 M KCl, the electrodes, with 1–2 μm tip diameter, showed Nernstian responses between pCa 3 and pCa 6, 12-20 mV between pCa 6 and 7, and 4–8 mV between pCa 7 and 8. Interferences from H^+, Na^+ and Mg^{2+} were negligible. High concentrations of CO_2 did not alter the electrode responses.

To be useful in the smaller cells, Ca^{2+} electrodes must have tips of sub-micrometre size, while retaining stable properties and ample resolution down to 10^{-7} M free Ca^{2+} or lower. For this purpose, Marban *et al.* [43] modified the sensor composition by replacing sodium tetraphenylborate by tetraphenylarsonium tetrakis(*p*-biphenyl)borate. This new PVC-gelled sensor offers various advantages over previous Ca^{2+}-selective microelectrode compositions: (i) mechanical stability is much improved over liquid sensors; the PVC gel very seldom backs up or escapes from the tip; (ii) electrode response is increased at low Ca^{2+} levels, giving greater resolution between 10^{-6} M and 10^{-7} M free Ca^{2+} than previously achieved in fine microelectrodes suitable for use in small cells; (iii) electrode lifetime and stability are greatly increased. These electrodes retain their responsiveness over periods of 3–6 h. Use of the electrodes is often limited not by ageing, but rather by blunting of the tip with repeated penetrations. The performance of the electrodes at low Ca^{2+} showed quantitative variations, even when they were constructed in a similar fashion from the same batch of sensor. Therefore, each electrode was individually tested.

The response is Nernstian (29.5 mV at 24°C) down to 10^{-6} M, with some 24 mV between 10^{-6} and 10^{-7} M, and 10 mV below 10^{-7} M, against a background of about 125 mM K^+ (Fig. 12).

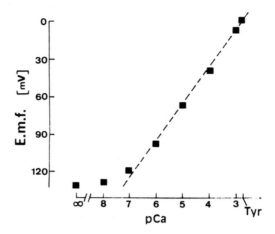

Fig. 12 — Calibration plots for a Ca^{2+} electrode with a bevelled tip of outer diameter *c*. 0.5 μm. The zero of the scale was defined as the potential in the Tyrode solution containing 1.8 mM Ca^{2+}. The dashed line indicates the theoretical (Nernstian) slope, i.e. 29.5 mV per pCa unit at 24°C. From [43].

At intracellular Ca^{2+} levels, while interferences from Mg^{2+} and H^+ can be neglected, Na^+ interference needs to be taken into account. At higher Ca^{2+} levels, above 0.1 μm, interfering cations have even less effect. The influence of temperature must also be considered, as the calibrations were normally done at room temperature, whereas the experiments were carried out at 35–37°C. Raising the temperature merely scaled the response of the e.m.f. by the ratio of absolute temperatures, in accordance with the Nernst equation. The response time of the electrodes is many seconds at micromolar levels of Ca^{2+}, so they are not expected to follow transients during single twitches. Attention was therefore focused on 'resting' levels and slow changes.

4.3   Experiments

In 1976, Brown *et al.* [35] conducted preliminary experiments involving penetration of barnacle muscle cells (100 μm) with Ca^{2+}-selective microelectrodes (1–2.5 μm) based on an alkylphosphate sensor. The same type of sensor has been used to measure Ca^{2+} activity in the giant neuron of marine molluscs, with a single-barrelled electrode with a 1 μm tip diameter [37].

Also in 1976, Nicholson *et al.* [17] used a double-barrelled microelectrode with a tip diameter of 2 μm, based on ETH 1001, to measure *in vivo* Ca^{2+} concentration in the extracellular space of cat cerebellum. At that time, most authors used this sensor, sometimes slightly modified, with single-barrelled electrodes to measure Ca^{2+} levels in the ventricular muscle of ferrets [43], in frog sartorius muscle fibres and ferret ventricular myocardium [31], in embryos of *Xenopus laevis* [31,42], in canine cardiac Purkinje fibres [44], and in sheep Purkinje fibres, sheep ventricular muscle, and frog skeletal muscle [40].

Others preferred the use of double-barrelled electrodes to measure Ca^{2+} levels extracellularly in the somatosensory cortex of cats (*in vivo*) [39], or intracellulary in single muscle fibres of *Balanus nubilus* [45], in rabbit ventricular muscle [46] or in salivary gland cells of snails [41]. These double-barrelled electrodes often have greater tip diameters than single-barrelled electrodes, except the last-mentioned [41] where a 0.5-μm tip diameter has been obtained.

4.4 Discussion and conclusion

It clearly appears that, though Ca^{2+}-selective microelectrodes with sufficient selectivity, sensitivity, and small tip diameter to make them suitable for intracellular measurements of Ca^{2+} have been reported, several problems still have to be solved.

On the basis of determinations of selectivity coefficients for various Ca^{2+} resins, Ca^{2+} electrodes cannot be expected to respond exclusively to Ca^{2+}. Thus, possible interferences, mainly from K^+ and Na^+, and secondarily from Mg^{2+}, must be minimized when working in intracellular conditions. This suggests that considerable errors might occur when calibration and measurements are performed in different ionic backgrounds. This problem may be partly overcome by using a constant interference calibration method. This requires the use of an internal solution whose cationic composition resembles the cytoplasm. Of course, the reliability of this approach depends on the assumptions made for the concentrations of the individual ions. For this, data from other ion-selective microelectrode measurements can be used.

It is also necessary to know accurately for the calibrating solutions the free Ca^{2+}, and so to determine the appropriate stability constant of the Ca^{2+} buffer.

Several authors have observed a super-Nernstian response for Ca^{2+}-selective microelectrodes between about pCa 4 and 6. Interestingly, this has not been a reported finding among the other groups who have used exactly the same Ca^{2+} cocktail. The reasons for this discrepancy remain unclear. Dagostino & Lee [15] claim to have seen this behaviour shortly after fabrication of the electrodes but not a few hours later. In contrast, the observations of Weingart & Hess [40] suggest that the entire calibration curve flattens with increasing time, and the non-linear behaviour persists. Tsien & Rink [31] originally proposed that the exchangeable inorganic cation in the cocktail (Na^+) might be responsible for this effect. However, the observation that the super-Nernstian response disappears with larger tip sizes would not support this hypothesis.

Tsien & Rink [47] have subsequently suggested that the reduction of shunt resistance through the wall of the glass might be responsible for the super-Nernstian behaviour. They report that embedding the Ca^{2+} sensor in a PVC matrix eliminates the extra slope. Moreover, this modification seems to improve the stability of the electrode and increases the response to low Ca^{2+} levels.

The performance of ion-selective microelectrodes is highly dependent not only on the membrane composition but also on the geometry of the glass pipettes used. The Ca^{2+}-selective liquid introduced by Oehme *et al.* in 1976 [38] has become one of the most reliable tools for determining intracellular as well as extracellular activities of free Ca^{2+} when used in electrodes with tip diameters larger than 1 μm. For such electrodes, the geometry (tip diameter, and thickness of the glass wall at the extreme tip) and the silanization technique used did not influence the electrode performance significantly. To measure free Ca^{2+} in small cells, electrodes with tip diameters below 1 μm had to be produced. Because of this reduction in tip size, the behaviour of the microelectrode cell assembly often deviated from the behaviour observed for larger tips. An over-Nernstian slope in the range 10^{-4}–10^{-6} M Ca^{2+} and a rather high detection limit were observed (yet an extremely low detection limit has been reported by Lee *et al.* [48] with the original membrane material).

Because problems of over-Nernstian slopes and/or high detection limits might be due to kinetic limitations in the diffusion of Ca^{2+} and interfering ions between the membrane phase and the sample solution, the sodium tetraphenylborate in the membrane was replaced by other salts. By incorporating tetraphenylphosphonium bis(1,3-diethyl-2-thiobarbiturate)trimethine oxonol, or tetraphenylarsonium tetrakis(*p*-biphenyl)borate, over-Nernstian slopes were apparently eliminated.

Limitations of this technique include the rather low speed of response at the low intracellular Ca^{2+} levels, and the requirement for simultaneous measurement of membrane potential. This problem can be overcome with the use of double-barrelled electrodes. The performance of these Ca^{2+}-selective microelectrodes indicates that they should be able to monitor Ca^{2+} levels inside cells. The performances of the smallest electrodes are only just adequate, and further improvements in electrode selectivity, speed, and stability are certainly desirable. Even if such technical improvements are achieved, two more fundamental problems will still have to be considered.

One is that a microelectrode takes a point sample of the Ca^{2+} level and can easily miss a localized change in Ca^{2+} concentration which could be functionally very important.

Another caveat is that unidentified constituents of the cytoplasm could perhaps disturb the electrode.

Also, all microelectrode recordings are susceptible to artefacts from impalement damage. Authors generally accept only those measurements which have satisfied the abovementioned criteria for good electrode sealing and cell viability, but some residual perturbation by penetration damage cannot be excluded.

5 Cl^--selective microelectrodes
Several types of microelectrodes have been used to measure intracellular chloride activity: those using silver chloride, which show a response close to the theory, but are either too large for use in most cells or difficult to make, and those based on a liquid ion-exchanger resin which have been widely used. The advent of microelectrodes sensitive to the chloride ion has enabled the internal Cl^- activity of a number of cell types to be directly investigated.

5.1   *Solid-state chloride microelectrodes*
Chloride microelectrodes have been made by coating a platinum wire protruding from the end of a glass micropipette with Ag-AgCl [49] or by depositing silver chloride inside the tip of a micropipette [50]. These electrodes all have the drawback of having a large ion-sensitive surface which must be introduced completely into the

cell, and their use is therefore limited to cells of the size of skeletal muscle cells or larger. While these electrodes may have a tip diameter of 1 μm or less, the sensitive area is of the order of 10 μm in length with a diameter of 5 μm or more at the top of this area. In addition to the problem of size, they are difficult to make.

To overcome these problems, Neild & Thomas [51] developed a new design for a chloride-selective microelectrode. This electrode used silver chloride and had a low resistance ($<$ 100 M Ω), a tip diameter of 1–2 μm, and a relatively slow response. In developing the new electrode the aim was to use the recessed-tip design of Thomas [52] with a Ag-AgCl wire in place of the sodium-selective glass. The finished electrodes (Fig. 13) give a response of 57 mV for a ten-fold change in chloride

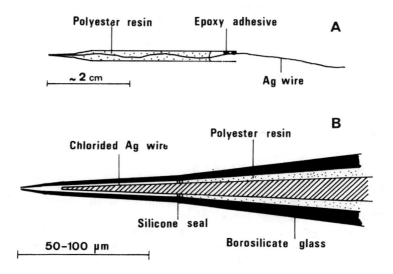

Fig. 13 — Construction of recessed-tip chloride-sensitive microelectrode. A, the complete electrode; B, enlarged view of the sensitive tip, with the diameter exaggerated relative to the length. From [51].

activity, and, depending on the tip diameter and volume of the recess, give complete responses within 1–2 min. This electrode is much more selective for chloride than for bicarbonate, and has a selectivity constant k_{Cl^-,HCO_3^-} of 3.4 $\times$ 10^{-4} [53].

5.2   Liquid Cl$^-$-exchanger microelectrodes

In 1970, Walker & Brown [54] described a single-barrelled microelectrode based on the liquid ion-exchanger Corning 476131 and filled with 1 M KCl, the length of the organic liquid column being 100–150 μm.

This electrode has an anion slope of 55–57 mV per tenfold change in Cl$^-$ activity at 25°C, independent of the cation in hydrogen, sodium, and potassium chloride solutions. The response of the electrode to two other anions, bicarbonate and isethionate, has been examined, because they may be present in appreciable quantities in intracellular fluid (Table 1). The resistance is about 10^{10}–10^{11} Ω, and the

Table 1 — Selectivity coefficients for a number of interfering ions determined with Cl$^-$-selective microelectrodes based on two different Corning sensors

Selectivity coefficient	Sensor	Refs.
$k_{Cl^-,bicarbonate} = 0.05\ (1.0\ M)$	Corning 476131	[54]
$k_{Cl^-,isethionate} = 0.2\ (1.0\ M)$		
$k_{Cl^-,bicarbonate} = 0.05\ (1.0\ M\ 0.1\ M)$		
$k_{Cl^-,isethionate} = 0.2\ (1.0\ M\ 0.1\ M)$	Corning 477315	3,55]
$k_{Cl^-,propionate} = 0.5\ (0.1\ M)$		
$k_{Cl^-,propionate} = 0.7\ (1.0\ M)$		
$k_{Cl^-,sulphate} = 0.03\ (0.1\ M)$	Corning 477315	[55]
$k_{Cl^-,sulphate} = 0.02\ (1.0\ M)$		

time constant is estimated to be 0.5 to 1.0 s.

The same year, Brown *et al.* [55] introduced a 150–200 μm length of Corning 477315 chloride ion-exchanger in a single-barrelled micropipette filled with 0.5 M KCl. Bolton & Vaughan-Jones [56] discussed the use of potassium sulphate- or potassium chloride-filled reference microelectrodes, and indicated that the latter gave the better results. These electrodes have a slope slightly less steep than predicted by the theory, being 55–57 mV for a ten-fold change in activity over the concentration range of 1.0×10^{-3} M–1.0 M at 20°C. This slope is also independent of the cation in hydrogen, sodium, and potassium chloride solutions. Table 1 reports the values of the selectivity constants for a number of interfering ions (bicarbonate, isethionate, propionate, and sulphate) determined in 1 M and 0.1 M solutions. These selectivity coefficients are quite similar to those mentioned above [54] (Table 1).

In another paper, Walker [3] announced a resistance in the range of 10^9–10^{10} Ω. If the resistance is too high ($> 10^{11}$ Ω), and the potential response to changes in Cl$^-$ activity is too sluggish to obtain meaningful calibration curves, the extreme tip can be bevelled with a suspension of aluminium oxide polishing powder stirred continuously. The resistance and the response time are then reduced [57]. Once the steady-state potential is attained, it remains constant to within 1 mV for several hours, even when several cell penetrations are made with the same electrode.

Fujimoto & Kubota [9] have studied in detail the effect of different parameters on the response of the Cl$^-$-selective microelectrode:

Interfering ions
Fig. 14 gives the e.m.f. of a Cl$^-$-selective microelectrode in test solutions containing Cl$^-$, HCO$_3^-$, SO$_4^{2-}$, H$_2$PO$_4^-$, or HPO$_4^{2-}$ at various concentrations at 25°C. The

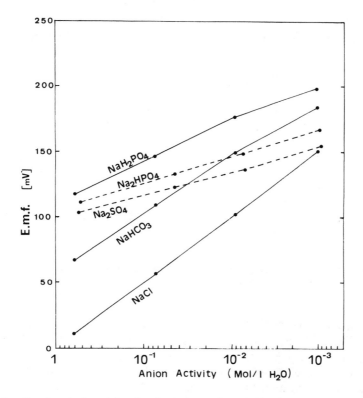

Fig. 14 — E.m.f. vs ionic activity plots for various anions obtained with double-barrelled Cl⁻ ion-selective microelectrodes at 25°C. The slope constants of the lines are nearly theoretical. From [9].

averaged slope obtained for the calibration line was in the range of 51.1–55.1 mV per ten-fold change in Cl^- activity from 1 to 10^{-3} M NaCl, which could be used in practical experiments. As can be inferred from Fig. 14, measurement of Cl^- activity may be difficult in an environment containing less than 5 meq/l of Cl^- because under such circumstances, interference of other coexistent anions is increased. However, since the measured intracellular Cl^- activity is usually in the order of 10 meq/l [58], it is possible that contributions of intracellular HCO_3^- and other anions to the e.m.f. of the Cl^- microelectrode are reduced to less than 9% of Cl^- activity. Therefore, it is feasible that measurements of intracellular Cl^- activity can be carried out with reasonable accuracy.

Response time
It usually averaged less than 1 s, ranging from 0.2–3 s [9]. The less the electrode resistance of the ionic barrel, the shorter the response time. It appeared that the electrical resistance of the ionic barrel showed a significant relationship to the length of the Cl^- exchanger column mounted at the tip. The electrical resistance in most of the ionic barrel was not more than a few multiples of 10^9 Ω. The shift of the electrode output was less than 1 mV for one hour.

Temperature effect
As shown in Fig. 15, temperature was found to have a profound effect on the e.m.f.

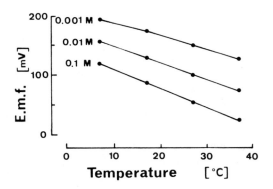

Fig. 15 — Effect of temperature on the e.m.f. of double-barrelled Cl⁻ ion-selective microelectrodes (concentrations refer to the standard KCl solutions used). From [9].

of the Cl^{2-} microelectrode. The vertical distance of two dots at a given temperature
indicates the size of the apparent slope constant for the relevant ion of that
microelectrode in the concentration range indicated. It is evident that the slope
constant of the microelectrode output increased with temperature. Values of the
slope constant of the microelectrode are shown in Table 2, where the n value (ratio of

Table 2 — Effect of temperature on double-barrelled Cl⁻ ion-selective microelectrodes. From [9]

Temperature (°C)	7	17	27	37
Nernstian slope [mV/p(X)]	55.5	57.5	59.5	61.5
Measured slope for Cl⁻ (mean ± standard deviation)	40.5 ± 4.3	45.8 ± 2.4	50.7 ± 1.6	55.0 ± 1.2
n	0.73	0.80	0.85	0.89
Reading of test solution (110 mM Cl⁻)	—	107 ± 5	112 ± 3	111 ± 2

the measured slope to the Nernstian slope), which represents a sub-ideal behaviour
of the Cl⁻-selective microelectrode with respect to the Nernstian response, is also
indicated. The n value increased almost linearly with temperature. Since the ion-

exchange properties are related to the oil-to-water partition coefficient of the ion and the association constant of the organic solute in the ion exchanger to the test ion, these factors are considered to be influenced by changes in temperature. The data suggest that so long as measurements are performed at the same temperature as in the calibrations or at a constant temperature, accurate results can be obtained.

pH effect
The effect of pH changes in the physiological range on the Cl^- microelectrode was negligible for a wide range of Cl^- concentrations (10–100 mM).

Protein effect
Bovine serum albumin (0–0.1 mM) exerted little effect on the e.m.f. independently of KCl concentrations, but 1 mM albumin produced an increasingly large deviation of the e.m.f. from that in the KCl standard solution with progressive dilution.

Effect of other substances
Fig. 16 shows the e.m.f. of Cl^--selective microelectrodes in response to various

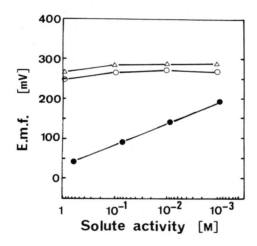

Fig. 16 — Effect of glucose or urea on double-barrelled Cl^- ion-selective microelectrodes. (●), e.m.f. in KCl solution; (○), e.m.f. in K^+-free glucose solutions; (△), e.m.f. in K^+-free urea solutions of different concentrations. From [9].

concentrations of standard KCl solutions, and to solutions of urea and glucose in redistilled water. Both urea and glucose exerted only a slight effect on the responses of the Cl^- exchanger microelectrode. This might be due to impurities in test materials (glucose or urea).

Spring & Kimura [59,60] have made a single-barrelled Cl^--selective microelectrode based on a chloride ion exchange resin (Orion), in which the reference filling solution was replaced by a chlorided silver wire inserted directly into the resin when the electrode was ready to be used. Its characteristics are presented in Table 3. The

Table 3 — Characteristics of the Orion Cl⁻-selective microelectrode

Selectivity coefficient	slope (mV/decade)	R (Ω)	response time (s)
$k_{Cl^-,HCO_3^-} = 0{,}1$ $k_{Cl^-,HPO_4^2} = 0{,}04$ $k_{Cl^-,SO_3^{2-}} = 0{,}2$	59	4×10^{10}	1

authors observed that sensitivity of the chloride ion exchanger to other anions introduced an artefact into measurements of intracellular Cl⁻ activity. A fraction of the apparent cell chloride activity was due to the electrode response to other intracellular anions.

Shindo & Spring [61] constructed, in 1981, single- and double-barrelled Cl⁻-selective microelectrodes also based on Orion Cl⁻ exchanger resin. The average slopes were 53.2 mV/decade and 47.0 mV/decade for the single- and the double-barrelled microelectrodes, respectively.

Cassola *et al.* [57] and Saito *et al.* [62] used the new more concentrated Corning 477913 Cl⁻ exchanger, in single- or double-barrelled microelectrodes. These electrodes were easier to make, and bevelling was not required as the resistance was in the order of 10^{10} Ω. The interference tests gave essentially the same results as those reported by Fujimoto & Kubota [9] ($k_{Cl^-,HCO_3^-} = 0.111$).

5.3 Experiments

The first experiments were carried out in large cells, such as the giant cell of the abdominal ganglion of *Aplysia californica* [54,55], or, extracellularly, in catfish cerebellum [63].

Bolton & Vaughan-Jones [56] studied chloride activity in frog Sartorius muscle, Macchia & Baumgarten [64] in toad semitendinosus muscle (*in vivo*), and Aickin & Vermue [65] in smooth muscle cells of the guinea pig. Several studies deal with chloride activity in renal proximal tubule of *Necturus* [59–61,66], of the bullfrog [9,58,67] or of the rat [57].

Measurements were also performed in epithelial cells of frog gastric mucosa [68], in *Necturus* gall-bladder [8], in salivary gland cells of *Chironomus* larvae [69], in mouse lacrimal acinar cells [62], and in frog skin epithelial cells [53,70,71].

5.4 Conclusion

Single- or double-barrelled chloride liquid ion-exchanger microelectrodes are available to measure steady-state chloride activity intracellularly. These electrodes have a slope quite near to the theoretical value, and the values of the resistance and of the response time are satisfactory.

The major disadvantage is that these electrodes are not very selective for Cl⁻ with respect to other anions present in the intracellular fluid such as bicarbonate and organic anions. Though the total sum of the organic anion concentrations within the cell cannot much exceed 0.1 M, the e.m.f. of the Cl⁻-selective microelectrode may possibly be affected in such a way that the anions react as if Cl^- concentration is larger than in reality. Judging from different studies, however, the size of such an anion effect on intracellular Cl^- estimation is presumed to be less than a few meq/l. Bicarbonate ion in the cell would also play a minor role in Cl^- reading, because its concentration is generally low (a few meq/l). Any overestimation of Cl^- due to intracellular HCO_3^- is likely to be less than 1 meq/l.

Despite many vulnerable points, the applicability of the Cl^--selective microelectrode to intracellular and steady-state or transient Cl^- activity measurements remains to be explored by technical improvements, since no other direct method for cell Cl^- determination has been developed so far.

6 Na$^+$-selective microelectrodes

Intracellular ionic activity measurements with Na^+-selective microelectrodes are important in many areas of biophysics and physiology. They are of great interest for the study of cell membrane transport of this ion, and of cellular functions related to the ion activities.

Because interpretation of membrane potential changes was often difficult or impossible, it was an old dream of electrophysiologists to construct ion-selective microelectrodes for measuring intracellular ion concentrations directly.

The first attempts were made with sealed-tip glass electrodes sensitive to Na^+, described by Hinke [72].

Then, in 1972, Thomas [73] reported a 'recessed-tip' Na^+-selective microelectrode having a diameter of only 1–2 μm, which allowed experiments lasting several hours to be undertaken. The major disadvantages of the recessed-tip Na^+-selective glass microelectrodes are the difficulty of the microelectrode construction and the slow response time [15].

Although selectivity of the glass microelectrodes for Na^+ over K^+ is better than that of the Na^+-selective microelectrodes made with available Na^+-selective liquid ion exchangers, these electrodes have several advantages and are quite reliable for intracellular Na^+ activity measurements.

Palmer & Civan reported in 1977 a Na^+-selective microelectrode based on a solution of potassium tetrachlorophenylborate as the sensor [69].

At about the same time, Kraig & Nicholson developed another Na^+-selective microelectrode based on a solution of the antibiotic monensin [74].

More recently, Steiner *et al.* [75] made a Na^+-selective microelectrode based on a new Na^+-sensitive neutral carrier (ETH 227) which allowed the measurement of Na^+ activity, free from K^+ interference over a range of activity ratios corresponding to those one would expect to encounter in cytoplasm.

6.1 Na$^+$-selective glass microelectrodes

6.1.1 Description

The first potentially usable electrodes were eventually made by the method of Hinke [72], the insulation and main body of the electrode being lead glass. The microelectrode tip is protruded outside the tip of the insulating glass pipette, and the response glass is directly in contact with a test solution [76].

Such electrodes gave excellent responses to Na^+ changes extracellularly, but for resistances less than $10^{11}\,\Omega$, at least $60\,\mu m$ of NAS 11-18 had to be exposed. Such electrodes proved very difficult to insert fully into the cell, because not only all the exposed NAS 11-18 but also the beginning of the insulation must be inside the cell. The cation-sensitive glass terminated in a spear tip [77] for easy insertion into the cell. The diameters of the tips varied from 5–$20\,\mu m$, and the lengths from 50–$200\,\mu m$. The response times to a ten-fold change in activity varied from a few seconds to about 2 min.

Then Thomas [52] describes a Na^+ recessed-tip microelectrode. In this design, the response glass is in contact with a test solution through a small open-tip of the insulating micropipette. The recessed-tip microelectrode was particularly designed for applications in relatively small cells [76].

With a tip diameter of about $1.0\,\mu m$, a complete response takes 1–2 min. This is slower than for previous designs, since the sodium ions have to pass through the tip. The cell membrane is penetrated only by the 1–$2\,\mu m$ diameter tip of the outer insulating glass. The resistance can be kept below $10^{11}\,\Omega$. A $100\,\mu m$ length Na^+-selective microelectrode has a resistance of about $10^{10}\,\Omega$ [73]. Electrodes are filled individually with 0.1 M NaCl, buffered to pH 8 with Tris-Cl. The volume of the dead space in the electrode tips was less than 10^{-9} ml.

6.1.2 Characteristics

6.1.2.1 Hinke-type electrode [78].
In mixtures of Na^+ and K^+, the Na^+ glass electrodes behave according to equation:

$$E_{Na^+}^\circ = E_{Na^+}^\circ + S_{Na^+}\log(a_{Na^+} + K_{Na^+,K^+}a_{K^+})$$

where $E_{Na^+}^\circ$ is the e.m.f. calculated for a given electrode from its response to a standard solution, a_{Na^+} and a_{K^+} are the ion activities, k_{Na^+,K^+} is the selectivity coefficient of Na over K, and S_{Na^+} is the e.m.f. for a unit log change in activity.

$E_{Na^+}^\circ$, k_{Na^+,K^+}, and S_{Na^+} are determined through calibration of the electrode.

At 18°C, S_{Na^+} is between 57 and 62 mV, and k_{Na^+,K^+} is consistently 0.005 or less. Dick & McLaughlin [77] give S_{Na^+} equal to 58 mV and selectivity coefficient k_{Na^+,K^+} varying from 0.01–0.02.

The activity coefficients, at 25°C, can be obtained from Robinson & Stokes [79].

The resistance of these microelectrodes is of the order of $10^9\,\Omega$. The electrodes have also been shown to be relatively insensitive to divalent cations and anions [80] as well as to ammonium, lysine, arginine, and albumin molecules [49]. It therefore seems reasonable to assume that none of the cytoplasmic constituents except the alkali metal cations affects the electrode readings significantly.

6.1.2.2 Inverted-type microelectrode
The response of an inverted-type microelectrode [81] with a tip diameter of 10–$20\,\mu m$ and an inverted section length of $c.$ $100\,\mu m$ is at least 50 mV per ten-fold change in Na^+ concentration in constant ionic strength solutions between 10–100 meq/l of Na^+. The slope is reduced at low Na^+ concentrations (below 10 meq/l) owing to competition with K^+.

Response time is of 1–3 min to reach 90% of the final value after a change of Na^+ concentration. This slow response time is probably due to the time taken by the fluid in the tip to equilibrate by diffusion with the calibration solution.

Resistances are of the order of 10^{11}–10^{12} Ω.

No value of the selectivity coefficient $k_{Na^+}^{}K^+$ has been found in the literature.

6.1.2.3   Recessed-tip microelectrodes. Little has been published about the characteristics of these electrodes, probably because they are similar to those described above.

Ellis [82] used Na^+ recessed-tip microelectrodes with tip diameters of about $0.4\ \mu$m. A ten-fold change in Na^+ concentration at constant ionic strength produced a Na^+-selective electrode response between 54 and 61 mV. The response of the electrodes was normally 90% complete within 90 s of a solution change.

Electrodes were accepted if they exhibited a selectivity of Na^+ over K^+ better than 0.014 [83]. The Na^+-selective microelectrodes must be carefully calibrated, since the selectivity of the microelectrodes depends on the ionic concentration [76]: it decreases with decreasing Na^+ concentrations (Fig. 17).

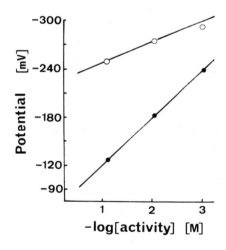

Fig. 17 — The relation between Na^+-selective glass microelectrode potentials and activities of Na and K ions. The potentials were measured in 1, 10, and 100 mM NaCl (●) and KCl (○) solutions. From [76].

Lee studied in detail the behaviour of these recessed-tip microelectrodes during ageing [76]. The results indicate that they should be used as soon after filling as possible for intracellular application.

Lee also compared the characteristics of Hinke-type electrodes and recessed-tip microelectrodes: Hinke-type electrodes have lower resistance, greater selectivity coefficients, and greater response time than the recessed-tip microelectrodes. He observed that microelectrodes with high resistance and short exposed-tip length had a lower specificity for Na^+ over K^+ than those with low resistance and long exposed-tip length.

The slow response time may be due to the recess space occupied by the fluid, which must be equilibrated with the test solution through a small tip of the insulating pipette. However, the microelectrodes with a long exposed-tip length have a slow response time since they have a long recess space. Lee concluded that the recessed-tip Na^+-selective microelectrodes require an optimum length of exposed tip for high sensitivity, low resistance, and fast response time. For intracellular application, the optimum lengths of the exposed tip may be between 40 and 80 μm.

The influence of various concentrations of divalent cations (i.e. Ca^{2+}, Ba^{2+}, Sr^{2+}, Mg^{2+}, Mn^{2+}) on the response of the electrodes was very small.

6.1.3 Experiments

Hinke [84] used the Na^+ glass microelectrode described above to measure Na^+ activities in the isolated giant axon of the squid *Loligo forbesi*, bathed in sea water or artificial sea water. The results obtained with the microelectrode were in good agreement with those obtained by flame photometry. Using the coefficient activity, the percentage of free intracellular sodium has been determined.

The same electrode was used to measure Na^+ activity into muscle cells from the propodite of crab and lobster, which have a 200–500 μm diameter [72].

The sodium activity in the cytoplasm of toad oocytes was also measured with these electrodes [77].

All these experiments have been performed in a chamber containing Ringer solutions.

Inverted-type glass microelectrodes have been used to study intracellular sodium changes in a large snail neurone (200 μm in diameter) during extrusion of injected sodium [78]. The preparation was placed in an experimental bath and the temperature was about 20–23°C. Others authors [81] describe large transient changes of Na^+ recorded with an inverted-type electrode in the cochlear duct endolymph of guinea pigs. These experiments were performed *in vivo*, except when prolonged anoxia brought about the death of the animal.

Thomas [73] used a recessed-tip microelectrode to study intracellular sodium activity and the sodium pump in snail neurones. The effective tip (the part that has to penetrate the cell) was only 1–2 μm in diameter, and it was possible to undertake experiments lasting several hours. The author determined the normal Na^+ concentration in snail neurones, and then investigated the effect on Na^+ concentration of Na pump inhibition by ouabain, of changes in external K^+ and Na^+ concentrations, and of an increase in membrane potential. These experiments were carried out in an experimental perfused chamber.

In another study [82], recessed-tip Na^+-selective microelectrodes were used to measure the internal activity of sheep cardiac Purkinje fibres. The effects of various factors that alter the internal Na^+ activity have been investigated, as changes in the external concentrations of Na^+, Ca^{2+}, K^+, and the effect of the cardiac glycoside ouabain, an inhibitor of the Na pump.

The effects of various external divalent cations on the intracellular Na^+ activity of sheep heart Purkinje fibres were studied [85].

An attempt has also been made [2] to determine the apparent activity coefficient of sodium, by measuring Na^+ activity in the papillary muscle cells of rabbit heart with glass microelectrodes. The total sodium concentration was determined with flame photometry, with sulphate space as a measure of the extracellular compartment.

More recently, these recessed-tip microelectrodes were used to study the K^+-activated Na pump current in sheep Purkinje fibres [86].

Measurements of intracellular sodium activities and of transepithelial transport were also carried out using these microelectrodes in cells of isolated gallbladders of *Necturus maculosus* under conditions of reduced and enhanced sodium entry across the luminal cell membrane [83].

6.2 Na^+-selective liquid ion exchanger microelectrodes

To alleviate the difficulties of Na glass microelectrode, Na^+-selective liquid ion exchanger microelectrodes have been made.

6.2.1 Monensin

Kraig & Nicholson [74] made a Na^+-selective ion exchanger microelectrode, which incorporates a 10% solution of monensin in nitrobenzene as the ion exchanger. Monensin is a biologically active compound that preferentially binds to Na^+ and has been proposed for use in macroelectrodes. The Na^+-selective microelectrode and a reference micropipette are glued with a tip spacing of less than 10 μm and connected.

Garcia-Diaz *et al.* [87] used as the sensing element a 10% solution of monensin in 3-nitro-*o*-xylene containing a small amount (about 1.0%) of potassium tetrachlorophenylborate in the ion exchanger solution, which significantly shortened the response time of the electrodes [8]. In 1980, Fujimoto & Honda [88] employed the same ionophore, but dissolved in Corning 477317, in a triple-barrelled microelectrode for simultaneous measurements of intracellular Na^+ and K^+ activities and membrane potential. The tips were made hydrophobic with a diluted solution in trichlorethylene. The internal solution was 0.49 M NaCl plus 0.01 M KCl. The pH of the internal solution was adjusted with 0.1 M citrate buffer to lower the pH sensitivity.

The characteristics of these monensin-based microelectrodes are summarized in Table 4. The linear regression line of the potential vs concentration curve, obtained at 25°C for Na^+ concentrations ranging from 25–250 mM, had a slope varying between 58.9 [74] and 56.0 mV [88] per decade change in Na^+. Fujimoto & Honda [88] note that the calibration curve of the Na^+-barrel exhibits a non-linear behaviour in the concentration range of Na^+ lower than 10 meq/l, representing a detection limit of Na^+ at approximately 3 meq/l.

The electrode was not sensitive to protons from pH 5 to 9. But a possible source of error is the interference of Mg^{2+}, since the selectivity coefficient is rather low ($k_{Na^+,Mg^{2+}} = 0.142$) [74]. This is a possible limitation for the monensin-based Na^+-selective microelectrode, especially when used for intracellular measurements.

To demonstrate the utility of the electrode in biological systems, Kraig & Nicholson [74] recorded changes in extracellular sodium level during spreading depression in the catfish cerebellum. It was the first direct evidence of a large Na^+

Table 4 — Characteristics of the monensin-based Na^+ microelectrodes

∅ (μm)	Slope at 25°C (mV/decade)	k_{Na^+,K^+}	Other selectivity coefficients	R (Ω)	Refs.
1–2	58.9	0.066 (0.1M) 0.077 (1.0M)	$k_{Na^+,Ca^{2+}}$ = 0.007 k_{Na^+,NH_4^+} = 0.0065 $k_{Na^+,Mg^{2+}}$ = 0.142 k_{Na^+,H^+} = 0.071	10^{10}	[74]†
<1	56.0 ± 11.0	0.167 (0.025–0.01 M) 0.08 (*c*. 0.1 M)		10^{10}	[88]‡

† Single-barrelled electrode
‡ Triple-barrelled electrode

shift in the extracellular space during spreading depression, provided by this Na^+-selective microelectrode.

Fujimoto & Honda [88] determined, *in vitro*, the Na^+ activity and the Na selectivity coefficient in biological samples (human serum and frog serum) with triple-barrelled Na^+- and K^+-selective microelectrodes, and flame photometry. These authors also measured Na^+ activity in proximal tubules and in Sartorius muscle of male bullfrogs, *in vivo*.

6.2.2   Tetraphenylborate
In 1977, Palmer & Civan [69] described a new liquid ion exchange microelectrode sensitive to Na^+. This microelectrode was made from a micropipette, as described above. The Na^+ exchanger was obtained by dissolving potassium tetrachlorophenyl-borate in triethylhexylphosphate to make a 1.5% solution. Kimura & Spring [89] used the same sensor, but potassium tetrakis-(*p*-chlorophenyl)-borate was dissolved in triethylhexylphosphate (5%). The shank was then filled with 0.5 M NaCl.

The latter authors insert a chlorided silver wire directly into the ion exchange resin instead of putting an electrolyte solution above the exchanger [60]. Electrode manufacture is greatly simplified by this approach, and large numbers of electrodes may be rapidly produced.

The electrodes were calibrated in NaCl solutions with or without KCl. The sodium activity was calculated taking into account the interference from the intracellular K^+, with k_{Na^+,K^+} of 0.38 for Palmer & Civan [69] and 0.28 to 0.5 for Kimura & Spring [89]. The slopes ranged from 61–66 mV [69] or 57.2–58.0 mV per decade [89].

Electrode resistances were about 10^{10} Ω.

These electrodes were used to study the Na^+ subcellular compartmentalization, especially the nucleus of *Chironomus* salivary gland cells placed in an experimental chamber [69]. Kimura & Spring studied the dependence of intracellular Na^+ activity on the electrical driving force across the luminal membrane of *Necturus* proximal tubule, placed in a chamber and perfused [60,89].

6.2.3 Neutral ligand ETH 227

Steiner *et al.* [75] first described a Na$^+$-selective microelectrode, based on synthetic electrically-neutral ion carriers. Among the Na$^+$-selective components suitable for liquid membrane electrodes, the neutral carriers 1 and 2 (ETH 227 and ETH 237, respectively) (Fig. 18) are particularly interesting. In Table 5, selectivities of micro-

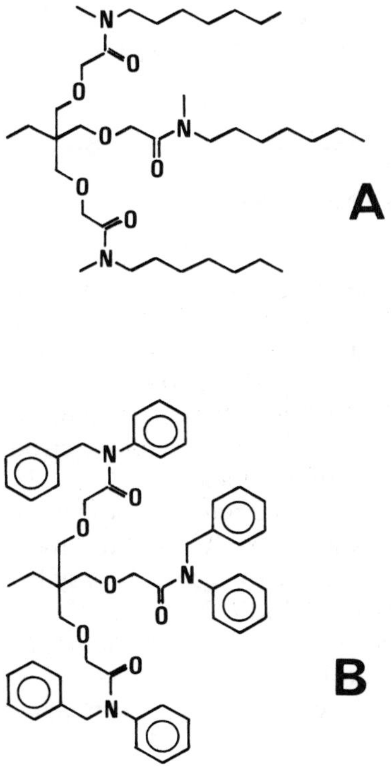

Fig. 18 — Na$^+$-sensitive neutral carriers ETH 227 (A) and ETH 237 (B). From [75].

electrodes based on these ligands are reported. Ligand 1 is shown to exhibit higher selectivities than ligand 2, except for Li$^+$. So only ligand 1, consisting of 1,1,1-tris[1-(2-oxa 4-oxo 5-aza 5-methyl)-dodecanyl]propane, has been retained and studied in greater detail.

To reduce the electrical resistance (from 2×10^{11}–1×10^{10} Ω) [5] and the slow response time of the microelectrode (5–10 min) [90], sodium tetraphenylborate was added to the ion-selective liquid, as described earlier [38] (Tetra-*p*-chlorophenylbor-ate can also be used [90,91]: see monensin-based microelectrode described above). As expected, and as shown in Table 5, there is some loss in selectivity with respect to different cations. Since the pivotal Na$^+$/K$^+$ selectivity is not affected, and simulta-

Table 5 — Selectivity factors $\log(k_{Na^+,X})$ for different Na^+ microelectrodes. From [75]

electrode based on	Li^+	K^+	Rb^+	Ca^+	Mg^{2+}	Ca^{2+}	Sr^{2+}	Ba^{2+}	H^+	acetyl-choline
				$\log k_{Na^+,X}$ for X =						
ligand 1 (ETH 227) without TPB$^-$†	0.4	−1.7	−2.0	−2.2	−2.7	−0.7	−1.1	−1.3	−0.9	
ligand 1 (ETH 227) with TPB$^-$	0.4	−1.7	−1.6	−1.8	−2.4	0.2	−0.2	−0.4	−0.8	−1.8
ligand 2 (ETH 237) without TPB$^-$	0.2	−1.1	−1.4	−1.6	−1.9	−0.1	−1.0	−0.9	−0.1	
Na-glass‡ (NAS$_{11-18}$)	−3.0	−3.0	−4.3	−3.0	−4.3				−3.0	

† tetraphenylborate
‡ Typical values for a Na^+-glass macroelectrode.

neously a reduction in resistance is obtained by adding sodium tetraphenylborate, this type of electrode is the most attractive for practical applications.

The most commonly used liquid ion exchanger was a 10% solution of the neutral ligand ETH 227 in 3-nitro-*o*-xylene or *o*-nitrophenyloctylether containing a small amount (about 1%) of tetrachlorophenylborate or tetraphenylborate [90,92–94].

Some authors used single-barrelled microelectrodes [15,90,94–97], but most of them used double-barrelled microelectrodes [71,75,92–94,98–100], as they succeeded in building this type of electrode with sufficiently fine tips. Kajino & Fujimoto used a double-barrelled microelectrode, twisted 180° [100].

The tip diameters of the microelectrodes vary from 0.15 μm [93] to 2 μm [75]. Most of them are less than 1 μm. Na^+-selective microelectrodes with fine tips (*c.* 0.15 μm) are constructed from Pyrex theta tubings (the method of construction has been described in Ref. [93]).

The characteristics of the electrodes are summarized in Table 6.

Table 6 — Electrical parameters of Na^+-sensitive theta-microelectrodes with tips 0.3 and 0.15 μm in diameter. Resistances (R), tip potentials (T.P.), changes in potential difference/decade (ΔV/dec.) are expressed as GΩ and mV, respectively. Conventional (reference) channels (Conv.) of Na^+ theta-microelectrodes were filled with 154 and 500 mM KCl saline: this accounts for the discrepancies in resistances in spite of same tip size. Also indicated is the yield (i.e. the percentage of correctly functioning electrodes out of the total production). From [75]

Channel	Tip size: 0.3 μm				Tip size: 0.15 μm			
	R	T.P	DV/dec	Yield	R	T.P.	ΔV/dec	Yield
Na^+	58 ± 11		−51.3 ± 0.6		84 ± 10		−50.3 ± 1.0	
Conv. (500 mM KCl)	0.2 ± 0.02	2.3 ± 0.4	0.8 ± 2	50%	0.4 ± 0.03	1.3 ± 0.9	−1.3 ± 0.3	38%

The selectivity coefficient of Na^+ over K^+ varies with the conditions used to determine it. Moreover, k_{Na^+,K^+} is dependent upon Na^+ and K^+ concentrations. So it is preferable to calibrate the electrodes under conditions approximating those to be expected in the cell interior, that is, a K^+ concentration of about 140 mM (a_{K^+} = 108 mM).

The sensitivity to Ca^{2+} is about 1.6 times higher than to Na^+ on a molar basis, except for two studies in which the electrodes exhibit a selectivity coefficient of 0.19 for 10 mmol/l of Ca^{2+} [71,98].

According to the data on the intracellular Ca^{2+} assessed by Ca^{2+}-selective microelectrodes, Ca^{2+} in the cell is in the order of 10^{-6}–10^{-7} M, a level 10^4–10^5 times lower than intracellular sodium activity [100]. Therefore, low intracellular Ca^{2+} would not induce a critical error in the estimation of intracellular Na^+ activities. But it is also preferable that calibration solutions contain a Ca^{2+} concentration equivalent to intracellular free Ca^{2+} [15].

The neutral carrier systems discriminate H^+ with respect to Na^+, as the selectivity coefficient of Na^+ over H^+ is about 0.1. Changes in pH of the sample solution in the range of pH 2–pH 10 do not heavily influence the e.m.f. if the concentration of Na^+ is kept low [75]. Therefore, pH changes over the physiological range will not interfere with the Na^+ determination.

Microelectrode resistances reported are high: 10^9–10^{11} Ω. Lewis & Wills [97] have identified an electrical artefact in certain liquid ion-selective microelectrodes. This artefact arises from the high electrical resistance of the ion-selective resin, which in some cases is comparable to the resistance of the microelectrode glass wall. This situation led to shunting of the exchanger potential and consequently artefactually high calculations of intracellular Na^+.

The response times vary from 0.2 s [71] to about 30 s [92] according to the authors. To improve the response speed, a silver shield can be painted on the shaft of the electrode (before baking) to which the output of the electrometer is connected, and, if necessary, the tip can be gradually bevelled in a spinning suspension of Al_2O_3 powder [99,101]. In addition, such shielded Na^+-selective microelectrodes, where any current loop between the tip and the glass wall is eliminated, show a better k_{Na^+,K^+} selectivity coefficient [97] and reduce the calculated intracellular Na^+ activity by approximately three-fold [101].

Dagostino & Lee [15] dealt with the stability of Na^+-selective microelectrodes. They showed that the slope, the selectivity coefficient, and the resistance are quite stable for a period of about 120 h ageing after construction. The microelectrode was kept in air or in 100 mM NaCl solution.

These electrodes have several applications:

In 1979, O'Doherty *et al.* [90] successfully impaled epithelial cells of *Necturus* small intestine with single-barrelled Na^+-selective microelectrodes.

Glitsch & Rasch [95] used these Na^+-selective microelectrodes to study the increase of the intracellular sodium activity during depolarizing action of noradrenaline on the cell membrane of quiescent sheep Purkinje fibres. Glitsch *et al.* [96] also measured the intracellular Na^+ activity in sheep cardiac Purkinje fibres to determine the ionic components of the pacemaker current. Daying & Achenbach [102] studied the sodium induced by activity, compared with the level of resting intracellular sodium in Purkinje fibres. Eisner *et al.* [103] studied the control of tonic tension activity in these fibres. The paper of Glitsch & Pusch [104] deals with the temperature dependence of the cardiac active Na^+ transport, still in sheep Purkinje fibres. All these experiments were performed in a perfused chamber.

Kajino & Fujimoto [100] used a double-barrelled Na^+-selective microelectrode to measure *in vivo* intracellular sodium activity in renal proximal tubules and Sartorius muscle of bullfrogs. Yoshitomi & Fromter [99] measured the intracellular Na^+ concentration of surface loops of proximal tubules in micropuncture experiments on rat kidney *in situ* and *in vivo*.

Na^+-selective theta microelectrodes with fine tips allowed Meyer *et al.*[93] to measure Na^+ activity of the epithelial cells of rabbit gall-bladder (cell diameter 5–10 μm).

Lewis & Wills [97,101] measured *in vitro* the Na^+ activity in the rabbit urinary bladder epithelium.

Experiments were also performed on isolated abdominal frog skin, at steady-state, and after exposure to amiloride [70]. Harvey & Kernan [71] measured with double-barrelled microelectrodes intracellular sodium activities in abdominal skin and isolated epithelia mounted in a chamber; the effect of amiloride was also studied.

6.3 Discussion and conclusion

The type of microelectrode system for measuring Na^+ activity and developed by Hinke, using NAS 11-18 glass, is characterized by a good selectivity for Na^+ over K^+. The reported values of the selectivity coefficient k_{Na^+,K^+} are considerably better than those of liquid ion-exchanger electrodes. The major disadvantages of these glass electrodes are difficulty in the microelectrode construction, high electrical resistivity, and slow response. The blocking of the microelectrode tip is also a problem, particularly in cases of impalement for a long period.

For the Hinke-type microelectrode, an appreciable length of Na^+-selective glass has to be incorporated into the cell. A length of 50–200 μm is commonly used, although it has been successfully reduced to 10–20 μm. Care must be taken that the entire length of the ion-selective glass is inserted into the cell. This is not difficult when large cells are used (squid axon, barnacle muscle, or toad oocyte); the problem is more serious with smaller cells.

Thomas designed an electrode made from the same NAS 11-18 glass, in which the ion-selective tip is recessed behind the insulating glass tip. This obviates the problem of positioning a finite tip length into the cell, but introduces a dead space between the insulating and ion-selective glass which must equilibrate with the contents of the cell. This again is not a problem with cells which are much larger than the dead space (approximately 0.24×10^{-12} l), so long as fast electrode response is not necessary; but smaller cells can have volumes comparable to the dead space, and the introduction of the electrode could significantly perturb their ionic composition.

An important drawback is that Na^+ changes of the order of some seconds cannot be recorded with a Na^+-selective glass electrode which requires minutes to respond. Only large transient changes can be recorded with these Na^+ glass electrodes. More conclusive than steady-state measurements are analyses of concentration transients. In small cells, such measurements require that the response time of the microelectrodes be sufficiently fast (of the order of 0.2–1 s), which is not easy to achieve with fine-tip, high-resistance electrodes, but can be accomplished if shielded electrodes with negative-feedback circuits are used.

Though some of these problems can, in principle, be overcome by advanced techniques of glass electrode miniaturization, the liquid ion exchanger electrodes are relatively simple to make, and are much more consistent and reliable. All the difficulties mentioned above have been alleviated by constructing a Na^+-selective liquid ion exchanger microelectrode.

Electrodes, based on monensin or potassium tetrachlorophenylborate, and then on the neutral ligand ETH 227, have been constructed. But it is obvious that the neutral carrier electrode shows a detection limit which is by about an order of magnitude lower than that of the monensin-based microelectrode. An analysis of the Nicolsky equation indicates that the detection limit of the monensin-based sensor is given mainly by the interference of K^+ and Mg^{2+} whereas the neutral carrier electrode suffers almost exclusively from K^+ interference.

In contrast to the monensin-based electrode, where the detection limit is just about the physiological Na^+ activity (10 mM), the neutral carrier electrode does make intracellular Na^+ activities accessible.

Whereas the classical Na^+-selective glass electrodes exhibit a preference for H^+ over Na^+ by a factor of about 10^3, the neutral carrier systems discriminate H^+ with respect to Na^+. Therefore, pH changes over the physiological range will not interfere with Na^+ determination.

It is apparent that the reported k_{Na^+,K^+} values vary from one investigator to another and differ somewhat between individual microelectrodes. Although the reasons for such variations are not completely clear, there are a few possible explanations. Firstly, poor silanization of the glass surface has been suggested. Secondly, an electrical shunt through the glass wall may affect ion-selective microelectrode potentials. Thirdly, poor contact (phase boundary) between test solution and ion-selective resin at the end of the tip is also possible. Dagostino & Lee [15] discuss each of these possibilities, without finding a satisfactory solution to the problem.

Single-barrelled ion-selective microelectrodes were the first to be constructed. Then, authors succeeded building double-barrelled Na^+-selective microelectrodes with tips fine enough to be used in small cells. Such double-barrelled electrodes allow the electrochemical potential to be measured simultaneously in the small cell, which is of great advantage if concentration transients are to be analysed. A triple-barrelled Na^+-, K^+-selective microelectrode has been constructed to measure the intracellular Na^+ and K^+ activities of a single cell, and its membrane potential. There are, however, a number of problems in the use of the multi-barrelled electrodes (tip size, mechanical strength, electrical resistance . . .). The triple-barrelled microelectrode capable of measuring not only the test ion (Na^+) but also one of the major interfering ions (K^+) can partly overcome the problem of the poor selectivity of the electrode for the test ion over the other interfering ions.

Until quite recently, the determination of Na^+ activity was restricted to large cells, and experiments were limited to steady-state measurements. During the 1980s, the use of the neutral ligand ETH 227 and the improvement of miniaturizing techniques have allowed the construction of Na^+-selective microelectrodes which are quite reliable, even though Mg^{2+}, Ca^{2+} and especially K^+ interferences must still not be neglected. These Na$^+$-selective microelectrodes are very useful tools for physiologists who could so improve their knowledge of Na$^+$ activity in cells and essentially of Na$^+$ transport across cell membranes, linked with the transport of other ions.

7 K$^+$-selective microelectrodes

The accurate determination of potassium for both clinical and research purposes is of recognized importance. Cation-specific glass microelectrodes have been employed for measuring K$^+$ concentrations and have provided valuable data relevant for electrophysiology of excitable nerve and muscle tissue. In the 1970s, a method was developed for measuring the K$^+$ activity, based on the liquid ion-exchanger principle. In contrast to ion-selective glass membranes where the selectivity coefficient for K$^+$ over Na$^+$ is high, some liquid ion exchangers exhibit a much better selectivity. This achievement is very advantageous for measurements in living systems, where both ions are often present together in concentrations of the same order of magnitude. Many designs of K$^+$-selective microelectrodes have been described : single-, double-, or even triple-barrelled microelectrodes, side-pore microelectrodes, coaxial microelectrodes . . .

Many of the potassium microelectrodes for investigative use have been made from tetra (*p*-chlorophenyl)borate; some others have been based on valinomycin, or on a bis (crown ether)-type ligand.

The development of these liquid ion-exchanger microelectrodes with fine tip has allowed intracellular K$^+$ activities to be measured in very small cells.

7.1 K$^+$-selective glass microelectrodes

In 1961, Hinke described [84] a method to obtain K$^+$-selective glass microelectrodes, based on NAS (Na$_2$O–Al$_2$O$_3$–SiO$_2$) 27-4 (Corning), with an exposed tip of 70–90 μm in diameter and 2–4 mm in length.

Later on, Dick & McLaughlin described [77] a K$^+$-selective glass microelectrode with tip diameters and lengths varying from 5–20 μm and 50–200 μm, respectively. The electrodes consisted of an outer tapering stock of insulating lead glass, with a protruding cation-selective glass terminating in a spear tip, for easy insertion into the cell.

Zeuthen *et al.* [11] used K$^+$-selective microelectrodes of the all-glass construction devised by Thomas [52,73]: the electrodes were of a recessed-tip design.

More recently, Lee & Fozzard [2] reported inner sealed micropipettes made from NAS 27-4 glass with tip diameters of less than 1 μm. The exposed tip lengths were less than 3 μm. The tip diameters of outer insulating micropipettes were about 1 μm.

The calibration of the K$^+$-selective microelectrodes in KCl–NaCl mixtures and the determination of the selectivity coefficient k_{K^+,Na^+} (0.1–0.5) showed that the microelectrodes were satisfactory only for solutions where the K$^+$ concentration was high or the Na$^+$ concentration was low [84].

The resistance of the microelectrodes was greater than 10^9 Ω immediately after they were made. This high resistance could be reduced to the order of 10^7 or 10^8 Ω by ageing the microelectrodes in 3 M KCl for a few days or more before insulation, which is also accompanied by a change in the selectivity coefficient [2]. The decrease in resistance was interpreted to indicate that the thin glass membrane at the microelectrode tip was completely hydrated and its resistance was much lower than that of the glass membrane of the microelectrode stem. This suggests that after sufficient ageing, the K^+-selective microelectrodes can be used without insulation, since with the low resistance microelectrodes, the potentials did not depend on the length immersed into a given test solution. The microelectrodes without insulation are restricted to extracellular applications; for intracellular use, insulation may still be necessary [76].

K^+-selective microelectrodes responded to some extent to H^+ activity [77,84]. They have also been shown to be relatively insensitive to divalent cations and anions, as well as to ammonium, lysine, arginine and albumin molecules [77].

7.2 K^+-selective microelectrodes based on ionized K^+ exchanger

To overcome the problem of the low selectivity for K^+ over Na^+, K^+-selective microelectrodes based on liquid ion exchangers have been developed. Moreover, these electrodes are easier to make, and very fine tip microelectrodes can be obtained.

The commercially available exchanger (Corning 477317) employs 1,2- dimethyl-3-nitrobenzene to dissolve potassium tetrakis(p-chlorophenyl) borate. Other solvents such as dioctylphtalate [91] or 3-nitro-o-xylene [89,105], were sometimes employed.

7.2.1 Single-barrelled microelectrodes

7.2.1.1 Classical single-barrelled microelectrodes. Walker first described [3] the miniaturization of ion-selective liquid ion-exchanger electrodes, for the express purpose of measuring intracellular ionic activities in living cells. Following this, many other authors described in detail how to make these electrodes [1,89,106–109].

To siliconize the inner wall of the micropipette, several methods, based on tri-*n*-butylchlorosilane in 1-chloronaphtalene [1,3,110], on dichlorodimethylsilane vapour or in CCl_4 [10,69,106,111,112], on trimethylchlorosilane [71,89,105,113], on a siliclad solution [11,108], or on silicone solution [8,10,109], have been used.

Kimura & Spring [89] used capillary tubing with an internal glass fibre, which readily conducted the ion-exchanger resin down to the tip.

After a small 200–500 μm column of K^+-exchanger resin had been introduced, the microelectrodes were generally filled with 0.5 M KCl as the internal filling solution.

Vyskocil & Kriz [1] attempted to find the most suitable shape parameters of glass microcapillaries for making K^+-selective microelectrodes. By setting the different parameters on the puller, several types of micropipette could be obtained which differ in shape (Fig. 19). They found that not all shapes are equally suitable for preparing K^+-selective microelectrodes. Types 1 and 2, depicted in Fig. 19, proved to be the most advantageous.

To measure K^+ activity in very small cells (10 μm), tip diameters were less than 1 μm; even tip diameters of 0.2 μm are reported [107,114,115]. Some microelectrodes with 1.6 μm diameters are described [10,63,106,108, 116], but they are used to measure extracellular K^+ activity.

The sensitivity of the electrode, as indicated by its slope in pure KCl solution, is most often excellent (Table 7).

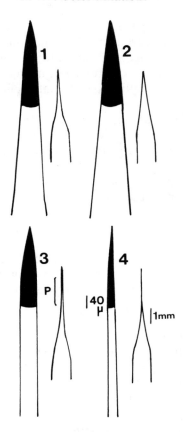

Fig. 19 — Different types of microelectrode shank and tip. Nos 1 and 2 are convenient for filling by exchanger and KCl solution, Nos 3 and 4 are not convenient. P: part of electrode shank with parallel walls. From [1].

Electrodes were generally calibrated in 0.001–1 M KCl and/or NaCl solutions, or in solutions of different concentrations of KCl with a standard background of 0.15 M NaCl [1]. Calibration curves of 6 different K^+-selective microelectrodes are shown in Fig. 20. Calibration curves were reproducible with a slope of 52 to 55 mV in the range above 10 mM KCl and a lower sensitivity for low concentrations of KCl because of interference by Na^+ [116].

The selectivity coefficients were determined either with the separate solution method (SSM) or with the fixed ion method (FIM) (Table 7). It was observed that the selectivity coefficient k_{K^+,Na^+} was concentration dependent [107]. Most of the values are in the 0.01–0.02 range, but higher values, up to 0.07, are reported. It can be seen in Table 7 that those microelectrodes which gave high k_{K^+,Na^+} values had small tip diameters. Kimura & Spring [89] noted that the selectivity deteriorated after the manufacture of electrodes, so they had to be used immediately.

The selectivities of the electrodes for other interfering ions with respect to K^+ have been determined for some interfering ions commonly found in biological

Table 7 — Characteristics of single-barrelled microelectrodes calibrated in pure KCl
solutions([a]) or in mixed solutions([b])

Slope (mV/decade)	k_{Na^+, Na^+}	ϕ (μm)	Refs.
058.5 ± 1.6[a]	0.02–0.066	0.2	[107]
53.0 ± 1.4[b]	id.	id.	[107]
59.5[a] (27°C)	0.016–0.07	0.2	[114,115]
58.9 ± 0.4[a] (37°C)	0.042	< 1	[111]
55 ± 1[a]	0.0114–0.0135		[105]
53.5 ± 1.5[b]	0.007	2.3	[116]
57.1 ± 0.4[b]	0.02	< 1	[89]
63.2 ± 2.6[b]	0.02–0.05	< 1	[8]
57.5 ± 2.5	0.015	< 1	[71]
57.7 ± 1.0[b]	0.017–0.014	1.8–2.0	[109]

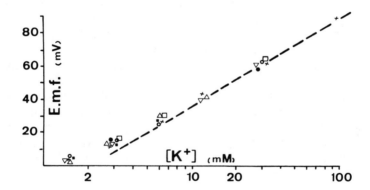

Fig. 20 — Calibration curves of 6 differents K^+-selective electrodes obtained in a series of
solutions containing 150 mM NaCl plus various concentrations of KCl. Slope is 53 mV/decade
for $[K^+]$ above 10 mM. From [1].

preparations [3,8,69]. The K^+-selective microelectrodes responded to tetraethylammonium chloride and other substituted amines (e.g. acetylcholine) [106] or quaternary ammonium compounds [63,111]. Palmer & Civan [69] determined a selectivity
coefficient for choline of 110: as little as 0.12 mM choline could create a 10% error in
the K^+ measurements. Choline, or other quaternary ammonium ions, are the only
ions which could interfere appreciably.

A change in pH from 6 to 8.5 did not produce a detectable voltage change
[69,116].

Other ions such as NH_4^+, Ca^{2+}, and Mg^{2+} do not contribute to the potential at all, because either their concentration is too small or the exchanger is not sensitive enough [107].

The maximum alterations in K^+ activity produced by temperature variations between 20 and 37°C corresponded to a limiting error equivalent to a 0.55 mM change in K^+ concentration [116]. Edelman *et al.* [107] have shown that calibration curves obtained at room temperature may be used for intracellular K^+ determinations at body temperature without introducing noticeable errors.

The resistance of the K^+-selective microelectrodes was about $10^9\,\Omega$, and even $10^8\,\Omega$ [10], which is a fairly low resistance for an ion- selective microelectrode [63]. When small tip microelectrodes are constructed, the resistance is approximately one order of magnitude higher ($10^{10}\,\Omega$), since the overall resistance of the microelectrode depends on two parameters: the inner diameter of the electrode tip and the length of the ion-exchanger column. Electrodes with extremely fine tips are not likely to function properly, because the resistivity of the ion-exchanger material is very high. So, to work with the smallest possible tips and the smallest possible tip resistance, authors sometimes prefer to work with single-barrelled K^+-selective microelectrodes instead of double-barrelled ones [107].

The time constant to ion change was less than 100 ms [63]. Variations in K^+ concentrations with a time course of 20 ms [10] or even 5–10 ms [106] were faithfully reproduced by the K^+-selective microelectrode. To obtain faster responses, electrodes with wider tips ($>3\,\mu$m) should be used.

The life span of the electrode was more than 1–2 weeks [1,10].

Single-barrelled K^+-selective microelectrodes have been used extracellularly to measure K^+ levels in *Aplysia* neurons [3], in exposed single neurons of snail [106], in cat [10,116,117], rat [11], and rabbit [118] cortex, in catfish cerebellum [63], and in guinea pig cochlea [108]. Intracellularly, these electrodes have been used in frog heart [119], in salivary glands of *Chironomus* [69], and of cat and dog [109], in proximal tubular cells of *Necturus* [89], and of rat [107], in gall-bladder epithelial cells of *Necturus* [8,105], of rabbit [114,115] and of guinea pig [115], in epithelial cells of rabbit gastric glands [111], and in abdominal skin and isolated epithelia of frog [71].

7.2.1.2   Side-pore microelectrodes. A different type of K^+-selective microelectrode has been developed for measurements of potassium levels in muscle tissue during contraction. The aim was to obtain a sensitive element, which does not disrupt muscle fibres, and, at the same time, is not itself mechanically damaged during intensive contraction.

The procedure for preparing this extracellular electrode is described by Vyskocil & Kriz [1] (Fig. 21). A side-pore channel of 3–5 μm was obtained, the tip of this electrode being 50–100 μm. The distal part of the channel was made hydrophobic and filled by the K^+-selective ion-exchanger, then the rest of the electrode was provided with a 0.5 M KCl solution. The resistance of the side-pore electrodes was found to be 2.5–$5.0 \times 10^8\,\Omega$. The electrode exhibits highly satisfactory specificity for potassium [1,120]. These electrodes have been used to measure changes in extracellular potassium concentration in rabbit and cat gastrocnemius muscles and in venous blood flowing from the cat gastrocnemius muscle [120]; K^+-concentration changes in

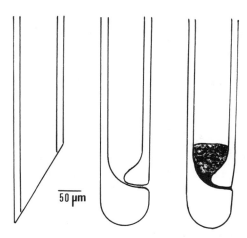

Fig. 21 — The procedure during preparation of a side-pore type electrode. From [1].

human muscles during volitional contractions have also been described [110].

7.2.1.3   Micro ion-selective field effect transistor. Haemmerli & Janata [113] described the construction of a potassium-selective micro ion-selective field effect transistor (K^+ μ-ISFET) and its characteristics. This device consists of a K^+-selective microelectrode (tip diameter < 1 μm) mounted directly on the gate of a pH-selective field effect transistor. The μ-ISFET approach allows the closest possible integration of the preamplifier with the microelectrode. The K^+-selective microelectrode was prepared as indicated in Ref. [3]. The electrode was made hydrophobic with hexamethyldisilazane and trimethylchlorosilane in 1-chloronaphtalene. The slope was determined with solutions containing different concentrations of K^+ and having a background of NaCl. The mean slope was 49–51 mV/decade change in K^+ activity. This slope is lower than the theoretical value because of sodium interference. Without sodium interference, the average sensitivity was equal to 53–57 mV/decade. The limit of detection was 1.4–2.2 mM and the selectivity coefficient k_{K^+, Na^+} was 0.019–0.031. The time constant (63% of the full response) was about 100–500 ms. The lifetime of the device was 10–15 days, and was limited by the slow evaporation of the hydrated gel. Most of the time, however, it was shortened by inadvertent breakage of the tip.

The performance of the K^+ μ-ISFET was compared with that of a K^+-selective microelectrode, on large photoreceptors from the barnacle *Balanus eburneus*. Both devices recorded similar quantitative changes as expected, since they had similar sensitivities. But the difference in quality of the recordings was striking. Compared to other systems commonly used, the K^+ μ-ISFET showed a very encouraging immunity to noise even though there appeared to be no significant gain in time response compared to the ion-selective microelectrode.

This probe has been used to monitor *in vivo* continuous K^+ levels in serum and interstitial fluid of dog [121].

7.2.2 Double- and triple-barrelled K$^+$-selective microelectrodes
7.2.2.1 Double-barrelled microelectrodes. Double-barrelled microelectrodes are
very useful for measuring intracellular ionic activity, since they allow simultaneous
recording of the potential due to the ionic activity plus the membrane potential with
the ionic barrel, and the membrane potential with the reference barrel. After
subtraction of these two potentials, the K$^+$ activity can be determined with
precision.

Several authors have described the construction of these microelectrodes
[1,7,9–11,93,122,123]. They were made either from glass pipettes with a middle
partition (theta microelectrodes) [10,92,93,122], or from two glass micropipettes
sealed together [1]. In the latter case, the tubes were sometimes twisted around their
axis by 180° [9] or by 360° [7,111,123] before they were pulled. Some of these glass
pipettes had an internal fibre [7]. To make the tip of the ionic barrel hydrophobic, the
same silicone-based solutions or vapours were used as for single-barrelled microelectrodes. For extracellular use, where small tip diameters are not necessary, the tips of
the electrodes were gently broken to adjust the final tip to 2.0–2.5 μm [123]. The
ionic barrel was filled with the K$^+$-exchanger resin on 200–500 μm and then with
0.1 M or 0.5 M KCl, since the reference barrel was filled with a NaCl solution
(0.15 M, 1 M, or 3.0 M), lithium acetate, or 1 M sodium acetate + 0.2 M KCl.

The characteristics of the electrodes are summmarized in Table 8. The slopes

Table 8 — Characteristics of double-barrelled K$^+$ ion-selective microelectrodes

Slope (mV/decade)	k_{K^+,Na^+}	ϕ (μm)	Refs.
40–56	0.01	0.3 (per barrel)	[11]
58	0.0143–0.025	<1	[12]
	0.01	2–3	[1]
56	0.007–0.011	1–3	[122]
52.6–57	0.0106	<1	[9]
46.4–53.4	0.02–0.066	0.5–1	[7]

were determined with different KCl solutions containing a constant amount of NaCl.
The selectivity coefficient for K$^+$ over N$^+$ ranged between 0.01 and 0.02; they were
concentration dependent.

Tip diameters were generally less than, or about, 1 μm. Meyer *et al.* [93] reported
a double-barrelled K$^+$-selective microelectrode with a tip diameter of 0.15 μm. *Juel*
[92] used double-barrelled microelectrodes with 25–50 μm tip diameter for extracellular use.

Authors reported resistance values from 1×10^8 [10,122–125] to 3×10^{10} Ω [7].
They are slightly higher than those of single-barrelled microelectrodes since, for the
same total tip diameter, the tip diameter of the ionic barrel of the double-barrelled
microelectrode is smaller.

Not all the authors employed the same definition for the rise time, or the time constant. Approximate response time constants of 0.5–3 ms were obtained by Lux [122]. Time constants of 1–2 ms [125] or less than 10 s [11] are reported. Fujimoto *et al.* [58] employed only electrodes with a response of less than 1 s. Teulon & Anagnostopulos [7] defined the rise time of the sensitive channel as the lag between the two channels at 90% of the steady-state response to cell impalement; this rise time was of the order of 0.5 s. Singer & Lux [124] reported rise times of about 2–10 ms, and Fujimoto & Kubota [9] reported rise times (from 10%–90%) ranging from 0.2–3 s. The shift of the electrode output was less than 1 mV for one hour. The electrodes were usually found to age in one or two days, although some were still usable after a week [9].

Selectivity coefficients for K^+ over interfering ions was determined, for example $k_{K^+,NH_4^+} = 0.20–0.25$ [9,12,58]. The K^+ selectivity to divalent cations, Ca^{2+}, or Mg^{2+}, was more than 1000: $k_{K^+,Ca^{2+}} < 0.001$ and $k_{K^+,Mg^{2+}} < 0.00099$.

The effect of pH changes in the physiological range on the K^+-selective microeoelectrode is negligible for a wide range of K^+ concentrations (up to 100 mM) [9].

Fujimoto & Kubota have reported [9] that 0.1 M bovine serum albumin exerted little effect on the e.m.f. of the microelectrode independently of K^{+2+} concentrations, but 1 mM albumin produced an increasingly large deviation of the e.m.f.; in addition, both urea and glucose had only a slight effect on responses of the ion-exchanger microelectrode. This might be due to impurities in the test materials [9].

As reported with single-barrelled K^+-selective microelectrodes, the ion exchanger used was sensitive to acetylcholine and to other quarternary ammonium compounds [122].

Values of the slope constant of the microelectrode at different temperatures are shown in Table 9. The n value (ratio of the measured slope to the Nernstian slope)

Table 9 — Effect of temperature on double-barrelled K^+ ion-selective microelectrodes. From [9]

Temperature (°C)	7	17	27	37
Nernstian slope [mV/p(X)]	55.5	57.5	59.5	61.5
Measured slope for K^+ (mean ± standard deviation)	49.2 ± 3.7	51.8 ± 3.4	53.9 ± 2.4	56.6 ± 1.6
n	0.89	0.90	0.91	0.92
Reading of test solution (4 mM K^+)	—	4.01 ± 0.07	4.03 ± 0.07	3.90 ± 0.06

increases almost linearly with temperature. This n value represents a sub-ideal behaviour of the ion-selective microelectrode with respect to the Nernstian response. These data suggest that so long as measurements are made at the same temperature as in the calibration or at constant temperature, accurate results can be obtained [9].

The many uses of double-barrelled K^+-selective microelectrodes include extra-cellular determination of potassium in cat [10,122] and rat cortex [11], in cat lateral geniculate nucleus [124], in the rat mesencephalic reticular formation [126], in the cat spinal cord [123,126,127], in the cat blood-brain barrier [128], and in the mouse soleus and extensor digitorum longus muscles [92]. These microelectrodes have been impaled in rat distal tubules [12], in renal proximal tubule of bullfrog [9,58], and in proximal tubular cells of *Necturus* kidney [7]. Experiments were also performed in epithelia cells of rabbit gastric glands [111] or rabbit gall-bladder [93], in abdominal skin or isolated epithelia of frog [71].

7.2.2.2 Coaxial microelectrodes. A procedure for preparing coaxial K^+-selective microelectrodes with a low longitudinal resistance of the liquid ion-exchanger sensitive barrel has been described by Ujec *et al.* [129]. The low resistance was attained by inserting another microelectrode (tip diameter of 3–5 μm) filled with 0.5 M KCl into the ion exchanger, so that the vertical distance between the inner and outer tip in the ion-exchanger barrel was 10–20 μm (Fig. 22). The lower longitudinal

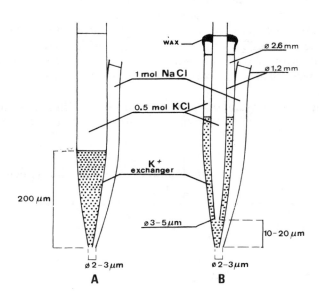

Fig. 22 — K^+-selective double-barrelled microelectrodes. A: conventional type. B: microelectrode with the inner coaxial micropipette. From [1].

resistance decreases the noise level and consequently increases the resolving power. This modification makes it possible to measure small and rapid K^+ concentration changes. The coaxial inserted micropipette lowered the longitudinal resistance of the ion-exchanger barrel by approximately one order of magnitude.

Moreover, while both the sensitivity and stability of the coaxial microelectrode are preserved, the lifetime was prolonged to several days. It was also found that this design can be of advantage in electrodes selective for Cl^- or Ca^{2+} in which the specific resistance is higher than that of K^+ ion-exchanger.

This new type of electrode was used for recording K^+ concentration changes in the frog dorsal root ganglion induced by stimulation of a nerve [130]. Transient concentration changes as low as 0.01 mM K^+ have been recorded.

7.2.2.3 Triple-barrelled microelectrodes. Fujimoto & Honda [88] described the construction of a triple-barrelled microelectrode for simultaneous measurements of intracellular N^+ and K^+ activities and membrane potential in biological cells.

Fig. 23 shows an example of the e.m.f. of the ionic barrels in solutions with

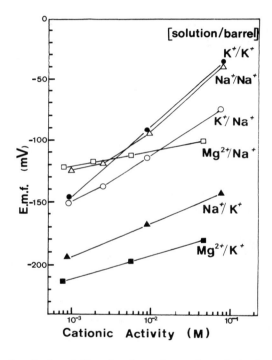

Fig. 23 — E.m.f. vs ionic activities plots for various cations obtained with a triple-barrelled Na^+, K^+-selective microelectrode. The e.m.fs. of the Na^+ barrel are presented by the open marks, whereas those of the K^+ barrel by the closed marks. The marks of triangles, circles, and squares are, respectively, Na^+, K^+, and Mg^{2+} ions in a series of calibration solutions with Cl^- as counter ions. From [129].

various cationic concentrations. The average slope for the K^+ barrel was 52.8–54.0 mV/decade, the average selectivity coefficient over Na^+ was 0.016–0.040 (concentration dependent), and the average electrical resistance of the K^+ barrel was 0.2–2.8×10^{10} Ω.

No pH effect was found on K^+ measurements when the pH changes ranged from 6.3–8.8.

Although these electrodes offer many adantages, there are, however, a number of problems in their use. The size of the tip of the electrode, in general, becomes larger as the number of barrels increases. However, the triple-barrelled microelectrode described by Fujimoto & Honda [88] had a tip less than 1 μm in outer diameter, which was small enough to impale most animal cells. Furthermore, the mechanical strength of the tip is increased. Another problem inherent in a high-resistance multi-barrelled electrode arises from cross-coupling among individual barrels.

These electrodes have been used to determine *in vivo* K^+ concentrations in single cells of the Sartorius muscle and renal proximal tubule of bullfrogs [88].

7.3 K^+-selective microelectrodes based on neutral carriers

7.3.1   *Valinomycin-based K^+-selective microelectrode*
In 1970, Frant & Ross [130] reported a new potassium macroelectrode based on valinomycin, which responded to potassium ion concentrations from 1 M to 10^{-6} M. The selectivity coefficient for K^+ over Na^+ was found to be equal to 7.6×10^{-5} at 0.1 M NaCl. A lower selectivity was observed when more concentrated NaCl solutions were used. This electrode has been used for the direct determination of potassium ion in human serum.

Later, Oehme & Simon [91] described the first microelectrode (tip diameter of about 2 μm) based on this neutral carrier. Dioctylphtalate was used to dissolve valinomycin and potassium tetra(*p*-chlorophenyl)borate. The role of the lipophilic charged carrier (i.e. K^+ tetra(*p*-chlorophenyl)borate) was to overcome the inherently high impedance of the exchanger.

Wuhrmann *et al.* [131] used double Pyrex tubings with an inner filament, instead of theta capillaries. To reduce the electrode resistance, the ion-sensitive component, based also on valinomycin and potassium tetra(*p*-chlorophenyl)borate, also contained 2,3-dimethylnitrobenzene dissolved in dibutylsebacate.

A double-barrelled microelectrode, whose tip had been broken at about 2 μm, was made, the ionic barrel being made hydrophobic with dichlorodimethylsilane in CCl_4 [91].

The response of the neutral carrier electrode was linear in the range 10^{-1}–5×10^{-6} M KCl (Fig. 24) giving a slope of 58.1–58.7 mV/decade [91] (56.9–58.3 mV/decade [131]). The resistance was about 10^{11} Ω, and the response time (10–60%) was 2 s [91] (10–90%: 30 s [131]). The e.m.f. of the microelectrode was constant within 0.3 mV over periods of 16 h.

Fig. 25 shows the outstanding selectivity of the electrode based on valinomycin, especially to H^+ and Na^+, in comparison with glass and other classical microelectrodes. It is obvious that the classical electrodes greatly prefer acetylcholine and other lipophilic cations such as tetrabutylammonium ions to potassium ions. Some of these cations were clearly discriminated from K^+ when valinomycin was used as the ion-exchanger component. The selectivity coefficients reported by Wuhrmann *et al.* [131] are quite similar to those given by Oehme & Simon [91]:

$$k_{K^+,Na^+} = 6.3 \times 10^{-4} \, , \; k_{K^+,Li^+} = 10^{-4} \, , \; k_{K^+,Mg^{2+}} = 10^{-5}$$

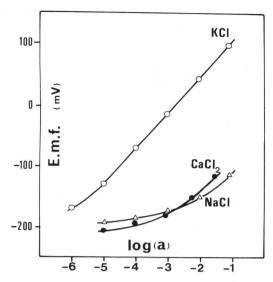

Fig. 24 — E.m.f. response curves of the K^+-microelectrode with neutral carrier solution (tip
diameter of 1.5 μm) to aqueous solutions of KCl, NaCl and CaCl$_2$. From [91].

$$k_{K^+,Ca^{2+}} = 3.16 \times 10^{-5} , \qquad k_{K^+,acetylcholine} = 3.16 \times 10^{-3} .$$

The inconvenience of a relatively high tip resistance and thus of a rather long
response time, is compensated by the otherwise excellent performance of the
valinomycin-based microelectrode, in terms of selectivity and slope. Wuhrmann
et al. [131] used this microelectrode to measure nuclear K^+ activity in salivary glands
of *Chironomus*. Nagy *et al.* [112] measured K^+ activity in different areas of the brain.

In 1983, however, Koch & Ladenson [132] reported that the commonly used
valinomycin-poly(vinylchloride) electrodes may be subject to interferences in biological fluids; thus the potassium data obtained from such electrodes should be
interpreted with extreme caution. Although such electrodes appear to work accurately in plasma, they do not work as accurately in urine, and may not be accurate at
all for red cell lysates.

7.3.2 Microelectrodes based on a new bis(crown ether)

A K^+-selective microelectrode with a tip diameter of 10 μm was described by Tarcali
et al. [4]. The ionophore used was 2,2'-bis[3,4-(15-crown-5)-2-nitrophenylcarba-
moxymethyl]tetradecane (BME 44) (Fig. 26) in a conventional mixture with sodium
tetraphenylborate, an ether and PVC.

Single-barrelled micropipettes with tips of 5–10 μm were filled to a length of
about 200 μm with the ion-exchanger solution.

The electrodes provided Nernstian responses (56–58 mV/decade) over the
10^{-5}–10^{-1} M range of potassium activity.

After the electrode had aged for 24 h, the 95% response time, at 10^{-3} M K^+, was

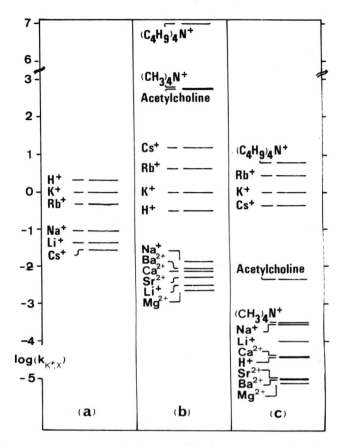

Fig. 25 — Comparison of the selectivity coefficients obtained by the separate solution technique with 0.1 M solutions of the chlorides (except for a 10^{-2} M solution of tetrabutylammonium iodide). (a), potassium-sensitive glass NAS_{27-4}; (b), K^+ microelectrode with classsical ion exchanger; (c), K^+ microelectrode based on neutral carrier. From [91].

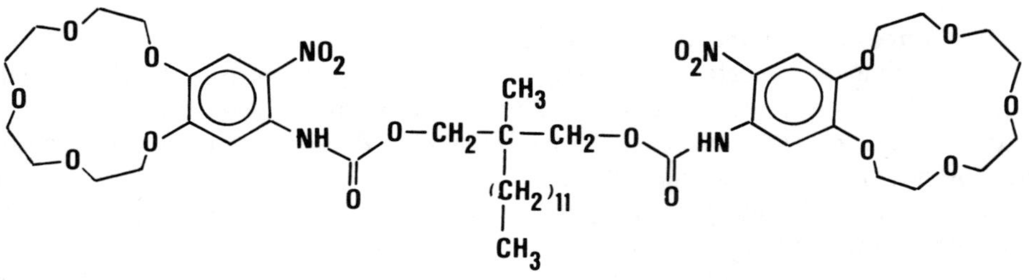

Fig. 26 — The ionophore BME 44. From [4].

in the range 1.5–2.5 s. It was found that this time was 10 s for freshly prepared electrodes. The decrease in the response time is most probably caused by structural rearrangements in the ion-selective membrane during solvent evaporation in the early stage of the electrode life.

The selectivity coefficients of these electrodes were evaluated by the mixed solutions method. Fig. 27 shows the selectivity coefficients obtained for the BME-44

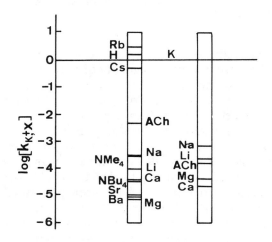

Fig. 27 — Comparison of the selectivity coefficients [$\log(k_{K^+,x})$] for the valinomycin and BME-44 electrodes. From [4].

electrode, compared with the selectivity data for a valinomycin-based electrode (measured by the separate solutions method) [131]. Because of the different methods of evaluation, direct comparison of the selectivities of the two types of electrode is questionable, but it appears that, even if the selectivity coefficient for K^+ over Na^+ for the BME-44-based electrode is slightly higher than that of the valinomycin-based electrode, the new electrode has adequate selectivity for application to *in vivo* measurements.

The stability of this microelectrode appeared to be good (3–4 weeks).

Changes in K^+ activity in different areas of the brain of rats were monitored *in vivo* using this type of microelectrode [111,112].

7.4 Conclusion

Several types of K^+-selective microelectrodes, based on sensitive glass, on ionized ion-exchanger systems, or on neutral carriers, are described in the literature and are available for measurements of K^+ activity. Single-, double-, triple-barrelled, or coaxial microelectrodes can be made. So far as the selectivity between K^+ and Na^+, the life-span and other electrical properties are concerned, no difference exists in electrochemical behaviour between single- and double-barrelled K^+-selective microelectrodes based on the same sensor and with the same tip diameter.

Very fine tip microelectrodes can record K^+ activity in very small cells; but they have rather high response time. And microelectrodes with tip diameters of 1 μm or more, with a response time of about 10 ms, can record K^+ activity changes extracellularly. Fast changes in K^+ activity in small cells cannot yet be measured.

Care must be taken with the presence of interfering ions, such as acetylcholine and quarternary ammonium ions for the K^+-selective microelectrodes based on Corning 477317, and the valinomycin-based microelectrodes seem not to be suitable in all physiological fluids.

8   pH-selective microelectrodes

Intracellular pH (pH_i) is an important regulator of cellular function.

The term pH can be conceptually difficult and misleading in biological systems because pH is subject to the influences of several simultaneously acting physico-chemical constraints. Indeed, the definition of pH, as the logarithm of the reciprocal of the hydrogen ion concentration, itself obscures the fact that H^+ is present in only submicromolar concentration in the brain. The overwhelming historical and current use of the term pH suggests that it would be better to recognize the limitations of the definition but retain its use for discussing the H^+ level in biological systems [133].

Most determinations of pH_i *in vivo* have been performed by using indirect methods (most often from the distribution of weak acids), since no suitable direct methods for reliable *in vivo* measurements have been available [134].

Until the introduction by Thomas of the recessed-tip pH microelectrode [135], microelectrodes could not be successfully applied to pH_i registrations in single mammalian cells. Double-barrelled recessed-tip pH microelectrodes have since been developed.

Other glass pH microelectrodes have been used, especially those with Hinke-type (or spear-type) design.

In 1980, Japanese workers miniaturized antimony microelectrodes and described their characteristics [136–138].

Then, as liquid ion-exchanger-based microelectrodes appeared for other ions, some authors tried to find a H^+-ion carrier suitable for pH microelectrodes. So Ammann *et al.* [139] developed a neutral carrier-based hydrogen-ion selective microelectrode suitable for extra and intracellular studies. In the 1980s, double-barrelled microelectrodes based on this sensor have been described [66,133,140–143].

The most useful and productive methods for measuring intracellular pH must be able not only to measure steady-state pH_i but also to monitor the dynamics of pH_i change [144].

8.1   Glass pH microelectrodes

8.1.1   Recessed-tip glass pH microelectrodes
8.1.1.1   Single-barrelled microelectrodes. Until 1974, when Thomas [135] developed a recessed-tip microelectrode available for intracellular measurements, all the pH-selective microelectrodes developed had the disadvantage that at least several

micrometers of the electrode tip had to be inserted into the cell to obtain a valid reading. The design of pH-selective microelectrode described by Thomas has the advantage that only the extreme tip needs to be inserted into the cell to be studied. This tip can be as small as 0.5 μm in diameter [145].

The design is based on that of the recessed-tip Na^+-selective microelectrode described earlier [73], and it has the pH-selective glass (typically code 0150 glass from Corning Glass Works) inside the tip of a borosilicate glass micropipette. The electrodes are perhaps rather difficult to make, but, once prepared, they can be used for many weeks and permit not only the direct measurement of the intracellular pH, but also the continuous recording of changes in pHi during experiments lasting for several hours [145].

The construction method is described in [135]. The electrode stem is filled with a solution containing approximately 100 mM NaCl and 100 mM citrate buffer (pH 6).

In contrast to the Na^+-selective microelectrodes described earlier, the pH-selective microelectrodes gave at first a small and noisy response to pH changes. Two or three weeks of soaking in water or chromic acid were needed before they would give their maximum response with least noise. One or two days after filling, a typical pH microelectrode was unstable, had a resistance of about $4 \times 10^{11}\,\Omega$, and gave a response of about 40 mV (instead of the theoretical 58 mV) to a pH unit change. After soaking for three weeks the resistance had fallen to $7 \times 10^{10}\,\Omega$, the response had increased to 55 mV/pH unit, and the noise level was less than 1 mV.

As expected from the properties of the code 0150 pH-selective glass, the electrodes gave linear responses to pH changes over at least the pH range of 2–9 and were not affected by changes in the concentration of other ions. With different pH microelectrodes the size of the response to pH changes varied slightly, ranging from 53–56 mV/pH unit, but with a given electrode the response remained the same over many days. The speed of the response, however, was often lower at the end of an experiment, presumably because the electrode tip tended to become partially blocked. Soaking for a few hours in chromic acid restored the original speed of response.

The response speed depended on the tip diameter and the volume of the recess, and was greatly affected by both the pH and the buffer concentration. Fig. 28 illustrates the response of a pH microelectrode to pH changes of approximatively one unit. In acid conditions, the response was relatively fast, but in the pH range 8–7, where the hydrogen ion concentration is less than micromolar, the response was very much slower, and appeared to depend principally on the buffer concentration.

Only electrodes with a reasonably fast response (less than 1 min to 90% response) have been used practically. It is difficult, if not impossible, to manufacture faster pH electrodes without sacrificing their sharpness and so causing more damage to the cell.

It is of interest to notice that the pH recorded is temperature dependent: the pH increases by 0.16 units between 28 and 37°C. The main disadvantage is that a reference electrode is necessary to measure the membrane potential, which is then subtracted from the potential recorded by the pH microelectrode. So, for intracellular measurements, the cell must be impaled with two microelectrodes.

Another disadvantage of the glass electrode is perhaps its very slow response in poorly buffered solutions. To overcome this problem, Thomas [146] modified the original method for making recessed-tip pH microelectrodes, allowing the recess

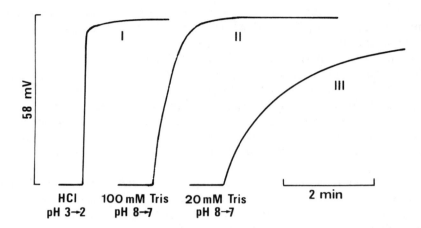

Fig. 28 — Pen-recordings of the responses of a pH-sensitive microelectrode to unit pH changes. The electrode tip was in a bath through which the test solutions flowed continuously. In the first recording (I) the test solution was changed from 10 mM HCl to 100 mM HCl; in the second recording (II) it was changed from 100 mM-tris maleate pH 8 to 100 mM-tris maleate pH 7; and in the third recording (III) it was changed from 20 mM-tris maleate pH 8 to 20 mM-tris maleate pH 7. From [135].

volume, and hence the response time, to be minimized.

Measurements of pH with these microelectrodes, have been performed in large neurones of snail [135,144,147], in mouse soleus muscle [148], and in sheep cardiac Purkinje fibres [145].

8.1.1.2 Double-barrelled microelectrodes. For small cells, the only practical method of inserting both the pH and the reference microelectrodes into the same cell is to use a double-barrelled pH microelectrode [149]. A double-barrelled design offers two important advantages: (i) since the membrane potential is simultaneously measured, the correct reference potential for the pHi measurement is easily obtained by subtraction ; (ii) the large change in potential upon cell penetration indicates when the electrode tip is intracellularly located [134].

Such a microelectrode combines a micropipette reference barrel to measure the membrane potential with the recessed-tip type of pH-selective glass microelectrode designed by Thomas [135].

De Hemptinne [150,151] applied a technique described by Zeuthen [152] to construct such double-barrelled pH microelectrodes with 0.5 μm tip diameter. Thus glass capillaries, with different diameters and containing a glass filament, are twisted 360° around one another and then glued. After pulling, the technique described by Thomas, with small adaptation, was applied to construct recessed-tip pH microelectrodes. Special care was taken to keep the recess volume small: the recess volume was less than 5–10% of the volume of the cell impaled. The pH-sensitive barrel was filled with a buffered solution containing 0.1 M NaCl buffered to pH 6 with 0.1 M citrate buffer. The reference barrel can be filled with 3 M KCl or another concentrated salt solution: for example, Vanheel & de Hemptinne [153] used 3 M KCl, 10 mM di-Na-EGTA, and 10 mM tri-K-citrate. The electrodes were found to be suitable for use after they were placed overnight in an oven at 75°C [151], to achieve hydration of the glass pH.

The sensitivity and response speed of the double-barrelled electrode are comparable to those of the single-barrelled electrode described by Thomas. The potential response was linear in the pH range 5–8 [149] and was 54–62 mV/pH unit; most electrodes measured 90% of the pH change within 2 min for Vanheel & de Hemptinne [153] and within 30 s or less for Cohen *et al.* [149]. These electrodes were used to measure pH in functionally isolated rat liver [149], in isolated sheep cardiac Purkinje strands [151,153] and in rat soleus muscle [150,151].

8.1.1.3 Coaxial double-barrelled microelectrodes. So far, double-barrelled electrodes have not been applied to intracellular pH measurements *in vivo.* Hagberg *et al.* [134,154,155] described a coaxial double-barrelled microelectrode. Single-barrelled recessed-tip pH microelectrodes were constructed. The pH pipette was inserted into an insulating glass pipette. Then, an outer pipette for the double-barrelled pH microelectrode was pushed into it [134] (Fig. 29).

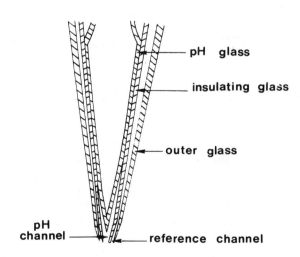

Fig. 29 — Schematic illustration of the design of the tip of the double-barrelled pH microelectrode. From [155].

The pH barrel was filled with 0.1 M NaCl buffered to pH 6 with a citrate buffer, and the reference barrel (between the outer and the insulating glass of the pH capillary) was filled with saturated K_2SO_4 with 50 mM KCl. The reference barrel was always easy to fill because the pH micropipette served as a filling filament. The present electrode construction has, however, the disadvantage that tip diameters less than 1–1.5 μm are difficult to make.

The response, response time, and stability of this coaxial double-barrelled pH microelectrode are similar to those previously described for single recessed-tip pH microelectrodes [135] or for double-barrelled recessed-tip pH microelectrodes [151].

This design has some advantages: (i) it is technically easy to make the coaxial double-barrelled microelectrodes from the single-barrelled ones; (ii) the electrode tips are resistant to breakage, which makes these electrodes ideal for *in vivo* measurements; (iii) owing to the shape of the tip, there are fewer problems with leaky punctures as compared with the parallel type of double-barrelled electrodes [154].

This microelectrode type has been used for *in vivo* pH measurements in rabbit skeletal muscle fibres [134,154,155].

8.1.2 Spear-type or Hinke-type microelectrodes
In the spear-type microelectrode [136,156,157], the pH-sensing tip protrudes from the end of a glass micropipette (Fig. 30).

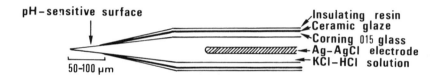

Fig. 30 — A spear type of glass microelectrode, whose tip is made to between 50 and 100 μm in length. Insulation is performed by the application of glazing technique with a supplement of insulating resin. From [136].

The important requirements for this type of electrodes are: (i) a sharp tip to minimize cell wall damage on penetration; (ii) a short pH-sensing length so that it will be completely within the cellular compartment being measured; (iii) a smooth transition between the pH-sensing portion and the insulated portion; (iv) a conical angle as small as possible for the portion of the microelectrode that will be in the cell wall: a small angle will make a better microelectrode to cell wall seal, thereby preventing leakage of intracellular fluid and shunting of the microelectrode; (v) finally, of course, a fast and reproducible response to pH and a design that can be easily cleaned and rejuvenated [156].

The sharpest tip can be obtained with the spear-type microelectrode, with tip diameters as small as 1 μm or less. On the other hand, it is necessary to have the complete sensing length inside the test fluid. Hebert [156] made these electrodes with sensing lengths of about 10 μm; sensing lengths of 50 μm were, however, made more commonly, these pH-microelectrodes having a response time of less than 15 s.

The stem of the electrode was filled with 0.1 N HCl solution.

For Hebert [156], the electrodes generally responded immediately with a response closely approximating the theoretical one, while, for Schneider *et al.* [157], they needed to be soaked for at least two weeks at room temperature in 0.1 N HCl before use.

The resistance of the electrodes varied between 10^8 and 10^9 Ω. The time between dipping the electrode into the buffer and the full response was between 1 and 15 s. The drift was maximally 2 mV/h. The electrodes were used for up to 15 months [157].

Fujimoto et al. [136] described the relationship between the slope constant of the glass microelectrode and the size of the pH-selective surface area. It is obvious that the shorter the pH-sensitive tip, the smaller the slope constant. This would suggest that manufacturing a smaller tip of less than 50 μm in length would inevitably be accompanied by a critical reduction of the pH response.

In applying the electrodes to biological systems, the main problem arises in the selection of a suitable filling solution for the reference electrode [157].

Another problem is the tip gelling of microglass pH electrodes, due to the extended time of contact between the electrode tip and the internal aqueous reference solution which leads to over-hydration of the glass membrane. Thus, the percentage of failure for making usable pH microelectrode was very high. So Savinell et al. [158] employed a non-aqueous solvent (0.1 M silver nitrate in acetonitrile), eliminating the water molecules from the standard inner electrolyte. The performances of macroelectrodes using a non-aqueous internal reference electrolyte or an aqueous one were compared. The authors found that the responses of uninsulated electrodes were quite sensitive to the quality of the tip profile. However, the responses of 11 electrodes studied were linear with Nernstian slopes in most cases, but the insulation with lead oxide was a failure [158].

To measure extracellular pH in human skin, Harrison & Walker [159] constructed a pH microelectrode with another design: a lead glass micropipette was fused internally to the pH glass approximately 150 μm above the tip, which can be made as small as 20 μm in diameter. Drift is less than 0.04 pH over a 16h period in phosphate buffer; the mean sensitivity is 55.5 mV/pH unit at 35°C, and it is accurate to 0.02 pH. The tip resistance is relatively high (10^{11} Ω), the response time is 80 s for a pH unit change, and 90% response occurs within 40 s.

In 1976, Pucacco & Carter [160] developed another technique to make single or double-barrelled spear-type microelectrodes: the tip of a high electrical resistance glass pipette was sealed with a thin membrane of H^+-selective glass.

Single-barrelled microelectrodes have been made with tip diameters ranging from 1.5–100 μm and double-barrelled microelectrodes with tip diameters from 2 to 28 μm. Two glasses were used in making the glass-membrane pH microelectrode: Corning 0150 glass was found to be satisfactory for large electrodes (greater than 50 μm tip diameter); however, for smaller ones, a uranium-containing pH-sensitive glass was used, because it has a lower specific resistance than Corning 0150 and because its working characteristics are better than 0150 glass. The composition in weight percentage is: SiO_2, 60; UO_2, 4; MgO, 8; and Na_2O, 28. The pH barrel was filled with 1 M magnesium acetate, and for the double-barrelled electrodes, the reference barrel was filled with 2.5 M KCl, 0.5 M KNO_3. The slope of all satisfactorily functioning glass-membrane pH microelectrodes was very close to the theoretical slope at 57–59 mV/pH unit at 22°C and 60–62 mV/pH unit at 37°C. The response time ranged from as fast as 15 s (10 μm single) to as long as 35 min (2 μm double). The relationship between response time and size was attributed to the increase in electrode resistance and apparent capacitance as the electrode size was reduced. The lifetime was variable, but in most cases was in excess of a week, and was, on average, 2–3 weeks.

Later on, Pucacco & Carter [161] detailed a method for thinning the pH glass in such a way that the manufacture of submicrometre glass-membrane pH microelectrodes was possible. The batch composition of the glass used was, in weight percentage: SiO_2, 58.1; CaO, 10.9; NaO, 27.0; and UO_2, 4.0. These electrodes had rapid response times (1–3 min) and maintained the desirable characteristics of all glass-membrane pH microelectrodes; that is, they were near the theoretical slope and had a well-defined sensing surface area. A 0.5 μm tip diameter electrode had a resistance of 2×10^9 Ω and a complete response time of 1.0 min. It appeared possible that even smaller electrodes could be manufactured. Moreover, it would appear possible to make these pH microelectrodes in the double-barrelled configuration, giving rise to other kinds of biological applications.

In conclusion, the advantages of the spear-type pH microelectrode, such as short response time, minimal drift, and long lifetime, would be compromised by further miniaturization, and the present tip size makes the electrode suitable primarily for extracellular measurements, as in the cortical sbarachnoidal space of cats, *in vivo* [157], in the luminal surface of the intestinal mucosa of guinea pig and rat, *in vivo* and *in vitro* [162], and in human skin [159].

8.2 Antimony microelectrodes

In 1980, several Japanese workers dealt with the miniaturization of an antimony electrode and studied the characteristics of this electrode [136–138,163]. The antimony microelectrode has certain advantages not only from technical but also from economic aspects, though it behaves non-ideally in several circumstances.

So single-barrelled Sb microelectrodes, with tip diameter of 1–2 μm, have been made, and also double-barrelled ones.

The design of the antimony microelectrode is shown in Fig. 31. It consists of a

Fig. 31 — The Sb microelectrode is constructed from a metal-filled glass capillary, whose tip is bevelled between 1 and 5 μm in outside diameter. From [136].

metal-filled glass micropipette electrode of a miniaturized 'Levin type' [136]. Fujimoto *et al.* [136] described its manufacture in detail.

Double-barrelled antimony microelectrodes have also been made [137,163]. Matsumura *et al.* [163] described the fabrication of the pencil-type Sb microelectrode

(Fig. 32). The outer diameter of the microelectrode tip was less than 1–2 μm. The reference electrode (the outer pipette of the double-barrelled type) was filled with Ringer solution or 0.1 M KCl solution. Special care was taken to avoid any cross-coupling between the reference and the antimony barrels. The response time of the microelectrodes, the linearity of the e.m.f. in response to various pHs, the stability of the e.m.f., and their biological applicability were studied by the Japanese team.

It is known that the Sb microelectrode, in general, is affected by physicochemical factors, including temperature, the ionic strength of the solution, the type of buffer, and the presence of organic solutes which are capable of forming complexes with the antimony metal. In short, the calibration of the antominy microelectrode should be done in tris-buffer rather than in phosphate buffer; the practical use of this electrode is limited to the test solutions whose ionic strength, as well as composition, is known; and the pH measurements are made at a constant temperature and partial pressure of oxygen.

The rise times of the Sb microelectrodes were within 10 s. This low response time is mainly due to the low resistance of the Sb microelectrodes, of the order of $(0.03–5) \times 10^8\,\Omega$, a range of values about 10 times lower than that of the glass microelectrode.

Standard calibration curves of both a glass microelectrode and a Sb microelectrode are shown in Fig. 33. Approximate linearities were observed over the pH range of 2–10. The average slope constant at 20°C was 44.3–59.5 mV/pH unit for the Sb microelectrode. The calibration curves were slightly different if the calibrations were carried out in phosphate buffers or in tris buffers.

The output voltages of the Sb microelectrode were not very stable. This might be due to the effect of oxidation and gradual dissolution of Sb metal into the test solution. It caused a considerable drift in the e.m.f. during a short period, frequently several mV an hour. But once a steady level had been reached, it tended to stabilize fairly well for a period sufficient to permit meaningful measurements.

The reproducibility of the pH data in six measurements of the same buffer was within the range of 0.0007. Since the e.m.f. of the Sb microelectrode tended to rise gradually by 1.0 to 1.5 mV in a series of successive measurements, the estimation of pH had to be made by frequent interpolations between two or more buffer calibrations before and after each test measurement [136].

Great care should be taken not to overlook the temperature effect.

The influence of the ionic strength of the test solution has been studied in detail by Matsumura *et al.* [137], who gave an empirical equation between the ionic strength and the deviation in pH readings.

Sakate *et al.* [138] dealt with the effect of protein on the Sb microelectrode for the pH measurement of biological fluids. It was found that bovine serum albumin, probably by binding to components in the electrode system, produced a marked deviation of both the e.m.f. and slope constant of the microelectrode. But the shift of the pH reading could be empirically predicted from the cubic function of protein concentration.

So, in applying the Sb microelectrode for biological experiments, it is desirable that calibration should be done with standard solutions which have almost the same components as test samples at the same temperature at which the biological measurements are made. By using tris-buffer, and with the appropriate correction for known changes of physicochemical factors, the accuracy of the Sb microelectrode

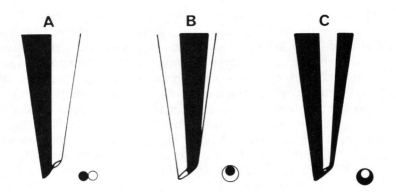

Fig. 32 — Three types of double-barrelled antimony microelectrodes: A, parallel; B, pencil; C, tube. From [163].

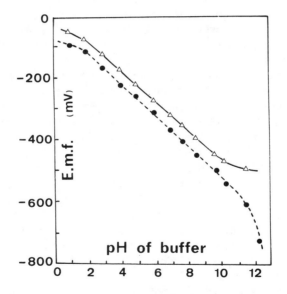

Fig. 33 — The pH response of pH-sensitive microelectrodes. $\triangle$, $\bullet$: data from the glass or the Sb microelectrode, respectively. From [136].

could almost attain the same level as that of the glass pH microelectrode.

The Sb microelectrodes were tested in blood samples obtained from bullfrogs, rats, and human subjects [136,138]. Using the double-barrelled type, the pH of intratubular fluid of bullfrog proximal tubule, of distal tubule, was measured *in vivo* [137,163]. Intracellular pH of frog Sartorius muscle has also been measured *in vivo* [163].

8.3   *Liquid ion-exchanger pH microelectrodes*

In contrast to glass pH microelectrodes, liquid membrane microelectrodes are easy to prepare, do not suffer from the disadvantages of tip geometry, and are less sensitive than the Sb microelectrodes to physicochemical factors.

In 1976, Le Blanc *et al.* [164] described a pH sensor suitable for continuous intravascular monitoring of blood pH. The H^+-carrier employed was *p*-octadecyl-*m*-chlorophenylhydrazonesoxalonitrile (OCPH), in an elastomeric polymer membrane.

Then Erne *et al.* [165] reported a synthetic H^+-selective ion-carrier which was suitable as a component for liquid membrane electrodes. A mixture of 3-hydroxy-*N*-octyl-picolinamide (20%), sodium tetraphenylborate (0.5%) and *o*-nitro-phenyl-*n*-octylether (79.5%) has been successfully used in microelectrodes with tip diameters of about 2 μm. This sensor had adequate selectivity both for intra- and extra-cellular studies, but tip diameters had to be above 2 μm to achieve membrane resistances below 10^{12} Ω [139].

Later, in 1981, Harman & Poole-Wilson [140] made a pH microelectrode with a tip diameter of 1 μm, based on the sensor previously used by Erne *et al.* [165]. The micropipette was made hydrophobic with dimethyldichlorosilane vapour, and liquid ion exchanger (OCPH in *n*-decanol) was then introduced. The shaft was finally back-filled with a sodium citrate buffer (pH 5) saturated with AgCl. The microelectrode was allowed to stabilize for 3 h before use.

The response of the microelectode was linear to changes of the pH in the range 5.8–8.0. No hysterisis to increasing or decreasing pH has been found. The response was not affected by osmotic pressure, CO_2, or changes (in mM) in Na^+ (4–150), K^+ (0–150), Ca^{2+} (0–10), or Mg^{2+} (0–10). The time to 90% response was less than 10 s.

The microelectrode was suitable for the study of rapid changes in intracellular pH in contracting tissues with small cells (right ventricular papillary muscle of the guinea pig) [140].

The electrode performance, unfortunately, was not fully specified. Limiting factors seem to be the rather long necessary conditioning time of at least 3 h and the high failure rate of the electrodes of approximately 40%.

Ammann *et al.* [139] described a H^+-selective microelectrode based on a neutral ion carrier (tri-*n*-dodecylamine: TDDA). The glass micropipette was made hydrophobic with *N*-(trimethylsilyl)dimethylamine [139,144] or with dimethyldichlorosilane [111]. The ion-selective liquid consisted of 10% TDDA and 0.7% sodium tetraphenylborate in *o*-nitrophenyloctylether. Sodium tetraphenylborate was added to the membrane phase to reduce anion interference and electrical resistance. Indeed, even electrodes with tip diameters of 0.8–1.0 μm have a resistance of about 10^{11} Ω. The internal filling solution was 40 mM KH_2PO_4, 23 mM NaOH, and 15 mM NaCl (pH 7). The e.m.f. measurements were performed at 22°C with a FET operational amplifier mounted on the top of the electrode. The response of this microelectrode to hydrogen ion additives for pH ranging from 5.5 to 12 was nearly Nernstian even in the presence of 60 mM Na^+. However, Kafoglis *et al.* [111]

reported that the relation between pH and potential was not log linear.

Fig. 34 shows the e.m.f. reponse of the microelectrode to different H^+ activities

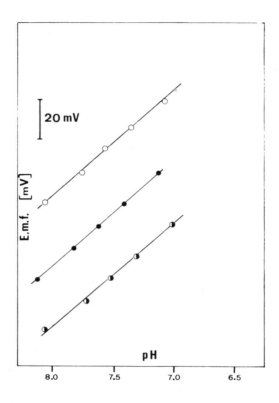

Fig. 34 — E.m.f. response of the microelectrode cell assembly to different H^+ activities in the physiological range at typical intra- and extracellular ion backgrounds. (○), H^+ response at constant extracellular ion background: Tris/HCl with 140 mM Na^+, 4 mM K^+, 0.6 mM Mg^{2+}, and 1.1 mM Ca^{2+}. (●), H^+ response at constant intracellular ion background: Tris/HCl with 10 mM Na^+, 200 mM K^+, 2 mM Mg^{2+}, and 0.01 mM Ca^{2+}. (●) H^+ response in pure tris/HCl buffer solutions. From [139].

in the physiological range at typical intra- and extra-cellular ion backgrounds. A selectivity factor k_{H^+,Na^+} lower than 2×10^{-13} can be estimated. No noticeable interference from ions is observed in the physiological concentration range. The addition of 24 mM sodium bicarbonate to solutions with extracellular background does not significantly influence the slope of the electrode response [139]. No interference was produced by inclusion of ouabain (10^{-4} M), histamine (10^{-4} M), or picoprazole (5×10^{-5} M) in the bathing media [111].

Owing to the relatively low electrical resistance of this neutral carrier-based microelectrode ($10^{11} \, \Omega$ for tip diameters of around 1 μm) the response is rather fast. The 90% response time is shorter than 5 s. The electrode shows an overall e.m.f. drift of 0.6 mV/h over at least 3 days.

Intracellular measurements were made in *Xenopus laevis* oocytes [139], in isolated rabbit gastric glands [111], and in snail neurons and crab muscle [144].

The use of double-barrelled ion-selective microelectrodes [133] allowed the simultaneous measurement of membrane potential and thus removed methodological uncertainties. The pH ion-selective microelectrode did not function in the presence of the proton-carrying metabolic blocker, dinitrophenol [133].

Yoshitomi & Fromter [166] described the manufacture of a double-barrelled microelectrode with a tip diameter of 0.2 μm. The microelectrode had Nernstian slopes (near 58 mV/pH unit) and a resistance of the order of 10^{12} Ω. By bevelling the tip in a suspension of Al_2O_3, and by painting a silver shield on the surface of the glass capillary, which covered both barrels from about 5 mm from the tip, the response time was lowered from 15 s to about 1 s.

So the improvements and modifications of the microelectrode design allowed membrane potentials and intracellular H^+ concentrations in small cells to be measured and the time course of intracellular pH changes to be followed directly [166].

Double-barrelled ion-selective pH microelectrodes have been used to measure pH_i in rat kidney *in vivo* [141,166], in the proximal tubule of *Necturus* kidney *in vivo* [66], and extracellular pH in rat brain during complete brain ischemia [133,142,167] and in R6-2 gliomas [143].

8.4 Conclusion

Several types of pH microelectrode are available for intra- or extra-cellular measurements. They do not have the same characteristics. The selection of the type of microelectrodes to use in a given experiment is best determined by the application, since each type of electrode has advantages as well as limitations. However, the recently developed single-or double-barrelled H^+-selective microelectrodes seem to be quite suitable for intra- or extra-cellular measurements and for monitoring rapid pH changes.

9 General conclusion

There is now no doubt that liquid ion-exchanger ion-selective microelectrodes have demonstrated their usefulness in physiological research during the past decade. Proven electrodes suitable for measuring Li^+, Mg^{2+}, Ca^{2+}, Cl^-, Na^+, K^+, and H^+ are now available, based either on organic ions or neutral carriers; these electrodes can be impaled in any cells, even the smallest, owing to their fine tips.

Ion-selective liquid ion-exchanger microelectrodes, however, do still have certain disadvantages and limitations, although these can be overcome, and improvements may be expected in the near future:

(i) Most of the microelectronics described are sensitive to interfering ions, so it is necessary to know the activity of these interfering ions in the cell impaled with the microelectrode, in order to take them into account in the determination of the activity of the principal ion. The construction of triple-barrelled microelectrodes, such as the triple-barrelled Na^+-selective microelectrode described above, could be an interesting solution.

(ii) Because of their relatively high resistance, most of the liquid ion-exchanger microelectrodes have a long response time. A shorter response time would allow intracellular transient changes in ionic activities to be measured. While the intracellular activity is being monitored, the preparation can be subjected to various manipulations that may affect intracellular activity. This would help to elucidate important aspects of the ionic mechanisms.

So, in spite of their present shortcomings, further improvements may be expected with the development of liquid ion exchangers with better selectivity coefficients, and the introduction of new technologies and modifications of fabrication.

In the future, microelectrodes sensitive to biologically active substances (e.g. enzymes, amino acids, transmitters) could help to extend the methodological possibilities of research. Until now, only electrodes with diameters of a few micrometres have been described; examples are a galactose-oxidase enzyme electrode to detect changes in raffinose concentration [168,169], an acetylcholine and choline ion-selective microelectrode [170,171], an enzyme microsensor for urea based on an ammonia-gas electrode [172], and a carbon fibre microelectrode using immobilized lactate dehydrogenase and sensitive to pyruvate [173].

References
[1] Vyskocil, F. & Kriz, N. (1972) *Pflugers Arch.* **337**, 265–276.
[2] Lee, C. O. & Fozzard, H. A. (1975) *J. Gen. Physiol.* **65**, 695–708.
[3] Walker, J. L. (1971) *Anal. Chem.* **43**, 89A–93A.
[4] Tarcali, J., Nagy, G., Toth, K., Pungor, E., Juhasz, G. & Kukorelli, T. (1985) *Anal. Chim. Acta* **1**, 231–237.
[5] Armstrong, W. McD. & Garcia-Diaz, J. F. (1980) *Fed. Proc.* **39**, 2851–2859.
[6] Fromter, E., Simon, W. & Gebler, B. (1981) In: Lübbers, D. W., Acker, H., Buck, R. P., Eisenman, G., Kessler, M. & Simon, W. (eds) *Progress in enzyme and ion-selective electrodes*. Berlin-Heidelberg-New York, Springer Verlag p. 35–44.
[7] Teulon, J. & Anagnostopoulos, T. (1982) *Am. J. Physiol.* **243**, F12–F18.
[8] Garcia-Diaz, J. F. & Armstrong, W. M. D. J. (1980) *J. Membr. Biol.* **55**, 213–222.
[9] Fujimoto, M. & Kubota, T. (1976) *Jap. J. Physiol.* **26**, 631–650.
[10] Lux, H. D. & Neher, E. (1973) *Exp. Brain Res.* **17**, 190–205.
[11] Zeuthen, T., Hiam, R. C. & Silver, I. A. (1974) *Adv. Exp. Med. Biol.* **50**, 145–156.
[12] Wright, F. S., Mc Dougal, W. S. & Yale, (1972) *J. Biol. Med.* **45**, 373–383.
[13] Walke, J. L. (1980) In: Koryta, J. (ed.) *Medical and biological applications of electrochemical devices*. New York, John Wiley p.109–128.
[14] Meier, P. C., Ammann, D., Osswald, H. K. & Simon, W. (1977) *Med. Progr. Technol.* **5**, 1–12.
[15] Dagostino, M. & Lee, C. O. (1982) *Biophys. J.* **40**, 199–207.
[16] International Union of Pure and Applied Chemistry (Analytical Chemistry Division — Commission on Analytical Nomenclature) (1976) *Pure Appl. Chem.* **48**, 127–132.
[17] Nicholson, C., Steinberg, R., Stockle, H. & Ten Bruggencate, G. (1976) *Neurosci. Lett.* **3**, 315–319.

[18] Ammann, D., Guggi, M., Pretsch, E. & Simon, W. (1975) *Anal. Lett.* **8**,709–720.

[19] Meier, P. C., Ammann, D., Morf, W. E. & Simon, W. (1980) In: Koryta, J. (ed.) *Medical and biological applications of electrochemical devices.* New York, John Wiley p.13–91.

[20] Vaupel, P., Gunther, H., Erdmann, W., Kunke, S. & Thews, G. (1972) *Verh. Dtsch. Ges. Inn. Med.* **78**, 133–136.

[21] Von Ardenne, M. (1976) *Z. Med. Labortechnik* **17**, 116–127.

[22] Sohtell, M. & Karlmark, B. (1976) *Pflugers Arch.* **363**, 179–180.

[23] Berman, H. J. (1974) In: Berman, H. J. & Hebert, N. C. (eds) *Ion-selective microelectrodes.* New York, Plenum Press p. 3–11.

[24] Guggi, M., Fiedler, U., Pretsch, E. V. & Simon, W. (1975) *Anal. Lett.***8**, 857–866.

[25] Zhukov, A. F., Erne, D., Ammann, D., Guggi, M., Pretsch, E. & Simon, W. (1981) *Anal. Chim. Acta* **13**, 117–122.

[26] Thomas, R. C., Simon, W. & Oehme, M. (1975) *Nature (Lond.)* **258**, 754–756.

[27] Lopez, J. R., Alamo, L., Caputo, C., Vergara, J. & DiPolo, R. (1984) *Biochim. Biophys. Acta* **804**, 1–7.

[28] Hess, P., Metzger, P. & Weingart, R. (1982) *J. Physiol. (Lond.)* **333**,173–188.

[29] Lanter, F., Erne, D., Ammann, D. & Simon, W. (1980) *Anal. Chem.* **52**, 2400–2402.

[30] Hess, P. & Weingart, R. (1981) *J. Physiol. (Lond.)* **318**, 14P–15P.

[31] Tsien, R. Y. & Rink, T. J. (1980) *Biochim. Biophys. Acta* **599**, 623–638.

[32] Ross, J. W. (1967) *Science* **156**, 1378–1379.

[33] Moody, G. J., Oke, R. B. & Thomas, J. D. R. (1970) *Analyst* **95**, 910–918.

[34] Ruzicka, J., Hansen, E. H. & Tjell, J. C. (1973) *Anal. Chim. Acta* **67**, 155–178.

[35] Brown, H. M., Pemberton, J. P. & Owen, J. D. (1976) *Anal. Chim. Acta* **85**, 261–276.

[36] Christoffersen, G. R. J. & Johansen, E. S. (1976) *Anal. Chim. Acta* **81**, 191–195.

[37] Owen, J. D., Brown, H. M. & Pemberton, J. P. (1977) *Anal. Chim. Acta* **90**, 241–244.

[38] Oehme, M., Kessler, M. & Simon, W. (1976) *Chimia* **30**, 204–206.

[39] Heinemann, U., Lux, H. D. & Gutnick, M. (1977) *J. Exp. Brain Res.* **27**, 237–243.

[40] Weingart, R. & Hess, P. (1984) *Pflugers Arch.* **402**, 1–9.

[41] Barber, A. (1986) *J. Exp. Biol.* **121**, 395–406.

[42] Rink, T. J., Tsien, R. Y. & Warner, A. E. (1980) *Nature (Lond.)* **283**, 658–660.

[43] Marban, E., Rink, T. J., Tsien, R. W. & Tsien, R. Y. (1980) *Nature (Lond.)* **286**, 845–850.

[44] Lee, C. O. & Dagostino, M. R. (1981) *Biophys. J.* **33**, 284a.

[45] Ashley, C. C., Rink, T. J. & Tsien, R. Y. (1978) *J. Physiol. (Lond.)* **280**, 27P.

[46] Bers, D. M. & McLeod, K. T. (1986) *Circ. Res.* **58**, 769–782.

[47] Tsien, R. Y. & Rink, T. J. (1981) *J. Neurosc. Meth.* **4**, 73–86.

[48] Lee, C. O., Uhm, D. Y. & Dresdnek, K. (1980) *Science* **209**, 699–701.
[49] Hinke, J. A. M. (1969) In: Lavallee, M., Shanne, O. & Hebert, N. C. (eds) *Glass microelectrodes*. New York, John Wiley p.349–375.
[50] Kerkut, G. A. & Meech, R. W. (1966) *Life Sci.* **5**, 453–456.
[51] Neild, T. O. & Thomas, R. C. (1973) *J. Physiol. (Lond.)* **231**, 7P–8P.
[52] Thomas, R. C. (1970) *J. Physiol. (Lond.)* **210**, 82P–83P.
[53] Harvey, B. J. & Kernan, R. P. (1981) *J. Physiol. (Lond.)* **318**, 54P–55P.
[54] Walker, J. L. & Brown, A. M. (1970) *Science* **167**, 1502–1504.
[55] Brown, A. M., Walker, J. L. & Sutton, R. B. (1970) *J. Gen. Physiol.* **56**, 559–582.
[56] Bolton, T. B. & Vaughan-Jones, R. D. (1977) *J. Physiol. (Lond.)* **270**, 801–833.
[57] Cassola, A. C., Mollenhauer, M. & Fromter, E. (1983) *Pflugers Arch.* **399**, 259–265.
[58] Fujimoto, M., Kubota, T. & Kotera, K. (1977) *Contr. Nephrol.* **6**, 114–123.
[59] Spring, K. R. & Kimura, G. (1978) *J. Membr. Biol.* **38**, 233–254.
[60] Spring, K. R. & Kimura, G. (1979) *Fed. Proc.* **38**, 2729–2732.
[61] Shindo, T. & Spring, K. R. (1981) *J. Membr. Biol.* **58**, 35–42.
[62] Saito, Y., Ozawa, T., Hayashi, H. & Nishiyama, A. (1985) *Pflugers Arch.* **405**, 108–111.
[63] Nicholson, C. & Kraig, R. P. (1975) *Brain Res.* **96**, 384–389.
[64] Macchia, D. D. & Baumgarten, C. M. (1979) *Pflugers Arch.* **382**, 193–195.
[65] Aickin, C. C. & Vermue, N. A. (1983) *Pflugers Arch.* **397**, 25–28.
[66] Anagnostopoulos, T. (1984) *J. Physiol. (Paris)* **79**, 401–405.
[67] Kubota, T., Honda, M., Kotera, K. & Fujimoto, M. (1980) *Jap. J. Physiol.* **30**, 775–790.
[68] Curci, S., Schettino, T. & Fromter, E. (1986) *Pflugers Arch.* **406**, 204–211.
[69] Palmer, L. G. & Civan, M. M. (1977) *J. Membr. Biol.* **33**, 41–61.
[70] Nagel, W., Garcia-Diaz, J. F. & Armstrong, W. McD. (1981) *J. Membr. Biol.* **61**, 127–134.
[71] Harvey, B. J. & Kernan, R. P. (1984) *J. Physiol. (Lond.)* **349**, 501–517.
[72] Hinke, J. A. M. (1959) *Nature (Lond.)* **184**, 1257–1258.
[73] Thomas, R. C. (1972) *J. Physiol. (Lond.)* **220**, 55–71.
[74] Kraig, R. P. & Nicholson, C. (1976) *Science* **194**, 725–726.
[75] Steiner, R. A., Oehme, M., Ammann, D. & Simon, W. (1979) *Anal. Chem.* **51**, 351–353.
[76] Lee, C. O. (1979) *Biophys. J.* **27**, 209–220.
[77] Dick, D. A. T. & McLaughlin, S. G. A. (1969) *J. Physiol. (Lond.)* **205**, 61–78.
[78] Thomas, R. C. (1969) *J. Physiol. (Lond.)* **201**, 495–514.
[79] Robinson, R. A. & Stokes, R. H. (1959) *Electrolyte solutions*. 2nd ed. London, Butterworths p. 230–238, 432–451.
[80] Eisenman, (1962) *Biophys. J.* **2**, 259–323.
[81] Sellick, P. M. & Johnstone, B. M. (1972) *Pflugers Arch.* **336**, 11–20.
[82] Ellis, D. (1977) *J. Physiol. (Lond.)* **273**, 211–240.
[83] Graf, J. & Giebisch, G. J. (1979) *J. Membr. Biol.* **47**, 327–355.
[84] Hinke, J. A. M. (1961) *J. Physiol. (Lond.)* **156**, 314–335.
[85] Deitmer, J. W. & Ellis, D. (1978) *J. Physiol. (Lond.)* **277**, 437–453.

[86] Glitsch, H. G., Pusch, H., Schumacher, Th. & Verdonck, F. (1982) *Pflugers Arch.* **394**, 256–263.

[87] Garcia-Diaz, J. F., O'Doherty, J. & Armstrong, W. McD. (1978) *Physiologist* **21**, 41.

[88] Fujimoto, M. & Honda, M. (1980) *Jap. J. Physiol.* **30**, 859–875.

[89] Kimura, G. & Spring, K. R. (1979) *Am. J. Physiol.* **236**, F295–F301.

[90] O'Doherty, J., Garcia-Diaz, J. F. & Armstrong, W. McD. (1979) *Science* **203**, 1349–1351.

[91] Oehme, M. & Simon, W. (1976) *Anal. Chim. Acta* **86**, 21–25.

[92] Juel, C. (1986) *Pflugers Arch.* **406**, 458–463.

[93] Meyer, G., Rossetti, C., Botta, G. & Cremaschi, D. (1985) *Pflugers Arch.* **404**, 378–381.

[94] Cremaschi, D., James, P. S., Meyer, G., Rossetti, C. & Smith, M. W. (1984) *J. Physiol. (Lond.)* **354**, 363–373.

[95] Glitsch, H. G. & Rasch, R. (1986) *Pflugers Arch.* **406**, 144–150.

[96] Glitsch, H. G., Pusch, H. & Verdonck, F. (1986) *Pflugers Arch.* **406**, 464–471.

[97] Lewis, S. A. & Wills, N. K. (1980) *Biophys. J.* **31**, 127–138.

[98] Harvey, B. J. & Ehrenfeld, J. (1986) *Pflugers Arch.* **406**, 362–366.

[99] Yoshitomi, K. & Fromter, E. (1985) *Pflugers Arch.* **405**, S121–S126.

[100] Kajino, K. & Fujimoto, M. (1982) *Jap. J. Physiol.* **32**, 997–1001.

[101] Wills, N. K. & Lewis, S. A. (1980) *Biophys. J.* **30**, 181–186.

[102] Daying, H. & Achenbach, C. (1985) *Pflugers Arch. (suppl.)* **403**, R25.

[103] Eisner, D. A., Lederer, W. J. & Vaughan-Jones, R. D. (1983) *J. Physiol. (Lond.)* **335**, 723–743.

[104] Glitsch, H. G. & Pusch, H. (1984) *Pflugers Arch.* **402**, 109–115.

[105] Hermansson, K. & Spring, K. R. (1986) *Pflugers Arch.* **407** (suppl. 2), S90–S99.

[106] Neher, E. & Lux, H. D. (1973) *J. Gen. Physiol.* **61**, 385–399.

[107] Edelman, A., Curci, S., Samarzija, I. & Fromter, E. (1978) *Pflugers Arch.* **378**, 37–45.

[108] Sellick, P. M. & Bock, G. R. (1974) *Pflugers Arch.* **352**, 351–361.

[109] Poulsen, J. H. & Bledsoe, S. W. (1978) *Am. J. Physiol.* **234**, E79–E83.

[110] Vyscocil, F., Hnik, P., Rehfeldt, H., Vejsada, R. & Ujec, E. (1983) *Pflugers Arch.* **399**, 235–237.

[111] Kafoglis, K., Hersey, S. J. & White, J. F. (1984) *Am. J. Physiol.* **246**, G433–G444.

[112] Nagy, G., Tarcali, J., Toth, K., Adams, R. N. & Pungor, E. (1985) *Anal. Chem. Symp. Ser.* **22**, 567–577.

[113] Haemmerli, A. & Janata, J. (1980) *Anal. Chem.* **52**, 1179–1182.

[114] Meyer, G., Henin, S. & Cremaschi, D. (1981) *Bioelectrochem. Bioenerg.* **8**, 575–579.

[115] Cremaschi, D. & Meyer, G. (1982) *J. Physiol. (Lond.)* **326**, 21–34.

[116] Prince, D. A., Lux, H. D. & Neher, E. (1973) *Brain Res.* **50**, 489–495.

[117] Singer, W. & Lux, H. D. (1975) *Brain Res.* **96**, 378–383.

[118] Futamachi, K. J. & Pedley, T. A. (1976) *Brain Res.* **109**, 311–322.

[119] Walker, J. L. & Ladle, R. O. (1973) *Am. J. Physiol.* **225**, 263–267.

[120] Hnik, P., Kriz, N., Vyscocil, F., Smiesko, V., Mejsnar, J., Ujec, E. & Holas,

M. (1973) *Pflugers Arch.* **338**, 177–181.

[121] McKinley, B. A., Saffle, J., Jordan, W. S., Janata, J., Moss, S. D. & Westenskow, D. R. (1980) *Med. Instrum.* **14**, 93–97.

[122] Lux, H. D. (1974) *Neuropharmacology* **13**, 509–517.

[123] Kriz, N., Sykova, E., Ujec, E. & Vyklicky, L. (1974) *J. Physiol. (Lond.)* **238**, 1–15.

[124] Singer, W. & Lux, H. D. (1973) *Brain Res.* **64**, 17–33.

[125] Sykova, E., Rothenberg, S. & Krekule, I. (1974) *Brain Res.* **79**, 333–337.

[126] Kriz, N., Sykova, E., Ujec, E. & Vyklicky, L. (1975) *J. Physiol. (Lond.)* **249**, 167–182.

[127] Lothman, E. W. & Somjen, G. G. (1975) *J. Physiol. (Lond.)* **252**, 115–136.

[128] Hansen, A. J., Lund–Andersen, H. & Krone, C. (1977) *Acta Physiol. Scand.* **101**, 438–445.

[129] Ujec, E., Keller, O., Machek, J. & Pavlik, V. (1979) *Pflugers Arch.* **382**, 189–192.

[130] Frant, M. S. & Ross, J. W. (1970) *Science* **167**, 986–988.

[131] Wuhrmann, P., Ineichen, H., Riesen–Willi, U. & Lezzi, M. (1979) *Proc. Nat. Acad. Sci. USA* **76**, 806–808.

[132] Koch, D. D. & Ladenson, J. H. (1983) *Anal. Chem.* **55**, 1807–1809.

[133] Kraig, R. P., Ferreira–Filho, C. R. & Nicholson, C. (1983) *J. Neurophysiol.* **49**, 831–850.

[134] Hagberg, H., Larsson, S. & Haljamäe, H. (1983) *Acta Physiol. Scand.* **118**, 149–153.

[135] Thomas, R. C. (1974) *J. Physiol. (Lond.)* **238**, 159–180.

[136] Fujimoto, M., Matsumura, Y. & Sakate, N. (1980) *Jap. J. Physiol.* **30**, 491–508.

[137] Matsumura, Y., Sakate, N. & Fujimoto, M. (1980) *Jap. J. Physiol.* **30**, 509–528.

[138] Sakate, N., Matsumura, Y. & Fujimoto, M. (1980) *Jap. J. Physiol.* **30**, 689–700.

[139] Ammann, D., Lanter, F., Steiner, R. A., Schulthess, P., Shijo, Y. & Simon, W. (1981) *Anal. Chem.* **5**, 2267–2269.

[140] Harman, M. C. & Poole-Wilson, P. A. (1981) *J. Physiol. (Lond.)* **315**, 1P.

[141] Henderson, R. M., Bell, P. B., Cohen, R. D., Browning, C. & Iles, R. A. (1986) *Am. J. Physiol.* **250**, F203–F209.

[142] Kraig, R. P., Pulsinelli, W. A. & Plum, F. (1986) *Am. J. Physiol.* **250**, R348–R357.

[143] Arnold, J. B., Kraig, R. P. & Rottenberg, D. A. (1986) *J. Cereb. Blood Flow Metab.* 6, 435–440.

[144] Szatkowski, M. S. & Thomas, R. C. (1986) *Pflugers Arch.* **407**, 59–63.

[145] Ellis, D. & Thomas, R. C. (1976) *Nature (Lond.)* **262**, 224–225.

[146] Thomas, R. C. (1978) *J. Physiol. (Lond.)* **277**, 14P–15P.

[147] Thomas, R. C. (1976) *Nature (Lond.)* **262**, 54–55.

[148] Aickin, C. C. & Thomas, R. C. (1977) *J. Physiol. (Lond.)* **267**, 791–810.

[149] Cohen, R. D., Henderson, R. M., Iles, R. A. & Smith, J. A. (1982) *J. Physiol. (Lond.)* **330**, 69–80.

[150] De Hemptinne, A. (1979) *J. Physiol. (Lond.)* **295**, 5P–6P.

[151] De Hemptinne, A. (1980) *Pflugers Arch.* **386**, 121–126.

[152] Zeuthen, T. (1976) *J. Physiol. (Lond.)* **254**, 8P–10P.

[153] Vanheel, B. & de Hemptinne, A. (1985) *Pflugers Arch.* **405**, 118–126.

[154] Hagberg, H., Larsson, S. & Haljamäe, H. (1985) In: Kessler, M., *et al.* (eds) *Ion measurements in physiology and medicine.* Berlin-Heidelberg, Springer Verlag p. 96–101.

[155] Hagberg, H. (1985) *Pflugers Arch.* **404**, 342–347.

[156] Hebert, N. C. (1974) *Adv. Exp. Med. Biol.* **50**, 23–38.

[157] Schneider, W., Wahl, M., Kuschinsky, W. & Thurau, K. (1977) *Pflugers Arch.* **372**, 103–107.

[158] Savinell, R. F., Liu, C. C., Kowalsky, T. E. & Puschett, J. B. (1981) *Anal. Chem.* **53**, 552–554.

[159] Harrison, D. K. & Walker, W. F. (1977) *J. Physiol.* **269**, 23P–25P.

[160] Pucacco, L. R. & Carter, N. W. (1976) *Anal. Biochem.* **73**, 501–512.

[161] Pucacco, L. R. & Carter, N. W. (1978) *Anal. Biochem.* **89**, 151–161.

[162] Rechkemmer, G., Wahl, M., Kuschinsky, W. & Von Engelhardt, W. (1986) *Pflugers Arch.* **407**, 33–40.

[163] Matsumura, Y., Kajino, K. & Fujimoto, M. (1980) *Membr. Biochem.* **3**, 99–129.

[164] Le Blanc, O. H., Brown, J. F., Klebe, J. F., Niedrach, L. W., Slusarczuk, G. M. J. & Stoddard, W. H. J. (1976) *Appl. Physiol.* **40**, 644–647.

[165] Erne, D., Ammann, D. & Simon, W. (1979) *Chimia* **33**, 88–90.

[166] Yoshitomi, K. & Fromter, E. (1984) *Pflugers Arch.* **402**, 300–305.

[167] Kraig, R. P., Pulsinelli, W. A. & Plum, F. (1985) *Brain Res.* **342**, 281–290.

[168] Geibel, J., Volkl, H. & Lang, F. (1984) *Pflugers Arch.* **400**, 388–392.

[169] Rehwald, W., Geibel, J. & Gstrein, E. (1984) *Pflugers Arch.* **400**, 398–402.

[170] Suaud-Chagny, M. F. & Pujol, J. F. (1985) *Analusis* **13**, 25–29.

[171] Jaramillo, A., Lopez, S., Justice, J. B., Salamone, J. D. & Neil, D. B. (1983) *Anal. Chim. Acta* **146**, 149–159.

[172] Joseph, J. P. (1985) *Anal. Chim. Acta* **169**, 249–256.

[173] Suaud-Chagny, M. F. & Gonon, F. G. (1986) *Anal. Chem.* **58**, 412–415.

2.3 HPLC WITH ELECTROCHEMICAL DETECTION
P. Blond

The dialysis technique originates from the push-pull cannula, described by Gaddum [1] and widely used later by Glowinski [2], for example in neurochemistry. It allows the local perfusion of tissue with artificial cerebrospinal fluid or physiological saline solution. In Gaddum's design, two concentric stainless-steel tubes are placed directly in the brain. The artificial perfusion fluid comes in through the central tube; the neurotransmitters and metabolites diffusing into the perfusion flow, are removed via the outer tube.

Ungerstedt *et al.* [3] have developed an intercerebral dialysis technique that involves continual flow in a closed system in which a narrow dialysis tube is inserted into the cerebral structure. This method is based on the principle that the perfusion fluid, which flows inside the dialysis tube, extracts from the surrounding tissue substances of low molecular weight which cross the dialysis membrane along a concentration gradient. Later, Zetterström *et al.* [4] combined this dialysis technique with sample analysis by HPLC and electrochemical detection (HPLC-EDT) to estimate endogenous extracellular levels of various neurotransmitters and metabolites in rat brain *in vivo*. Simultaneously, Justice *et al.* [5,6] described an on-line monitoring system for chromatographic analysis of dialysed perfusate in freely moving animals.

Successively, we shall describe the dialysis probe, the equipment used and the results of biological compound analysis by HPLC-EDT.

1 Dialysis procedure
Several assemblies are used for *in vivo* studies of neurotransmitter metabolism using a dialysis procedure.

1.1 *Transcerebral dialysis*
The transcerebral dialysis by linear hollow fibre inserted transversally has been developed by Imperaro & Di Chiara [7,8]. Fig. 1 shows schematically the position of the dialysis tube in the caudate nucleus. The dialysis tube is an acrylic copolymer hollow fibre with a molecular weight (m.wt) cut-off of 50 000 (Vita fibre type 3×50, AMICON, Lexington, USA).

The implantation is executed on anaesthetized male rats whose body temperature, blood pressure, and cardiac frequences are controlled. A hole is drilled on each side of the temporal bone at the level of the selected nucleus according to a stereotaxis atlas [9,10]. The dialysis tube, into which is inserted a tungsten wire (0.15 μm diameter), is fastened in a transverse position. The surface of the dialysis tube is covered with super epoxy resin, except for two zones, 4.0 mm wide, separated by a central zone of 2 mm wide also covered by super epoxy (corresponding to the caudate of each side). Then the dialysis tube–tungsten wire assembly is inserted and positioned according to the calculated reference point. Each end of the dialysis tube is connected to a needle inserted into an Eppendorf tip and secured in a vertical position to the parietal bone with dental cement. The rats are allowed 24 h to recover before any experiment is performed. The dialysis tube is connected to perfusion apparatus which delivers a constant flow of 2 μl/min throughout the experiment.

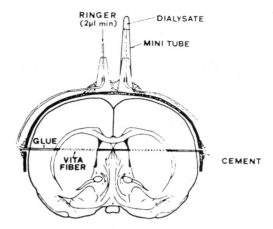

Fig. 1 — A trans-striatal dialysis preparation in an awake rat. The drawing shows a frontal section of rat brain and the transversally inserted dialysis tube. Note that the tube is covered with super-epoxy resin in its extrastriatal portions. From [8].

1.2 The dialysis loop

The miniature dialysis loop shown in Fig. 2 was achieved by Clemens & Phebus [11],

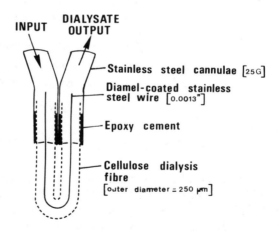

Fig. 2 — The *in vivo* brain dialysis probe proposed by Clemens & Phebus [11]. This probe was stereotaxically lowered into the corpus striatum and cemented into place with dental cement.

Sharp *et al*. [12], Lerma *et al*. [13], and other workers. It was made by bending one or two centimetres of miniature cellulose dialysis tubing into an open-ended loop. Fine stainless-steel wire was incorporated into the loop to provide the necessary stiffness.

The ends of this loop inserted into the corpus striatum are stainless-steel cannulae and are glued into place with cement.

The dialysis loop is used frequently because it is easy to implant unilaterally in certain cerebral structures. The major drawback of loops is that their outer diameters are often greater than 500 μm.

1.3 The dialysis cannula

The assembly reported by Hudson *et al.* [14] (Fig. 3) consists of the above guide

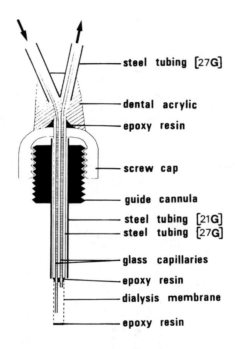

Fig. 3 — Dialysis probe and guide cannula according to Hudson *et al.* [14].

cannula and a dialysis probe. The guide cannula is commercially available (C 313 DC from Clarke Electromedical, London, UK). This had to be modified and equipped with a tubular dialysis membrane (approximately 300 μm outer diameter). Hudson *et al.* give precise details of the assembly construction.

As early as 1983, Justice *et al.* [6] constructed a dialysis cannula (Fig. 4) from Spectrapor HF hollow fibre (Spectrum, Los Angeles, USA) with a m.wt cut-off of 5000 and a diameter of about 200 μm. Kapoor & Chalmers [15] use only one central glass tube inside the hollow fibre (Fig. 5).

Recently, Combe [16] used a commercially-available dialysis cannula (Carnegie Medicin, Solna, Sweden) with an outer diameter of 500 μm and a m.wt cut-off of 15 000. Results show that this probe has a much higher relative recovery than other hollow fibres for the catecholamines (e.g. adrenalin, noradrenalin, . . .).

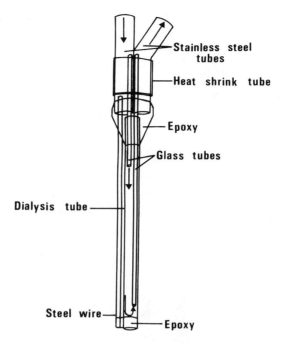

Fig. 4 — Details of the dialysis cannula (not drawn to scale) constructed by Justice *et al.* [6]. Diameter of the cannula is about 200 μm. The length depends upon the structure to be perfused. The six-pin connector which surrounds the top of the cannula is not shown. Arrows show the direction of flow. From [6].

1.4 *Dialysis membranes*

Dialysis membranes are shaped like hollow fibres. They are either cellulosic membranes or, more often, synthetic polymer fibres which are usually included in clinical dialysers for human kidney dialysis. The outer diameter of these fibres is generally between 200 and 500 μm, and the m.wt cut-off between 1500 and 50 000. The most frequently used fibres are:

- Dow 50, flexible cellulose tubing (Dow Chemical Co., USA) with a small outer diameter of 250 μm and a m.wt cut-off of 5000 [4];
- Hollow fibre C-Dak regenerated cellulose (Japan Medical Supply) used by Kato *et al.* [17]: 250 μm outer diameter and 90% m.wt cut-off of 5000;
- Vita Fibre 3×50 (Amicon): 350 μm outer diameter (acrylic polymer) and m.wt cut-off of 50 000 [4,8];
- Diaflo tube (Amicon): 200 μm internal diameter and m.wt cut-off of 10 000. Diaflo tube is less fragile and has a smaller cut-off value [19] than Vita fibre;
- Crupaphan hollow fibre of 200 μm internal diameter with which Lerma *et al.* [13] constructed a U-shaped loop (Disscap hollow fibre dialyser from Hospal, Colorado, USA);
- Clarins TE 07 hollow fibre (Terumo, Japan; 200 μm internal diameter and m.wt cut-off of 1500) used by Kapoor & Chalmers [15] to build a dialysis cannula.

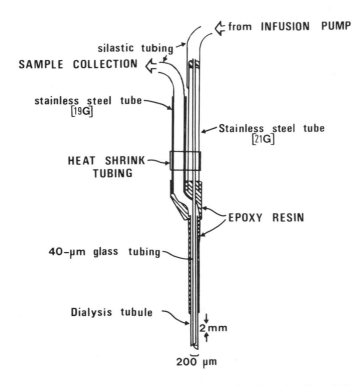

Fig. 5 — Details of the Kapoor & Chalmers dialysis probe. From [15].

1.5 *Perfusion assembly*

The probe cannula inlet is connected to a microperfusion pump via polyethylene tubing. The dialysis tubes are perfused with a physiological saline solution (Ringer solution), whose composition is 147 mM Na^+, 2.3 mM Ca^{2+}, 4 mM K^+, 155.6 mM Cl^- (pH = 6–6.5). Sometimes, the saline solution also includes Mg^{2+} ions [20].

These solutions are continuously perfused in a dialysis probe. The constant flow rate is most often 2 μl/min [21,22]. For direct injection into a microbore column, the perfusion rates used by Wages *et al.* [20] ranged from 0.08–0.21 μl/min. Various pump models can be used: Harvard perfusion model 2274 equipped with Hamilton syringe (Harvard Apparatus, South Natick, USA); peristaltic pumps with a polyethylene tube of small internal diameter (0.1 mm): Model 12000 Varioperpex from L.K.B. (Bromma, Sweden), Model P3 from Pharmacia (Uppsala, Sweden); micro-infusion system CMA/100 (Carnegie Medicin).

The perfusates are collected every 20 min in plastic microtubes mounted over outlet cannula. Various procedures can be observed, depending on the collected compounds: precautions are not always necessary, for amino acids for example. Nevertheless, for catecholamine studies in extracellular tissues, each collecting tube contained 10 μl of 1 M perchloric acid to prevent air oxidation of monoamines [19,23].

The dialysate samples (24 μl) containing 5-hydroxyindoleacetic acid (5-HIAA) and tryptophan collected by Hudson *et al*. [14] were immediately made up to 50 μl with 0.2% cystein, and 40 μl was taken for analysis.

Sometimes, the samples are specially prepared before the analysis by HPLC-EDT; one example is the organic phase extraction described by Kapoor & Chalmers [15].

After completion of the collection, samples are injected, both directly or after centrifugation, into HPLC column for analysis. Alternatively, the samples can be immediately frozen on ice or dry ice while awaiting analysis [19].

As a rule, at the end of an experiment, the authors verify the localization of the dialysis probe by histological control.

1.6   Histological control

On completion of some experiments during which the dialysis tube was continuously perfused with Ringer solution, rats were perfused intracardially with formaldehyde at 10% [8,19]. The brains were removed and frozen, and 40 μm coronal and sagittal brain sections were obtained after cresyl violet staining. The site of the dialysis tube was observed histologically. A degree of local trauma appeared and could be judged from the appearance of neurones and myelinated bundles in the vicinity of the dialysis probe. Histological control performed on a section of cerebral cortex [19] revealed a normal neuronal morphology around the probe except for a narrow zone adjacent to the dialysis tube (Diaflo from Amicon, 200 μm internal diameter) that showed neuronal pyknosis (Fig. 6).

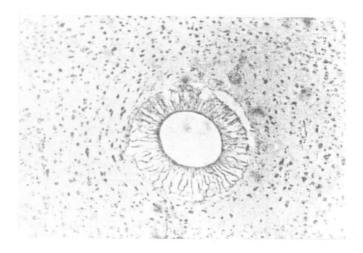

Fig. 6 — Histological control performed on a section of cerebral cortex ($\times$75) of an animal implanted with a dialysis tube and perfused for 8 h. Cresyl violet stain. From [19].

1.7   In vitro *recovery testing probes*

In vitro experiments are carried out to test the recovery of each biological compound (e.g. dopamine, DA; 3,4-dihydroxyphenylacetic acid, DOPAC; homovanillic acid, HVA; 5-hydroxytryptamine, 5-HT; ascorbic acid, AA; amino acids; . . .) through the dialysis membrane. Each probe is perfused with artificial solutions (usually Ringer solution, previously passed through a 0.5 μm filter) at the selected flow rate; they are placed in a beaker containing the biological compound at standard concentrations in the same Ringer solution. Several perfusate samples are collected at intervals of 20 min and analysed by HPLC-EDT. Recoveries in the perfusate are often expressed as a percentage of the standard concentration.

The recovery of various compounds across the dialysis membrane is a function of both the temperature and the perfusion flow rate. For example, the DOPAC recovery at 37°C is considerably greater than at 23°C [20].

Wages *et al.* [20] propose the following simple equation to calculate the amount of material recovered at a particular perfusion flow rate:

$$A = C P R \tag{1}$$

where A equals the amount of material recovered per minute (pg/min), C is the surrounding concentration (pg/μl), P is the perfusion flow rate (μl/min), and R is the recovery at that flow (%).

Considering the case of laminar flow in a hollow cylinder with semipermeable walls, Jacobson *et al.* [24] give the following equation:

$$C/C_o = 1 - \exp(-K_o A/F) \tag{2}$$

in which the diffusion in the direction of flow is ignored. A equals the area of the semipermeable membrane, C is the bulk concentration in the tube, C_o is the uniform concentration outside the tube, F is the flow rate, and K_o is the average mass transfer coefficient. K_o is nearly constant, independent of flow rates in a restricted flow domain, which validates the data previously obtained from various model dialysis experiments.

The recovery factors are thoroughly reviewed by Wages *et al.* [20] in their discussion of the on-line microbore liquid chromatographic analyses of brain dialysate. The effect of the perfusion flow rate over a wider range of values (between 0 and 10 μl/min) is illustrated in Fig. 7. The relative recovery is defined as the amount in the perfusate per time unit. It is zero at zero flow, reaches a broad maximum at about 2–3 μl/min, and declines at higher flow rates.

Nevertheless, the recovery from brain tissue is expected to be lower than from the *in vitro* standard solution, owing to reduced mass transport to dialysis cannula in the

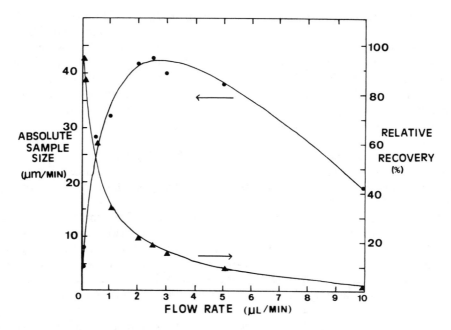

Fig. 7 — Amount of sample collected per minute as a function of perfusion flow rate [equation
(1)] using dialysis (●). Calculations were made assuming an external concentration of 1.0 μM
and adjusting for the recovery (▲) across the dialysis membrane at each perfusion flow rate.
From [20].

brain. This discrepancy can lead to underestimating extracellular concentrations.
This error can be reduced by perfusing at very slow rates (40–100 nl/min) where
recovery approaches 100%.

2 HPLC assembly

2.1 *Conventional HPLC-EDT*
For the analysis of dialysate samples collected in the Eppendorf tube at the dialysis
probe outlet, all authors describe a conventional HPLC device which consists of a
high-pressure pump with flow control and a 250×6 mm stainless-steel column
packed with a reverse phase (C8 or C18) or, if necessary, a cation-exchange column
[22].

The mobile phase consists of buffer solution adjusted to exact pH value,
supplemented with some methanol. If a counter-ion is necessary, sodium octane
sulfonate is chosen.

The biological compounds are separated and determined against an external
standard solution, using an electrochemical detector in which the working electrode
is a carbon paste electrode, a vitreous carbon electrode, or a carbon fibre brush
electrode [25].

Fig. 8 shows a chromatogram obtained by Imperato & Di Chiara in 1984 [7] under basal conditions to separate and identify the substances present in the dialysate: uric acid, DOPAC, HVA, 5-HIAA, and also DA provided that the highest sensitivity is used; so far, these experiments have been carried out only on tissue. The identification of the compounds in the dialysate in comparison with reference compounds is based on the following criteria: retention time, co-elution, and identification of the electrochemical responses (the exact chromatographic conditions followed by Imperato & Di Chiara are given in the key to Fig. 8).

2.2 Perfusion systems connected on-line to analytical systems

Justice *et al*. [6] were the first to construct an on-line monitoring system for chromatographic analysis of dialysis perfusate. The outlet dialysis tube is connected to the HPLC injection valve (Rheodyne injector Model 7413; Rheodyne Inc., Cotati, USA). A 100 μl sample loop is used for sample introduction. Direct introduction is facilitated because the sample is prepurified of any protein which could degrade column performance. With this system, the analytical column can be a conventional column (4.6×100 mm) [6] or a microbore column (1×100 mm) [20] with 3 μm C18 packing. The flow rate of the solvent is, according to the column, 1.2 ml/min for a conventional column and from 80–100 μl/min for a microbore column. The electrochemical detector (Bioanalytical Systems, West Lafayette, USA) is equipped with a glassy-carbon electrode held at a potential of +0.75 V vs. a Ag/AgCl reference electrode. The system is illustrated in Fig. 9.

In the system described by Wages *et al*. [20], the injection valve is in line with the dialysis cannula and the pull syringe of the perfusion pump. With this arrangement, the perfusion fluid from the cannula continuously fills the sample loop. Samples are automatically injected at preset intervals and the valve returned to the load position to begin filling with the next sample while the previous sample is eluting.

Damsma *et al*. [22] and Westerink & Tuinte [26] have also used a perfusion system and an on-line connected analytical system. But to measure choline and acetylcholine in perfusate samples, an enzymatic post-column reactor containing acetylcholinesterase (EC 3.1.1.7) and choline oxidase (EC 1.1.3.17) covalently bonded to CN-Br activated Sepharose 4B (Pharmacia) was inserted between the analytical column and the electrochemical detector. The platinum working electrode detects the hydrogen peroxide generated by the enzymatic reaction. The absolute detection level for acetylcholine is 50 fmol/injection.

3 Neurochemical results

To study the monitoring of neurotransmitter catechols and indols and to account for specificity of each metabolite, neurochemists use many drugs which are often enzyme inhibitors.

The reader may refer to Appendice A1 of Chapter 2.1 (Part 1) for information on the main metabolic lines for catecholamines.

3.1 Pharmacological investigations

Many careful studies on the pharmacological effects of neuroleptic drugs have been undertaken by Zetterström *et al*. [4,18,27,28]. First, the basal extracellular concentration (that is, before drug administration) of neurotransmitter amines, DA and 5-

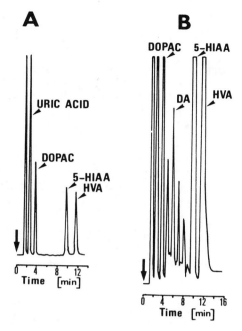

Fig. 8 — Chromatogram of 40 μl of dialysate (20 min sample) mixed with 5.0 μl of 0.1 N HCLO$_4$. A: tracing obtained at the lowest sensitivity on a two-pen recorder showing the peaks of DOPAC (58 pmol, 50 nA full scale), 5-HIAA (60 pmol, 20 nA full scale), and HVA (25 pmol, 20 nA full scale). B: tracing obtained at the highest sensitivity showing the peak of DA (0.25 pmol, 0.2 nA full scale).

Analytical conditions:
— HPLC: Dual piston pump Model M4S from Waters Associates (Milford, USA); PD-2 pulse damper from Bioanalytical Systems; Reverse phase column (250×4.6 mm), octadecyl silica particles of 5 μm Supelcosil LC-18-DB (Supelco, Bellefonte, USA); Pellicular guard column type LC-18 Supelco; Mobile phase: 0.23 M sodium acetate, 0.015 M citric acid, 100 mg/l EDTA, pH 5.5; Flow rate: 2 ml/min.
— EDT: Amperometric controller (Bioanalytical Systems); Flow cell equipped with a silicon carbon paste working electrode (type TL-3 from Bioanalytical Systems) at a potential of +0.65 mV (oxidation) vs. a Ag/AgCl reference electrode.
— Two-pen recorder.

($\downarrow$): injection. From [7].

HT, and their metabolites in the striatum and nucleus accumbens of anaesthetized or freely moving rats have to be estimated. The effects of several neuroleptic drugs on extracellular neurotransmitter levels can then be investigated. For example, the effects of amphetamine [4], apomorphine [18], sulpiride, haloperidol, and cisflupenthixol [27], p-chloramphetamine [28] on the extracellular levels of DA and its metabolites in striatum of the conscious rats have been studied.

Four neurotransmitters are determined with HPLC-EDT in collected striatal perfusates, i.e. DA, DOPAC, HVA, and 5-HIAA. All drugs modify the basal level of DA and metabolites, but the time course and dose–response magnitude of effects on each individual metabolite differ markedly, depending on the neuroleptic used. Zetterström *et al.* suggest that neuroleptic drugs have two separate actions on the

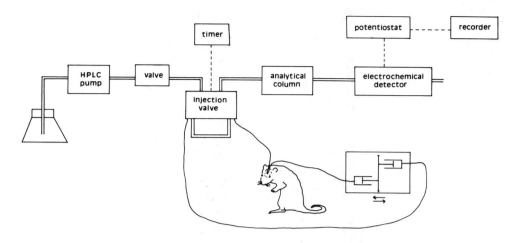

Fig. 9 — On-line monitoring system for HPLC-EDT analysis of brain dialysis perfusates.

DA neurone *in vivo*, one causing an increase in DA release (DA recovered in striatal perfusate) and another producing an increase in DA metabolism, which is probably a consequence of increased DA intraneuronal synthesis.

The same workers have recently investigated the *in vivo* DA release and metabolism in various brain structures [12]. The dialysis loops were implanted unilaterally in striatum, nucleus accumbens, and medial prefrontal cortex. Levels of DA and its acid metabolites were determined both in perfusate samples and in tissues extracted from killed rats. The results show regional differences in basal DA release and metabolite levels (Table 1).

Two recent studies by Sharp *et al*. [29,30] compared the effects of several drugs, i.e. amphetamine, *p*-chloramphetamine, and a mixture of amphetamine and sulpiride, on DA release in the striatum and nucleus accumbens, and highlighted induced behaviours of individual animals, for example locomotive activity, repetitive head and forepaw movements, and sniffing.

Rollema *et al*. [31] have used brain dialysis of DA and its metabolites to study the effects of MPTP (1-methyl 4-phenyl 1,2,3,6-tetrahydropyridine) and its active metabolites. Thus, the metabolite MPP^+ (1-methyl 4-phenylpyridinium) has a profound and instantaneous effect on the release of striatial DA and 3-methoxy dopamine, a metabolite of DA.

Di Chiara & Imperato [32] have established the temporal relationship between DA release in the nucleus accumbens and hypermobility after low doses of ethanol (0.5 g/kg) in freely moving rats.

Ruggeri *et al*. [23] have used these intracerebral microdialyses with U-shaped cannula to analyse how cholecystokinin (CCK peptides) and neurotensin may modulate DA release and metabolism in the rostral and caudal nucleus accumbens of halotane anaesthetized male rats.

Table 1 — Levels of DA and its acid metabolites in regional/brain perfusates and whole tissue extracts. From [12]

	DA	DOPAC	HVA	DOPAC/HVA	DOPAC/DA	Da perfusate/DA tissue (relative to striatum)
Striatum						
Tissue (nmol/g)	97.29±9.80 (8)	14.76±1.46 (8)	16.54±1.64 (8)	0.89	0.15	
Perfusatre (pmol/40 μl)	0.23±0.03 (11)	23.21±2.70 (11)	20.91±1.57 (11)	1.11	101	1.00
Nucleus accumbens						
Tissue (nmol/g)	55.10±6.27 (7)	13.81±2.45 (7)	9.01±0.88 (7)	1.53	0.25	
Perfusate (pmol/40 μl)	0.15±0.02 (9)	40.73±4.59 (9)	20.41±1.80 (9)	1.99	272	1.15
Frontal cortex						
Tissue (nmol/g)	1.28±0.14 (8)	0.42±0.07 (8)	1.24±0.15 (8)	0.32	0.33	
Perfusate (pmol/40 μl)	0.03±0.01 (10)	0.63±0.07 (10)	2.24±0.31 (10)	0.28	21	10.59

DA, DOPAC and HVA in perfusates or tissue extracts were measured simultaneously, using HPLC-EDT. Statistical analysis of the dialysis measurements revealed no correlation between basal DA levels and basal levels of either DOPAC or HVA. Numbers in parentheses indicate number of animals. Mean values ± standard deviation are indicated. From [12].

Routledge & Marsden [21] have used intracerebral dialysis to investigate auto-regulation of 5-HT release in the frontal cortex and to study the adrenergic involvement of hypothalamic regulation of blood pressure.

3.2 Comparative studies
This section describes some results of intracerebral dialysis studies using unilaterally-implanted cannula to perform comparative determination of neurotransmitter levels.

Thus, Holman [33] determined the concentrations of DOPAC, HVA, and 5-HIAA at the end of each experiment, (that is, *post mortem*) in the caudal nucleus of each cerebral hemisphere both with and without a cannula. In perfusion experiments where no drug was administered, it appeared that the formation of DOPAC and HVA, but not that of 5-HIAA, can be weakly but significantly increased in the caudal nucleus containing the dialysis cannula.

Zetterström *et al.* [34,35] used the intracerebral dialysis technique *in vivo* to show the spontaneous DA release from grafted nigral neurone. The dialysis loops were implanted both into the re-innervated transplanted region of the grafted host striatum and, in a similar position, into the contralateral non-lesioned side of the grafted rats. The results provide additional evidence that the grafted DA neurones exert their functional effect through a continuous active transmitter release from their newly established terminal in the host target. This study has also shown that an acute unilateral lesion of the substantia nigra causes an increase in the release of newly synthesized DA in contralateral intact striatum and that it probably represents a permanent compensatory phenomenon.

Sharp *et al.* [36] have published a very interesting comparison of drug-induced changes in extracellular levels of DOPAC and 5-HIAA measured by intracerebral dialysis and differential pulse voltammetry (DPV) with carbon fibre electrodes. A pretreated carbon fibre electrode [37] was implanted in one striatum and a dialysis loop in the contralateral striatum. The results (Fig. 10) show that *in vivo* DPV using

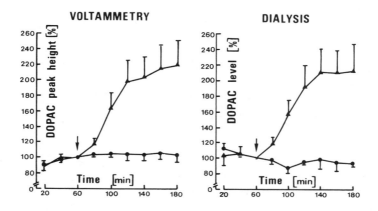

Fig. 10 — A comparison of the effect of haloperidol (0.5 mg/kg) on DOPAC peak height of the voltammogram from rat striata and DOPAC levels in dialysis perfusates from the contralateral side. Haloperidol (▲) or saline (●) injected (↓) 60 min after the start of the experiment. Results given as percentage of the pre-injection (60 min) value ± standard error (each value is the average of 5 pieces of data). From [36].

carbon fibre electrodes gives similar results to intracerebral dialysis for monitoring DA metabolism (DOPAC). However, care should be taken in interpreting changes in the 5-HIAA oxidation peak as some other electroactive compounds, probably uric acid, may contribute to the peak.

Workers are well aware of this difficulty of several electroactive species oxidizing at similar potentials. Nevertheless, pharmacological manipulation using enzyme inhibitors of neuronal metabolism facilitates the interpretation of the voltammetric signals 'in vivo'. Another validity test of graphite electrode signals has been described by Rice *et al.* [38]. Simultaneous voltammetric and chemical sampling of DA release are carried out, the first by using a carbon fibre electrode with a nafion anionic polymer film, the second by using a microalumina extraction probe positioned next to the top of the recording electrode to absorb released catecholamines on coronal slices of rat striatum. Dopamine was desorbed into a microvolume of acid and analysed by a very sensitive HPLC-EDC assembly.

4 Conclusion

The use of intracerebral dialysis has provided a relatively simple and sensitive method for determination of *in vivo* extracellular amine levels. The study by Sharp *et al.* [36] provides the first direct comparison, in the same animal, of changes in the proposed DOPAC and 5-HIAA voltammetric peaks (obtained with a carbon fibre electrode) with changes in the extracellular levels of these metabolites *in vivo* (obtained with brain dialysis HPLC-EDT). The pharmacological manipulation of the signals obtained by DPV with carbon fibre electrodes gives the assignment of voltammetric peaks and the accuracy of DOPAC determination for both methods.

It has been shown previously that the best working conditions for intracerebral dialysate analysis involve many factors, some of which are contradictory: concentration of neurotransmitters in brain tissues, dialysis recovery (slower perfusion rates give higher recovery, but also smaller sample volumes per time unit). The operating parameters for HPLC analysis are dependent on the experimental device. But the major biological constraint is minimizing perturbation of the neuronal environment. One of the goals of current research is miniaturizing the microdialysis cannula. The use of currently available cannulae precludes the use, on-line, of HPLC systems with microbore columns and capillary columns, with the concomitant increase in sensitivity. Knecht *et al.* [39] have detailed the construction of an electrochemical glass cell which uses a single graphite fibre as the working electrode. Using an open tubular column, they have demonstrated that the detection limit is of the order of a femtomole. The linearity of the detector response was verified. It should be possible to use this apparatus in a variety of situations.

In vivo voltammetry and microdialysis have both been aimed at the monitoring of biogenic amines and metabolites in brain research, and it is probable that this type of application is now possible in the various situations likely to be encountered. Not many studies, however, have developed their application to *in vivo* mesurements of electroactive species in other biological fluids or tissues. Wang *et al.* [40] describe the design and evaluation of long voltammetric microelectrodes for *in vivo* monitoring of acetaminophen in primate species. Other applications of microdialysis with electrochemical detection should soon be described.

References

[1] Gaddum, J. H. (1961) *J. Physiol.* **155**, 1–2.

[2] Glowinski, J. (1981) *Fed. Proc.* **40**, 135–141.

[3] Ungerstedt, U., Herrera-Marschitz, M., Jungnelius, U., Stahle, L., Tossman, U. & Zetterström, T. (1982). In: Kohsaka, M., Shohmore, T., Tsukada, Y. & Woodruff, G. N. (eds) *Advances in dopamine research*. New York, Pergamon Press p. 219–231.

[4] Zetterström, T., Sharp, T., Marsden, C. A. & Ungerstedt, U. (1983) *J. Neurochem.* **41**, 1769–1773.

[5] Johnson, R. D. & Justice, J. B. (1983) *Brain Res. Bull.* **10**, 567–571.

[6] Justice, J. B., Wages, S. A., Michael, A. C., Blakely, R. D. & Neill, D. B. (1983) *J. Liquid Chromatogr.* **6**, 1873–1896.

[7] Imperato, A. & Di Chiara, G. (1984) *J. Neurosci.* **4**, 966–977.

[8] Imperato, A. & Di Chiara, G. (1985) *J. Neurosci.* **5**, 297–306.

[9] Konig, J. F. R. & Klippel, R. A. (1970) *The rat brain: a stereotaxis atlas*. New

York, R. E. Kreiger.

[10] Paxinos, G. & Watson, C. (1982) *The rat brain in stereotaxic coordinates*. London, Academic Press.

[11] Clemens, J. A. & Phebus, L. A. (1984) *Life Sci.* **35**, 671–677.

[12] Sharp, T., Zetterström, T. & Ungerstedt, U. (1986) *J. Neurochem.* **47**, 113–122.

[13] Lerma, J., Herranz, A. S., Herreras, O., Abraira, V. & Del Rio, R. M. (1986) *Brain Res.* **384**, 145–155.

[14] Hudson, P. H., Sarna, G. S., Kantamaneni, B. D. & Curzon, G. (1985) *J. Neurochem.* **44**, 1266–1273.

[15] Kapoor, V. & Chalmers, J. P. (1987) *J. Neurosci. Methods* **19**, 173–182.

[16] Combe, P. (1987) *Pharm. Thesis*, No. 137079, University of Lyon 1 (Claude Bernard), France.

[17] Kato, T., Dong, B., Ishii, K. & Kinemuchi, H. (1986) *J. Neurochem.* **46**, 1277–1282.

[18] Zetterström, T. & Ungerstedt, U. (1984) *Eur. J. Pharmacol.* **97**, 29–36.

[19] L'heureux, R., Dennis, T., Curet, O. & Scatton, B. (1986) *J. Neurochem.* **46**, 1794–1801.

[20] Wages, S. A., Church, W. H. & Justice, J. B. (1986) *Anal. Chem.* **58**, 1649–1656.

[21] Routledge, C. & Marsden, C. A. (1985) *Biochem. Soc. Trans.* **13**, 1058–1060.

[22] Damsma, G., Westerink, B. H. C., De Vries, J. B., Van Den Berg, C. J. & Horn, A. S. (1987) *J. Neurochem.* **48**, 1523–1528.

[23] Ruggeri, M., Ungerstedt, U., Agnati, L. F., Mutt, V., Härfstrand, A. & Fuxe, K. (1987) *Neurochem. Intern.* **10**, 509–520.

[24] Jacobson, I., Sandberg, M. & Hamberger, A. (1985) *J. Neurosci. Methods* **15**, 263–268.

[25] Blond, P., Ponchon, J. L., Masoero, G. & Buda, M. (1981) *Journées d'Electrochimie 1981, Brussels, 2–5 June 1981*. Abstract book p. 38.

[26] Westerink, B. H. C. & Tuinte, M. H. J. (1986) *J. Neurochem.* **46**, 181–185.

[27] Zetterström, T., Sharp, T. & Ungerstedt, U. (1985) *Eur. J. Pharmacol.* **106**, 27–37.

[28] Sharp, T., Zetterström, T., Christmanson, L. & Ungerstedt, U. (1986) *Neurosci. Lett.* **72**, 320–324.

[29] Sharp, T., Zetterström, T., Ljungberg, T. & Ungerstedt, U. (1987) *Brain Res.* **401**, 322–330.

[30] Sharp, T., Zetterström, T., Ljungberg, T. & Ungerstedt, U. (1986) *Eur. J. Pharmacol.* **129**, 411–415.

[31] Rollema, H., Damsma, G., Horn, A. S., De Vries, J. B. & Westerink, B. H. C. (1986) *Eur. J. Pharmacol.* **126**, 345–346.

[32] Di Chiara, G. & Imperato, A. (1985) *Eur. J. Pharmacol.* **115**, 131–132.

[33] Holman, R. B. (1985) *Neurochem. Intern.* **7**, 177–183.

[34] Zetterström, T., Brundin, P., Gage, F. H., Sharp, T., Isacson, O., Dunnett, S. B., Ungerstedt, U. & Björklund, A. (1986) *Brain Res.* **362**, 344–349.

[35] Zetterström, T., Herrera-Marschitz, M. & Ungerstedt, U. (1986) *Brain Res.* **376**, 1–7.

[36] Sharp, T., Maidment, N. T., Brazell, M. P., Zetterström, P., Ungerstedt, U.,

Bennett, G. W. & Marsden, C. A. (1984) *Neuroscience* **12**, 1213–1221.

[37] Gonon, F. G., Fombarlet, C. M., Buda, M. J. & Pujol, J. F. (1981) *Anal. Chem.* **53**, 1386–1389.

[38] Rice, M. E., Oke, A. F., Bradberry, C. W. & Adams, R. N. (1985) *Brain Res.* **340**, 151–155.

[39] Knecht, L. A., Guthrie, E. J. & Jorgenson, J. W. (1984) *Anal. Chem.* **56**, 479–482.

[40] Wang, J., Hutchins, L. D., Selim, S. & Cummins, L. B. (1984) *Bioelectrochem. Bioenerg.* **12**, 193–203.

Index